Muntjac and Water Deer

Muntjac and Water Deer

Natural History, Environmental Impact and Management

Arnold Cooke

PELAGIC PUBLISHING

Published by Pelagic Publishing
PO Box 874
Exeter
EX3 9BR
UK

www.pelagicpublishing.com

Muntjac and Water Deer
Natural History, Environmental Impact and Management

ISBN 978-1-78427-190-9   *Paperback*
ISBN 978-1-78427-191-6   *ePub*
ISBN 978-1-78427-192-3   *PDF*

A CIP record for this book is available from the British Library

Cover photographs: Pair of muntjac (Kevin Loader)
                        Water deer (Kevin Robson)

Contents

Acknowledgements

The following people have kindly provided photographs for this book and are identified by their initials: Marc Baldwin, Norma Chapman, Michael Clark, Rosemarie Cooke, Lynne Farrell, Josh Hellon, Kevin Loader, Matt Lodge, Mike McKenzie, Guy Pilkington, Kevin Robson and Judith Wakelam. The remaining photographs are my own, with camera trap images being indicated where appropriate. The following images were taken at Whipsnade Zoo: Figures 3.9a–c, 3.12a and 15.1; Figure 3.6b was taken at the British Wildlife Centre.

I am grateful to the following who have allowed me to include unpublished information: Alan Bowley, Roy Butters and the Abbots Ripton Deer Management Group, Peter Green, Martin Guy, Matt Hamilton, Ron Harold, Shelley Hinsley, Tony Mitchell-Jones and Peter Walker. The British Deer Society gave me permission to use the latest distribution maps of the two species and to reproduce material from *Deer* magazine. The distribution maps were prepared by Ryan Burrell of the Game & Wildlife Conservation Trust. Weather data in Chapter 8 were extracted from records of the Royal Meteorological Society's station at the National Institute of Agricultural Botany in Cambridge.

Preparation of the manuscript was considerably aided by comments on sections and chapters received from Marc Baldwin, Alan Bowley, Richard Broughton, Roy Butters, Norma Chapman, Graham Downing, Lynne Farrell, Peter Green, Keith Kirby, Tony Mitchell-Jones and Charles Smith-Jones. A number of people have helped me try to keep up with the literature, but Endi Zhang, Norma Chapman and Marc Baldwin deserve a special mention. Statistical advice was received from Tim Sparks, Ken Lakhani and Peter Rothery.

National Nature Reserve staff who have been willing sources of advice and practical help were: Jeremy Woodward, David Massen, Chris Gardiner, Tim Sutton, Gordon Mason, Ron Harold, Alan Bowley, Chris Evans, Katy Smith, Roger Boston, John Robinson, Roy Harris, Jim Frith, Dave Hughes, Andrew Mason and Bryan Nelson. Botanical studies were underpinned by the expert guidance of Lynne Farrell, who also greatly contributed to surveillance of deer over several decades. I also received help in the field from my wife Rosemarie and son Steven. At an administrative level, I have received assistance from the Nature Conservancy Council, English Nature, the Institute of Terrestrial Ecology, Natural England, Bedfordshire County Council and the Wildlife Trust for Cambridgeshire.

In addition to those above, it has been good to discuss the reserves and/or deer with: Henry Arnold, Trevor Banham, Harry Barnett, Henry Bexley, John Brown, Steve Burgoyne and the stalkers at Marston Thrift, Richard Champion, Raymond

Chaplin, Donald Chapman, Min Chen, Kath and Mick Claydon, John Comont, Jamie Cordery, Oliver Dansie, Brian Davis, Michael and Jum Demidecki, Jack Dempster, Barry Dickerson, John Eastman, Rob Fuller, Robin Gill, Emma Goldberg, Nick Greatorex-Davies, John Heathcote, David Hooton and members of the East Anglian Deer Forum, Don Jefferies, Jochen Langbein, Bob Lawrence, Richard Lawrence, Dave Leech, Carol Long, Maurice Massey, Brenda Mayle, Roger Mitchell, Ed Mountford, David Nobbs, Roger Orbell, Andy Papworth, George Peterken, Ernie Pollard, Steph Powley, Rory Putman, Oliver Rackham, Mark Ricketts, Sharon and Graham Scott, Bob Smith, Robert Smith, Terry Smithson, Brian Staines, Ray Tabor, Andrew Tanentzap, Callum Thomson, Alastair Ward, Peter Watson, Terry Wells, Ian Wyllie and those people credited in the text with providing views and information. This project has rambled on for more than 40 years – apologies if I have forgotten any names along the way.

Finally, I wish to thank Nigel Massen at Pelagic Publishing and Chris Reed of BBR Design for their help, encouragement and patience in the production of this book.

Abbreviations

BDS   British Deer Society
BRC   Biological Records Centre
BTO   British Trust for Ornithology
Defra   Department for Environment, Food and Rural Affairs
DMG   Deer Management Group
DVC   deer–vehicle collision
ITE   Institute of Terrestrial Ecology
IUCN   International Union for Conservation of Nature
JNCC   Joint Nature Conservation Committee
MAFF   Ministry of Agriculture, Fisheries and Food
NBN   National Biodiversity Network
NNR   National Nature Reserve
PTES   People's Trust for Endangered Species
RSPB   Royal Society for the Protection of Birds
SSSI   Site of Special Scientific Interest
VC   Vice County

CHAPTER 1

Introduction

1.1   The species

The cold winter sun throws long shadows from the alders across the fen field. Fringing reeds twitch and an animal moves slowly and deliberately out from cover along a well-used trail through the tall grass. Its back gleams in the late afternoon sunlight, which also reflects off its dagger-like canine teeth. The animal oozes menace, an impression enhanced by its torn ears. Suddenly, and without warning, it makes an eerie clicking noise and hurtles forwards through the grass towards a previously unseen and unaware rival. The second animal is too late in appreciating the danger, and a cloud of hair puffs into the air as tusk hits rump before the sanctuary of the reeds can be reached. Another head appears above the grass as a muntjac's grazing is disturbed by the commotion. Although she has a young fawn hidden snugly nearby, neither she nor it has anything to fear from the aggressive male water deer. He wants to drive away the second male but is totally indifferent to the presence of her and her fawn. He paws the ground and marks the patch with dung to warn off other intruding males of his kind.

This is a book about two extraordinary species of deer and the scene above provides a couple of examples of how different they are from the other species of deer in this country and in the rest of the continent. For example, the male water deer *Hydropotes inermis* lacks antlers, but instead has long canine teeth for fighting. It could in some situations be mistaken for a feral big cat (Figure 1.1a), an error that has been made in my part of England. The male muntjac *Muntiacus reevesi*, meanwhile, has an equally bizarre appearance with its smaller tusks and large facial glands (Figure 1.1b). This species has also retained its ancestral feature of breeding at all times of year, which means that fawns can be seen even in midwinter. Both species have a number of other unusual features which are dealt with in this book.

The two species are still frequently confused, probably because they are of similar size and are found together in some parts of England. The muntjac is much more common and better known, so the usual mistake is to label a water deer as a muntjac. They are similar in some ways but differ in others – especially when environmental impact is considered.

Figure 1.1   Males (bucks) of the two species: (a) an aggressive water deer with long canines poised to attack a rival (MM); (b) a muntjac licks the gland in front of its eye, causing its jaw to move sideways (KR). See Section 2.7 for an explanation of such behaviour.

Although they are now at home in this country, both species originate from the Far East – the water deer from China and Korea, and the muntjac from China and Taiwan. There is concern worldwide about environmental problems caused by introduced, invasive species. The Department for Environment, Food and Rural Affairs (Defra) has set up a Non-Native Species Secretariat to give advice on such issues, and the Secretariat's website lists descriptions of both of these deer species among its 'popular portal searches'. Invasive plants and animals in Britain are also the subject of a comprehensive field guide (Booy et al. 2015).

An alternative view of introduced species was put forward by Fred Pearce (2015), who argued that the case against alien invaders had been overstated. He pointed out that problems often occur in ecosystems that have been degraded by man; and new invasive species may be able to exploit and reinvigorate natural systems. This is a credible argument in many circumstances, but can it apply to the case of muntjac causing impacts in ancient woodland in England? It is true that the structure and species composition of such woodlands have in the past been heavily influenced by us, while limited deer grazing and browsing has helped to shape conservation interests. However, if the right balance of deer browsing is a good thing, should we not encourage native roe deer *Capreolus capreolus* rather than muntjac? Things start to become more complicated when it is realised that the roe deer roaming southern England have originated from foreign stock brought here in the 1800s.

Furthermore, muntjac have become habituated to people and are common in many suburban areas where they may be viewed fondly by those who do not have tasty vegetables and flowers to protect in gardens or allotments. In this way, they have become the people's deer in some parts of lowland England – a species of deer that can be easily and frequently seen outside the confines of zoos.

Views on muntjac therefore can be polarised towards those who love them and those who hate them. Water deer, on the other hand, are much rarer and consequently less well known – but again views are polarised. The name, Chinese water deer, advertises their foreign origins and this leads to suspicion in many minds. However, those who know something about them may appreciate that the jury is still debating whether they pose an environmental threat and regard them more favourably than muntjac.

In this book, I use the name 'water deer' rather than 'Chinese water deer' because there is just one species and the deer come from Korea as well as China. Similarly, the species of muntjac that is living here is called Reeves' or the Chinese muntjac. I do not refer to it as the Chinese muntjac as it is also found in Taiwan, and several other species of muntjac occur in China. I refer to Reeves' muntjac when I need to be specific, but otherwise it is simply called 'the muntjac'.

But why write a book about two unrelated species of deer? Their lives in their native ranges and here in England were and are curiously interwoven. In parts of China, it is still possible to find them together. They were brought to this country at about the same time and their establishment here owes much to the same aristocratic family. In East Anglia, they have lived in the same landscapes, and have been confronted by the same threats and opportunities. They have colonised and interacted over the last 50 years in the area where I live – and I do not feel I can deal with one without considering the other. In a sense, it is my story as well as theirs. The book draws particularly on what has happened in this country, but includes information from elsewhere, especially the Far East.

1.2   Origins of the deer in this country

Reeves' muntjac was named after John Russell Reeves who first imported it to this country in 1839. He worked for the British East India Company and was also interested in botany. One subspecies, *Muntiacus reevesi reevesi*, comes from the Chinese mainland and another, *M. r. micrurus*, occurs in Taiwan. Reeves' muntjac were introduced into the park at Woburn Abbey in Bedfordshire in the 1890s, where they bred but did not thrive (Chapman et al. 1994a). The place of origin of the Reeves' muntjac is unclear because records of transactions are incomplete. They may have been from mainland stock or from Taiwan, or be of mixed parentage. Genetic evidence initially indicated that Woburn animals were not the only sources of introduction to this country (Williams et al. 1995), but a more recent study concluded that there was a single founding event involving only a few females (Freeman et al. 2016).

The water deer was first described in the west by Robert Swinhoe (1870), who was an eminent English biologist working as a diplomat in China and Formosa (now Taiwan). Swinhoe proposed its Latin name, which translates as 'horn-less water-drinker' after its lack of antlers and its fondness for marshy ground. The species was imported to this country in the late nineteenth century, being first kept in London Zoo in 1873 and with 19 being introduced to Woburn Abbey between 1896 and 1913 (Chapman 1995). Genetic studies by Richard Fautley (2013) suggested that the imported water deer came from the Chinese mainland and that this ancestral population had become extinct. Two subspecies of water deer have been recognised: *Hydropotes inermis inermis* in China and *H. i. argyropus* in Korea. However, a genetic study by Koh et al. (2009) revealed two distinct forms, with one occurring in both countries and one solely in Korea.

1.3   Recent events on a global scale

This section outlines the species' changing fortunes in their native ranges, where they are not faring well, and summarises their occurrence elsewhere in the world. The International Union for Conservation of Nature (IUCN) reassessed the status of Reeves' muntjac as of Least Concern in 2016 (Timmins and Chan 2016). There is little hard information, however, and the IUCN authors recommended that its position should be assessed again when new information becomes available. Its Chinese population was estimated at more than 2,000,000 by Sheng (1992a) but declines were believed to have occurred because of hunting and habitat loss (Timmins and Chan 2016), and it has been assigned to the category of Vulnerable on the Chinese Red List (Jiang et al. 2015). Timmins and Chan (2016) suggested that 'the relative lack of concern for this species may stem from its abundance in the UK'. There is no estimate of population size in Taiwan but these authors considered the population to be stable; a ban on hunting and extinction of its main predator, the Formosan clouded leopard *Neofilis nebulosi brachyura* (Chiang 2007), have presumably helped to offset losses for other reasons, such as habitat degradation. The muntjac's southern range in China was illustrated as down to latitude 22° north by both Sheng (1992a) and Jiang et al. (2015), but extensive use of camera traps in recent years put its southern limits at 25° north (Timmins and Chan 2016). This contraction is reflected in the map in Figure 1.2.

Historically, the water deer was a common animal that roamed over much of the wetlands of eastern China, as far north as latitude 42° (Sheng and Ohtaishi 1993; Zhang 1996). During the twentieth century, its range and numbers contracted markedly. These authors presented maps indicating the range during the 1990s was between latitudes 24° and 34° north and between longitude 110° east and the Chinese coast. Within this area, though, there were only three principal centres of population and these were widely separated (Figure 1.2). Estimates of the total number in China were about 10,000 (Sheng 1992b) or in the range 10,000–30,000 (Sheng and Ohtaishi 1993). As Helin Sheng (1992b) had estimated the annual hunt total at 10,000, it is not difficult to understand why the population was decreasing,

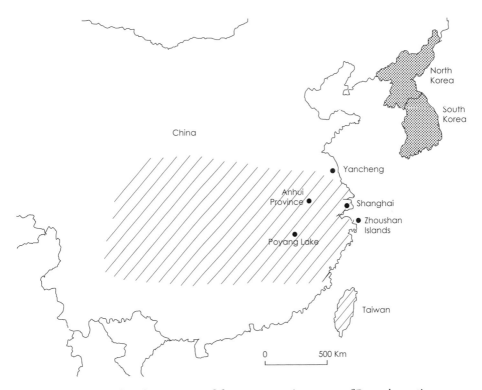

Figure 1.2   A distribution map of the current native ranges of Reeves' muntjac
and water deer based on information in Sheng (1992a), Kim et al. (2011), Harris and
Duckworth (2015), Jiang et al. (2015), Chen et al. (2016) and Timmins and Chan (2016).
Diagonal hatching shows the overall extent of the range of the muntjac in China and
Taiwan. Similarly, stippling indicates the overall range of the water deer in North
and South Korea. In China, dots show locations where water deer may be found,
the principal ones being Poyang Lake, the Zhoushan Islands and Yancheng.

especially as habitat loss was continually exacerbating the situation (e.g. Xu et al.
1998; Chen et al. 2009; 2016; Harris and Duckworth 2015). Min Chen provided an
estimate of less than 5,000 in 2011 (Fautley 2013). There have been conservation initi-
atives in recent years, such as captive breeding and release of animals at Shanghai
(Chen et al. 2016; He et al. 2016) and a level of protection in several reserves.

The Korean water deer has suffered from habitat loss and excessive trapping and
hunting, but has benefited from loss of natural predators; high numbers can still be
found in some areas of South Korea (Won and Smith 1999; Harris and Duckworth
2015; Jung et al. 2016). Despite a recent estimate for that country of 500,000–
700,000, there is some concern at the current high level of persecution and road
deaths (Chun 2018). The IUCN has categorised the water deer as Vulnerable since
1994 (Cooke and Farrell 1998), and the species has the same assignation on the
Chinese Red List (Jiang et al. 2015). A vulnerable species on the IUCN Red List is
one that is likely to become endangered unless the factors threatening its survival
improve. Until the early 1990s, the water deer had been graded as Rare by the IUCN,
a category of lesser concern.

In addition to its population in Britain, Reeves' muntjac now occurs in a number of other countries outside its native range, becoming particularly well established in Japan (National Institute of Environmental Studies 2017). Escapes occurred from zoos in southern Chiba and Izuoshima in Japan during the 1960s and 1970s; significant populations have developed and a range of agricultural and horticultural crops have been damaged. Import, transport and keeping in captivity are all prohibited in Japan. An introduction to France failed (Cooke 1999a), but there have been considerable concerns about the appearance of muntjac in the wild in Ireland (Dick et al. 2010; Carden et al. 2011; Freeman et al. 2016), the Netherlands (Hollander 2015), Belgium (Baiwy et al. 2013) and Denmark and Germany (European Union 2017).

Muntjac have been bred in captivity throughout Ireland and reported sightings in the wild have increased in recent years. The first definite record was of a buck shot in County Wicklow in the Republic in 2007, while the first in Northern Ireland was a traffic casualty in County Down in 2009. Since then there have been a number of other sightings (Figure 5.3). People are urged to report occurrences, and measures are taken to try to limit and control releases. In the Netherlands, reports of muntjac date from the late 1990s in the Veluwe area, although most of the recent sightings are from the south in the province of Noord-Brabant, and are thought to be escapes or deliberate releases for the purpose of hunting. There is also concern that deer may be entering the Netherlands from Belgium. Provisions are in place to reduce the likelihood of new releases. Muntjac are held in zoos and parks in Belgium and incidental reports of the species in the wild in Kempen and West Vlaanderen are likely to be due to escapes and also to deliberate releases. In addition there is a possibility that establishment in the Netherlands may result in colonisation from across the border. Muntjac will not respect political borders and will move in either direction! The difficulty in eradicating or controlling populations is recognised. Preventative measures have been outlined, but these cannot be guaranteed to stop deliberate or accidental releases to the wild.

In contrast, France was the only country apart from England where water deer survived for a while. They were introduced in 1954 (Cooke 1999b) but, after living near Limoges for several decades, the population died out (Gérard Dubost, personal communication).

Both muntjac and water deer are kept in captivity in other parts of the world. Marc Baldwin has informed me that he knows of at least 49 zoos in 11 countries that keep, or recently kept, muntjac, and nine zoos in four countries that have water deer. A simple Internet search demonstrates that muntjac and water deer populations are maintained on ranches in Texas for the purpose of commercial hunting. Doubtless muntjac, in particular, are widely kept under unrecorded circumstances – and escapes or releases to the wild in other countries will almost certainly occur.

1.4   Growth of knowledge about these species

It is worth briefly outlining the landmark studies on these two species, beginning with the muntjac. In Section 1.5, I will describe how my interest in deer evolved and fitted in with this accumulating knowledge base.

The first person to study muntjac long-term seems to have been Eileen Soper (1969) in her garden in Hertfordshire, beginning in 1959. In the 1960s, Oliver Dansie and Michael Clark were beginning their studies, also mainly in Hertfordshire; and Donald and Norma Chapman became interested in muntjac as well as fallow deer *Dama dama*. Among publications in the 1970s and 1980s were: Dansie's general accounts for the British Deer Society (BDS) in 1970 (with a second edition in 1983) and for *The Handbook of British Mammals* (1977); and Clark's *Mammal Watching* (1981), in which he used the muntjac as a model for what could be learned by close observation in the field and in captivity. During the 1970s, several detailed articles had been published on the behaviour of captive Reeves' and Indian muntjac *M. muntjak* (Dubost 1971; Barrette 1977a; 1977b). Two papers by the Chapmans in 1982 and 1983 showed that our muntjac were all of the Reeves' species. A postgraduate study on the social organisation, diet and habitat use of Reeves' muntjac in an Oxfordshire wood and at Whipsnade Zoo in Bedfordshire was undertaken during the early 1980s by Stephan Harding (1986).

In 1979, Donald and Norma Chapman began to organise a major study of muntjac and other deer in the King's Forest in Suffolk, the most southerly block of Thetford Forest. This study went on into the 1990s and involved a large number of people, including two research students working on deer ecology: Peter Forde in the 1980s and Jonathan Keeling in the 1990s. In a study area of about 200 ha, work included estimating the population size and numbers in each sex/age class over a period of years, mapping home ranges (Figure 1.3), recording food items and defining habitats used (e.g. N.G. Chapman et al. 1985; 1993; Claydon et al. 1986; Harris and Forde 1986). Members of the same team, led by Norma Chapman,

Figure 1.3   A muntjac with a radio collar in the King's Forest in Suffolk, where a major study on muntjac took place during the 1980s and 1990s (JW).

documented the early history of the muntjac in Britain, described how it spread, much abetted by man, and plotted its expanding distribution up to the early 1990s (Chapman et al. 1994a).

Books reviewing deer species in China were published in 1992 and 1993. A chapter in the former book on Reeves' muntjac (Sheng 1992a) revealed it to be a numerous species that was heavily hunted, but little studied. Hunting was not undertaken on any scientific basis. A year later, a meeting was held in Cambridge to discuss the biology, impact and management of muntjac in Britain, and a slim volume was published, edited by Brenda Mayle of the Forestry Commission. Later, in 1996, a booklet on muntjac was written by Norma Chapman and Stephen Harris – this was the first of a series on British deer published by the Mammal Society and the BDS.

An important study undertaken by staff of the Institute of Terrestrial Ecology (ITE) and Forest Research and funded by the Ministry of Agriculture, Fisheries and Food (MAFF) was begun in the mid-1990s (Staines et al. 1998). This reviewed distribution, status and habitat requirements of deer in lowland England and Wales, and undertook studies on the ranging behaviour of fallow deer and muntjac. This was followed in the early years of the twenty-first century by studies on population size, habitat associations and landscape-scale management of muntjac and roe deer in Thetford Forest, led primarily by staff of the University of East Anglia (e.g. Hemani et al. 2005; 2007; Waeber et al. 2013).

In Taiwan, McCullough et al. (2000) reported on home range, activity patterns and habitat relations. Po-Jen Chiang studied the now-extinct Formosan clouded leopard for many years, and submitted a doctoral dissertation in 2007 on the ecology and conservation of the leopard, its prey and other carnivores in Taiwan. Reeves' muntjac was thought to be an important prey species so various aspects of its ecology were studied.

Summaries of the natural history of the species in Britain were included in two books on muntjac management (Smith-Jones 2004; Downing 2014). The first of these authors augmented his considerable field knowledge by visiting Whipsnade Zoo to observe the behaviour of the muntjac population confined within the perimeter fence.

From this brief description, it will be clear that far more of the basic research on the muntjac has been done in this country rather than in the species' native range. The first person to map the two species in Britain was Kenneth Whitehead in 1964. Since then there have been several schemes to update their distributions. For instance, the National Biodiversity Network (NBN) Gateway maintains up-to-date maps and the BDS has organised a series of surveys between 1972 and 2016.

When Lynne Farrell and I began our study on water deer in 1976, there was little written material to digest. Kenneth Whitehead's 600-page volume on deer in the British Isles in 1964 had a few pages on the fate of various introductions and basic details about appearance. The first descriptive account about the habits of water deer was by Harris and Duff (1970). Although evidently based on observations of captive deer at Whipsnade Zoo and the park at Woburn Abbey, it was very instructive and has generally stood the test of time. This was followed by Raymond Chaplin's book on deer in 1977. He had kept water deer and studied them in the field, so had many first-hand observations to contribute. In the late 1980s, Stefan

Stadler, a research student from Germany, studied behaviour and social organi-sation of water deer at Whipsnade; his thesis was finished in 1991 and is still a mine of information.

Meanwhile, Chinese researchers had been much more interested in water deer than in muntjac, and Helin Sheng (1992b) reviewed results and observations on habitat, abundance, reproduction, population structure, home range, scent-marking and status. In an article in the same volume, Helin Sheng and Endi Zhang discussed the conservation of all species of deer in China. Endi Zhang then came to Britain as a research student to follow in the footsteps of Stefan Stadler and work further on the behavioural ecology of water deer at Whipsnade (Zhang 1996). He returned to China to lead a research team investigating in particular the ecology and conservation of water deer in that country. This developed into a programme, led by Min Chen, to reintroduce water deer to the Shanghai area, from which they were evidently lost in the early twentieth century (Chen et al. 2016). Another recent strand of work has centred on attempts to unravel and understand the genetics of the two subspecies of water deer in China and Korea (e.g. Koh et al. 2009; Fautley 2013).

The French mammalogist, Gérard Dubost, led studies on a population of water deer at Branféré Zoological Park between 2002 and 2005, producing a series of papers relating to reproduction, growth and behaviour (e.g. Dubost et al. 2008). The European studies listed here have all focused on captive animals, whereas Chinese investigations have largely been on the few remaining wild populations. Not surprisingly, there has been an emphasis in China on conserving the species, whereas in Britain this sentiment has been largely absent. Indeed, both the water deer and the muntjac are on Schedule 9 of the Wildlife and Countryside Act, 1981, which prohibits their release without an appropriate licence.

1.5   My interest in the deer

My interest in deer first developed in the 1960s when I was a research student in Reading studying the composition and formation of birds' egg shells. My course was laboratory based but I was keen to enjoy the nearby countryside as much as possible. I often made excursions in the evening to places such as Knowl Hill and Bramshill Forest, and became more familiar with crepuscular creatures such as woodcock *Scolopax rusticola*, nightjar *Caprimulgus europaeus* and fallow deer – and I saw my first wild muntjac.

Encounters around Reading made me want to develop a personal, spare-time project on deer when, in 1968, I joined the Nature Conservancy at Monks Wood Experimental Station in Cambridgeshire to work on the effects of toxic chemicals on wildlife. In 1976, after a few false starts, I asked my friend and colleague, Lynne Farrell, if she would like to join me in determining whether a project on water deer at Woodwalton Fen might be viable. She agreed without hesitation and I still remember the first evening visit when we saw four deer – and the project developed into a three-year study to discover as much a possible about the natural history of the species. From 1979/80 we scaled it down into a winter programme to follow the fortunes of the population. Because this was nothing to do with our official work, we could be flexible and the project had no defined end point. The surveillance programme has rolled on for more than 40 years.

From 1973, the Nature Conservancy was reorganised into the Nature Conservancy Council (NCC) and ITE. I opted to stay at Monks Wood with ITE, but in 1978 joined the NCC, based at nearby Huntingdon. My responsibilities were to be adviser for Britain on (1) the effects of toxic chemicals on fauna and flora and (2) the conservation of amphibians and reptiles. Lynne and our colleagues, Don Jefferies and Tony Mitchell-Jones, carried out post-mortem investigations on any water deer casualties that were found, but I was already dealing with enough dead creatures, so declined to join them. By the mid-1980s, our group within NCC had moved to Peterborough. This meant my base had changed three times in 12 years, but I was still living and working in much the same area.

In 1985, I was told by a friend at ITE that muntjac had damaged coppice regrowth in Monks Wood. I also discovered that the warden of the wood was leaving; up to that point he had been supplying me with records of muntjac. As a consequence, Lynne Farrell and I began to survey the muntjac (and water deer) in Monks Wood. In the meantime, muntjac had started to colonise both Woodwalton Fen and nearby Holme Fen in 1980. I began surveillance at Holme Fen in 1987 – now being involved with deer at three National Nature Reserves (NNRs) that had a combined area of about 6 km².

A further reorganisation occurred in the early 1990s, when NCC was divided into the three national conservation bodies and the Joint Nature Conservation Committee (JNCC). I eventually ended up as adviser on toxic chemicals for both English Nature and the JNCC. By now Lynne Farrell had moved away but continued to help when needed. Around that time, deer damage in woodland was becoming an important issue. Surveys of coppice damage were beginning, but little had been published. English Nature was aware of the poor condition of Monks Wood and I persuaded senior staff to second me back to ITE to spend a year studying muntjac impact in the wood. This took place from May 1993 until April 1994 and allowed me not only to document impact and the effectiveness of management, but also to derive methodology and baseline data for future monitoring.

Between 1995 and 1998, I was able to collaborate with the researchers on the MAFF-funded project on fallow and muntjac deer in Monks Wood and Rockingham Forest (Staines et al. 1998), and in turn obtain information from the project. For instance, I helped to catch and put radio collars on muntjac in Monks Wood, and this assisted me by having individuals in the wood that could be readily identified, located and tracked by their coloured collars and transmitted signatures (see Figure 1.3 for a photograph of a muntjac with a radio collar). During the period 1993–1998 I visited many other woods, learning much about muntjac, and found that people were turning to me for advice on deer impacts and how to monitor and manage them. From that time on, I worked closely with managers of woods and other types of reserve, as well as with deer stalkers. In 1998, Lynne and I were responsible for a booklet on Chinese water deer, the second to appear in the series on deer published by the Mammal Society and the BDS.

That year, 1998, marked the beginning of great change. In June, I took early retirement from English Nature and was promptly diagnosed with a life-changing illness. Once I was over the initial effects of the condition and the treatment, however, I found that I was able to continue with my work on deer due to the degree of flexibility afforded by my recording and by informal meetings. Another landmark event for me was the start of stalking within Monks Wood in the autumn of 1998; and, from 2000, culling began in other local conservation woodlands. The next few years up until 2010 marked a period when monitoring the effect of control on muntjac populations formed a significant part of my studies. The Deer Initiative had been set up in 1995 to facilitate the management of sustainable deer populations in England and Wales, and an East Anglian Deer Forum was instigated in 2004. The Forum kept me in regular contact with a range of individuals and representatives with an interest in deer in our region.

Despite an emphasis in my studies on muntjac management, the population of water deer at Woodwalton Fen had been demanding more of my attention. After 20 years of relative stability, the population increased in size from the mid-1990s because of landscape changes inside and outside the reserve. By 2010, browsing impacts in the reserve had become serious and information was required on the relative contributions of water deer and muntjac. Over a period of several years, the use of camera traps helped to reveal the culprits. Acquisition of these devices opened up a whole new phase of work, particularly on the behaviour of both species. Some of these behavioural studies are reported for the first time in this book.

Piecing together the book also stimulated fieldwork that otherwise would probably not have been done. It was written over a period of more than five years and during that time I tried to write during the appropriate season. So that when, for instance, I compiled the text about muntjac grazing on ground flora, I was able to go to a wood and check a fact or fill a gap. I also endeavoured to acquire a more complete knowledge of deer distribution, both through the county and the region.

I see my main contributions to our knowledge about muntjac as being long-term surveillance of populations and observations on impacts. Surveillance has demonstrated how long populations take to reach a plateau and has also given pointers to factors that affect population size. The impact work at Monks Wood and other fragments of ancient woodland illustrated the breadth and extent of impacts; and I later monitored, using novel methods, how remedial management affected both the muntjac populations and their impacts. The work in Monks Wood was described as 'pioneering' by Fuller and Gill (2001), while a review by Defra said it was 'probably the best documented example of deer damage to conservation woodland and the steps taken to deal with it' (Wilson 2003). My observations on impacts were a factor in the decision to add muntjac to Schedule 9 of the Wildlife and Countryside Act, 1981.

Surveillance of the water deer at Woodwalton Fen with Lynne Farrell apparently remains the only long-term study on a wild population of this species conducted anywhere in the world. In the first three years it produced basic information enabling us to contribute an account of the species for the BDS (Cooke and Farrell 1983) and provide entries for the species in the third and fourth editions of *Mammals of the British Isles*. The study has also demonstrated that an isolated colony of water deer can survive for more than half a century and has helped to explain how and why fluctuations in numbers have occurred. More recent investigations, often with camera traps, have shown that water deer are capable of causing certain types of impact in a wetland reserve, but these are so far slight compared with those inflicted by muntjac in woodlands.

1.6   Main study sites

The bulk of the work on muntjac and water deer has been undertaken in Natural England's three large NNRs close to my home. Some brief details about the sites are given in this section, together with Figure 1.4 which shows their locations relative to one another and other features. It is worth stating again that there has been a series of reorganisations of the national conservation bodies; and the Nature Conservancy, the NCC and English Nature managed these reserves before Natural England took over.

Monks Wood NNR is 157 ha in size and is one of the largest ancient woods in Cambridgeshire. It is on the Oxford Clay escarpment in the south-west corner of the Fens, and ash *Fraxinus excelsior* and oak *Quercus robur* are the dominant tree species (Figures 1.5 and 1.6).

Woodwalton Fen NNR is a 208 ha fenland remnant, most of which was cut in the past for peat. It did, however, survive the widespread drainage that has occurred since the seventeenth century, and is now a patchwork of mixed fen, marsh, reed

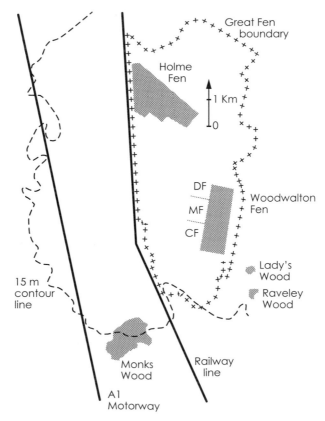

Figure 1.4   A map of the south-west corner of the Fens, with shading showing the reserves most frequently referred to in this book, together with other selected topographical features. The locations of farms to the west of Woodwalton Fen are indicated by their initials: Darlow's Farm (DF), Middle Farm (MF) and Castlehill Farm (CF). The boundary of the Great Fen is demarcated by small crosses.

Figure 1.5   The main ride of Monks Wood in September 2015.

Figure 1.6   Fields outside the southern boundary of Monks Wood have been used for
different types of ecological research since the 1960s. Here, the fields have just been mown
in summer, 1993. The edge of the wood is on the left, buildings of the research station
are in the background and one of the hedges used for experiments is on the right.

bed, meres, scrub and woodland (Figures 1.7, 1.8 and 4.3a). Locations of farms to the
west of the reserve are indicated in Figure 1.4.

Holme Fen NNR lies on the peat immediately to the south-west of the former
Whittlesey Mere, which was the largest of the old fenland meres. After drainage
in the nineteenth century, part of the area was used for crops but later abandoned.
Most of the site developed into birch woodland, and, at 266 ha, it is the finest
example of this type of wood in lowland Britain – and the largest woodland in
Cambridgeshire. The Great Fen Project is a partnership between Natural England,
the Wildlife Trust, Huntingdonshire District Council and the Environment Agency.

It is restoring more than 3,700 ha of mainly arable land back to fenland. The project will connect Woodwalton Fen and Holme Fen NNRs (Figure 1.4) and allow much better control of the water regime. Views of parts of the southern and northern edges of Holme Fen are shown in Figures 1.9 and 1.10 respectively.

Two other small fragments of ancient woodland, Raveley Wood (6 ha) and Lady's Wood (7 ha), are also included in Figure 1.4. These are owned and managed by the Wildlife Trust. Muntjac and their impacts have been recorded in both woods for more than 20 years.

Figure 1.7   A water deer in the drier south of the reserve at Woodwalton Fen. For a view of this wetland habitat, see Figure 4.3a.

Figure 1.8   A view from the western flood-bank of Woodwalton Fen looking in a northerly direction with Holme Fen on the north-western skyline. The closest fields are part of Middle Farm with Darlow's Farm beyond. The photograph was taken in 2002 before these arable fields were converted to grassland as part of the Great Fen Project (see Figure 8.8b).

Figure 1.9   The southern edge of Holme Fen in November 2006, photographed soon after arable fields had been converted to grassland.

Figure 1.10   A view taken in March 2016 by a drone above the north-west corner of Holme Fen. The drone was looking in a south-easterly direction with birch woodland on the right and recreated fenland on the left (JH).

1.7   The book

The book aims to pull together published and unpublished information on natural history, environmental impact, how the deer can be managed and whether management works, in order to be a reference source for anyone requiring basic or detailed information on either or both species. It is a personal account, concentrating especially on topics of particular interest to me.

Virtually all statistical testing is omitted. However, where it is stated that, for example, some variable has changed over time, this statement should have been made on the basis of a statistical test. Often it has not been possible to determine with confidence why something has changed; or whether, when a change in one variable is related to a change in another, that this truly represents cause and effect. There is scope for further work on some of the observations presented here.

The first part of the book covers various aspects of natural history. The following two chapters draw together basic information on appearance, signs, sounds and senses, habitats, food, mortality and breeding gathered from this country and the species' native ranges. Remaining chapters on natural history relate to other aspects of their lives and presence here: locating deer in the field; colonisation of Britain, the detailed colonisation of a county and colonisation of specific sites; how and why unmanaged populations have changed; how deer species interact; and density and numbers. Many of these chapters concentrate on my local deer in Cambridgeshire where large populations have built up. Although no patches of land are exactly the same, my observations should be of relevance to situations elsewhere. Thus, what I have seen and recorded should help other people to understand how the species might settle and colonise their areas.

The second part of the book is on environmental impact and its management. My perspective is very much from the point of view of nature conservation, although other forms of impact are touched on. Muntjac deer have been a major recent driver of changes in woodland structure and species composition in lowland England, and many of the principles relating to such woodland impacts are also applicable to the activities of other species of deer. My studies on muntjac have largely focused on Monks Wood in Cambridgeshire, where muntjac density was one of the highest ever documented, meaning that the wood was probably as badly damaged by muntjac as any in the country. However, fallow deer were absent and the impact caused there by muntjac was probably no worse than that caused by fallow and a combination of deer species in many woods in this country. However, interest in environmental impacts is not solely restricted to woodlands. The highest densities of water deer occur in wetlands, where there is some potential for conflict. The focus on water deer in the book is on whether issues occur. The paradox of the water deer is that it is a Red Data Book species that may now be more abundant here than in China, yet is still viewed with suspicion by many British conservationists and land managers who think it may become 'another muntjac'.

The final chapter is a brief summary of current knowledge and attitudes and what might happen or be done in the future.

The structure of the book means that certain topics can crop up in several chapters. This can lead to problems of the chicken-and-egg variety and also to some repetition. I have tried to minimise repetition while keeping each chapter as self-contained as possible.

CHAPTER 2

Basic natural history

2.1 Introduction

This chapter provides information from the Far East, Britain and France on size, appearance, antlers and tusks, ageing methods, signs, senses, communication, social behaviour, habitats, food and mortality. Other aspects of natural history are discussed in later chapters in the first part of the book.

The scientific classification of deer (the family Cervidae) has taken several twists and turns over the last century; initially, appearance, skeletal structure and geographical distribution held sway, but then genetic studies indicated different taxonomic classifications. Throughout this time, however, there has been general agreement that the water deer is an unusual species that is not closely related to the muntjac. The most recent classification (Gilbert et al. 2006) has the water deer, along with roe deer, in the tribe Capreolini in the subfamily Capreolinae. Reeves' and other muntjac species are in the tribe of Muntiacini in the subfamily Cervinae. It is believed that ancestors of the water deer once had antlers, but that these have been lost (Randi et al. 1998; Gilbert et al. 2006).

Species of muntjac are unusual in that males have long bony face-ridges and pedicles on which their simple antlers grow (Dansie 1983; Chapman and Harris 1996). They are an ancient group that has retained the primitive feature of long canines in the upper jaw. The water deer has even longer canines and has a lifestyle that is more akin to a large rodent than a deer, with rapid growth and development of both individuals and populations but with a relatively short lifespan (Dubost et al. 2011a).

2.2 Body weight and size

These are small deer, but in specifying exactly how small, some care is needed. They continue to grow for roughly two years, so that an individual in the field may appear to be a full-sized adult but could still have some growing to do. In addition, size and weight vary between and within populations and between the sexes. To illustrate some of these statements, average body weights are given for Reeves' muntjac in several places in England (Table 2.1). Deer in the Breckland Forests had

Table 2.1   Average whole body weights of Reeves' muntjac in some
English populations. Sample sizes are given in brackets.

Location	Age	Average whole body weight, kg (number in sample)		Source
		Males	Females	
Breckland Forests	At least 2 years	14.9 (41)	13.3 (30)	Chapman (2008)
Monks Wood	50–80 weeks	10.5 (15)	9.8 (30)	Abbots Ripton Deer Management Group
	At least 2 years	12.7 (39)	11.0 (43)	
Holme Fen	50–80 weeks	12.7 (13)	10.9 (11)	Abbots Ripton Deer Management Group
	At least 2 years	13.6 (19)	12.2 (16)	

been culled or were road casualties, while those from Monks Wood (2006–2010) and Holme Fen (2009–2012) were shot. Stalking started in these two reserves in 1998 and 2005 respectively. In Table 2.1, analysis of body weight data focused on periods several years after stalking started in an attempt to reduce or eliminate any effect of high deer density on body weight. Muntjac produce fawns at all times of year, so season is of no help in estimating age. State of pregnancy will affect body weight in females.

The average increase in weight from yearling to older deer varied between 7% for males at Holme Fen and 21% for males at Monks Wood; the increase was 12% for females at both sites. The forest deer were on average heavier than the older deer at Monks Wood by 17% for males and by 21% for females. Those from Holme Fen were intermediate; yearlings from Holme Fen were virtually the same weight as older deer from Monks Wood. Sexual dimorphism is well known for this species and males were consistently heavier than females, despite the fact that most of the does were probably pregnant. Helin Sheng (1992a) provided average weights for three areas of China: 12.7–13.9 kg for males and 11.1–12.2 kg for females. These were presumably animals that had been shot as the species was widely harvested for skins, but age range was not given. If the samples included variable proportions of yearlings, then it becomes difficult to compare them with information from England, although they fall within the ranges in Table 2.1. The heaviest muntjac reported for England by Norma Chapman (1992a) were a buck at 18 kg and a doe at 16 kg, while Charles Smith-Jones (2004) and Graham Downing (2014) refer to animals up to 20 kg or more.

Reeves' muntjac is one of the smallest of the muntjac tribe, with the Taiwanese subspecies being smaller than the Chinese subspecies; for *M. r. micrurus*, adult males weigh about 12 kg and females 8 kg (McCullough et al. 2000). Adult bucks of Reeves' muntjac in England are less than 10% of the weight of mature red deer stags *Cervus elaphus* living in the Quantock Hills in Somerset (Staines et al. 2008). In red deer, weight and size vary according to habitat and density. With the weights of muntjac in Table 2.1, it was a surprise to find that both sexes were significantly heavier in birch woodland at Holme Fen than in oak-ash woodland at Monks Wood, a few kilometres away. Muntjac were reported to be heavier in the population at Kineton in Warwickshire than in that at Yardley Chase in Northamptonshire (Gough 1999);

the suggested reason was 'frequent and sustained daytime and night-time disturbance' at the latter site leading to reduced food intake and stress. Such factors are
difficult to quantify, but seem unlikely to explain the differences between muntjac
at Monks Wood and Holme Fen. Neither population would have been expected to
be unduly stressed by lack of food at the times the samples were taken.

Average shoulder heights of adult muntjac shot in the Breckland Forests were
50 cm for males and 47 cm for females (Chapman 2008). Sheng (1992a) provided
information for two areas in China, but again ages of the deer were not specified: 45
and 49 cm for males; 40 and 45 cm for females.

Water deer tend to be taller and heavier than Reeves' muntjac. Information has
only recently become available on the timing of tooth eruption that can help age
water deer (Dubost et al. 2008; Seo et al. 2017) and some samples that have been
weighed are composed of animals of unknown ages. First-year deer are lighter
in weight than older individuals (Middleton 1937; Chaplin 1977; Cooke and
Farrell 1998; Dubost et al. 2008), so a high proportion of young deer in a sample
would be expected to depress the average weight. Average weights of samples
from different populations are collated in Table 2.2. Animals on the Woodwalton
farmland had been shot, whereas those at Branféré had been weighed alive and
the Cambridgeshire deer were found dead. Procedures for collecting the other
samples were unclear. Apart from the Whipsnade samples, ranges of average
weights for both males and females were small despite the different collection
procedures. Average weight for females was consistently higher than that for males.
The figure of 15.2 kg for females at Branféré Zoo was significantly higher than the
weight for males, and was for non-gestating deer, so removing any effect of foetus
weight (Dubost et al. 2008). Thus, females tend to be marginally heavier than males
irrespective of whether they are pregnant. The heaviest deer recorded locally in
Cambridgeshire, either found dead or shot, were 19 kg for both males and females,
but, as with muntjac, I have been told about animals of more than 20 kg being shot
in this country. The very low weights at Whipsnade for 'adult' deer were blamed on

Table 2.2   Average whole body weights of water deer in different
locations. Sample sizes are given in brackets.

| Location | Age | Average whole body weight, kg (number in sample) | | Source |
		Males	Females	
China	Unreported	15.9 (4)	16.3 (2)	Sheng 1992b
Zhejiang, China	Unreported	14.8 (18)	15.2 (21)	Zhang 1996
Whipsnade Zoo	Unreported	11.1 (12)	11.7 (7)	Zhang 1996
Branféré Zoo, France	At least 2 years	14.0 (43)	15.2 (62)	Dubost et al. 2008
	Yearlings	14.4 (9)	14.8 (5)	
Cambridgeshire	First year and older	14.3 (11)	15.6 (7)	Cooke and Farrell (2008)
Woodwalton Fen farmland	First year and older	14.8 (27)	15.6 (16)	M. Guy (pers. comm.)

nutritional factors (Zhang 1996; Section 2.9) – Middleton (1937) had reported similar low weights roughly 60 years earlier.

Average shoulder height of water deer ranged from 47 cm for males and 49 cm for females at Whipsnade (Zhang 1996) to 56 cm for males and 54 cm for a solitary female in China (Sheng 1992h). For deer found dead in Cambridgeshire, average heights were 52 cm for males and 50 cm for females (Cooke and Farrell 2008). The roe deer can be confused with water deer in the field, but is larger, being roughly 65–70 cm at the shoulder and weighing 18–30 kg (Fawcett 1997).

To conclude on body weight, muntjac bucks can vary in size between populations but some are as heavy as a big water deer, although they may not be as tall. It is different for the does, with water deer being appreciably heavier than muntjac and taller too. While in the muntjac the buck is much bigger than the doe, in the water deer the doe is slightly larger. It is unusual to find that the female is larger in any species of deer (Dubost 2016); typically, for other deer species, males are bigger at birth and reach full size later, but the reverse is true for the water deer. Why buck water deer are not larger is open to speculation. It may possibly be related to the way they fight, as they do not have to rely on sheer strength to fight rivals using antlers. Perhaps small size helps agility and confers an advantage when fighting with tusks. Also, maintaining a permanent territory would be more costly energetically for a larger male (Stadler 1991).

2.3   Appearance

Table 2.3 has been compiled to describe the main features of adult muntjac, water deer and roe for comparison. Coat colours are fairly similar, although to my eye the other species cannot match the rich dark brown summer coat of some muntjac; while water deer are generally paler. Some water deer can appear very pale, but become much redder in the light of the setting sun. Water deer readily shed tufts of hair in winter; it is mainly white with shades of brownish banding near the tip. Loss of hair while fighting in the midwinter rut is described in the next chapter (Figure 3.8). Deer often look scruffy during the spring moult. The final six features in the table are particularly useful in distinguishing these species in the field. Females lack tusks and antlers and so can be slightly more difficult to identify; females of the three species are shown in Figure 2.1.

2.4   Antlers and tusks

Antlers are bony structures growing from pedicles on the skulls of males of nearly all deer species, including muntjac. Antlers are cast and regrown each year. Adult buck muntjac have a regular cycle, casting antlers during May–July and then regrowing them during the following months (Chapman and Harris 1996; Chapman 2008). The growing antlers are covered in a skin called velvet. When the antlers have attained their final size, blood stops flowing and the velvet is rubbed off by the deer. These antlers are then retained until the following summer. As muntjac breed throughout the year, a young buck has to synchronise his antler cycle to the seasonal cycle of adults. This is achieved by growing the first set of antlers from around 8–11 months of age and keeping these for variable amounts of

Table 2.3   A comparison of the appearance of adult muntjac, water deer and roe, based on information in Chapman and Harris (1996), Fawcett (1997), Cooke and Farrell (1998), Harris and Yalden (2008) and personal observations.

Feature	Muntjac	Water deer	Roe
Winter coat colour	Dark grey brown, with sandy head for male	Pale brown or peppery grey brown	Grey brown
Time of spring moult	April–June	April–May	March–May
Summer coat colour	Red brown, buff beneath	Pale red brown	Reddish brown
Time of autumn moult	September–October	September–October	September–October
Tail and rear	Long tail (about 15 cm), white underside, raised when alarmed	No distinct markings, very small tail	White or buff caudal patch, tuft of long hair in centre for female
Antlers on male	Present on long pedicles, mainly in velvet May–August	None	Present, in velvet November–February
Tusks on older males	Small	Prominent	None
Faces	V-shaped black stripes in male, black kite-shaped pattern on forehead in female, prominent facial glands, white chin	Eyes and nose resemble three black buttons in winter, faces grey, brown or black, often white band around nose	Often mainly grey, large eyes, black nose and black around nose to corners of mouth, white chin
Ears	Stick out, mobile. Gap between ears greater than width of an ear	Upright, close together, hairy. Gap between ears less than width of an ear	Large, mobile. Gap between ears similar to width of an ear
Back and stance	Body sometimes hunched; short-necked	Back concave when standing with head up, rump higher than shoulder	Straight back when standing with head up

time before casting during May–July. Thus a buck born in March may cast for the first time in his second summer, whereas a deer born a month later may drop his first set of antlers a year later.

The first set of antlers consists of simple spikes with no coronet at their base. The coronet appears in subsequent sets and the end of each antler usually curves backwards and inwards. Maximum length seldom exceeds 10 cm (Chapman 2008). The International Council for Game and Wildlife Conservation has a system of trophy evaluation; for muntjac bucks, whether a trophy head is of bronze, silver or gold quality depends on the size, shape and uniformity of antlers.

The muntjac has other weapons in the form of relatively long pointed canine teeth (tusks) in its upper jaw. Most deer species either lack canines completely or have small ones which are relics from the time when ancestral species had long canines that they used for fighting (Chaplin 1977). For a buck muntjac, the tusks are his main weapons, against both other muntjac and predators (Chapman 1997a). These tusks have a sharp cutting rear edge and are designed to slash through skin and muscle; it is not unusual to come across bucks showing such injuries. However,

Figure 2.1   Adult females for comparison: (a) muntjac (GP); (b) water deer (KR); (c) roe (MB).

damage to tusks can lead to loss of territory and reduced chances of breeding. Norma Chapman recounted the story of one particular buck in her study area in the King's Forest in Suffolk that had broken both tusks when he was attacked and ousted from his territory by a rival. His injuries included a slash on his shoulder and two deep wounds on his rump. He survived but thereafter lived in suboptimal habitat, with no easy access to females.

The permanent canine erupts at about five months of age and is about 20 mm around the curve of the tooth after 14 months (Chapman and Harris 1996; Chapman 1997a; 2007). At between three and five years, average length is 29 mm with a maximum of 38 mm. There is some movement of the tooth in the gum. Aitchison (1946) suggested that was to allow sideways movement of the lower jaw when chewing the cud. In females the permanent upper canine begins to erupt much later, around one year of age (D.I. Chapman et al. 1985), but remains small and never shows below the lip.

It is, though, the water deer which is famous for having long canines (Figure 2.2a). Apart from using the front legs to kick rivals (Stadler 1991), these are a buck water deer's only weapons. The rear edge of a tusk remains sharp through friction with the lower lip, the inside surface of the tooth being polished (Harris and Duff 1970). Unlike the muntjac, however, water deer do not seem to use their tusks against predators, only against rival bucks. These tusks are even less firmly implanted than those of the muntjac and a muscular mechanism allows a buck to snarl with facial muscles bringing his tusks forwards and inwards (Aitchison

1946; Figure 1.1a). Their flexible movement also permits sideways chewing and the tusks can be held back when the animal is grazing close the ground (Figure 2.2b). Loose attachment may dampen impacts sustained during fights or other activities. Despite this modification, broken tusks are not rare among older bucks (Figure 4.6). Females have small permanent canines which are not visible on the live animal in the field.

In order to follow early growth of the tusks I have liaised with Martin Guy who stalks over the expanse of farmland to the south-west of the Woodwalton Fen reserve. During three recent winters he measured tusk length in bucks that had been shot. The measurement was a straight line from tip to gum. Open season for shooting is November–March and results are shown in Table 2.4. Bucks with tusks of at least 30 mm during November–December or at least 40 mm during January–March were assumed to be older deer. Ranges of tusk lengths for the young deer were 0–22 mm in November–December and 15–32 mm in January–March. Only two of the nine young bucks had erupted canines when examined in November, aged about five months. Eruption can, however, occur earlier as evidenced by one found dead on the farmland in October 1977. Tusks were about half-grown by March (Tables 2.4 and 2.5).

At what age, then, can tusks be seen on young bucks in the field? When Lynne Farrell and I started our study at Woodwalton Fen in 1976, Raymond Chaplin told us that tusks are obvious and well grown by one year of age; any deer noticed with tusks in winter is at least 18 months old. This statement was generally borne out by

Figure 2.2 Water deer tusks: (a) the skull of a mature buck;
(b) a grazing buck with the tusks held back (KL); (c) a first year buck in
early April with a tusk just visible (camera trap photograph).

Table 2.4   Tusk length of young male water deer culled on the farmland close to the reserve at Woodwalton Fen, 2011–2016. Number of deer examined is given in brackets. No bucks were shot during February. Information supplied by Martin Guy.

Month	Average tusk length (number of deer)
November	1 mm (9)
December + January	17 mm (7)
March	25 mm (9)

Table 2.5   Average tusk length in samples of adult buck water deer found freshly dead or shot in Cambridgeshire. Tusks on the shot deer were measured by Martin Guy.

	Average tusk length (number of deer)
Found freshly dead in Woodwalton Fen	60 mm (9)
Found freshly dead elsewhere in Cambridgeshire	49 mm (6)
Shot on farmland close to Woodwalton Fen	47 mm (7)

our experiences. Tusks just visible below the upper lip were most obvious during the period April–June (Cooke and Farrell 1981). We used a telescope in the early years of our study at Woodwalton Fen, but failed to see many short, just-visible tusks prior to spring, despite recruitment being good in some winters. The deer in Figure 2.2c was photographed on the Fen in early April 2014. On the other hand, researchers at Whipsnade have been able to detect tusks on young bucks in late winter, possibly because it is easier there to observe more closely and for longer periods. At Poyang Lake in China, Sun and Dai (1995) used a telescope to identify first-winter bucks aged 5–8 months by their short tusks (aided I suspect by size and behaviour of the deer).

Average tusk length in a sample of 20 adult males in China was reported as 53 mm (Sheng 1992b). Records from Cambridgeshire have gradually accumulated over the years (Table 2.5). Animals found freshly dead in the reserve at Woodwalton Fen had longer tusks than those shot on the farmland outside the reserve or found dead elsewhere. The latter group included a road casualty and dead deer from farmland and from other local reserves. Although few deer have been measured, the evidence indicates that bucks with the biggest tusks were in the best habitat inside the reserve at Woodwalton Fen. Maximum tusk length was 72 mm. At least two of the bucks found dead elsewhere or shot had lost a tusk. Average tusk length for a sample of bucks at Whipsnade in the 1990s was only 44 mm (Cooke and Farrell 1998). Average length of canines in females found dead in Cambridgeshire was 5 mm, with a maximum of 8 mm.

Examination of tusks from 13 of the deer found dead showed that all were close-rooted apart from those in one deer. The exception was a buck found inside the reserve in November 1977; it had tusks measuring 51 mm. This suggests that by or during their second winter, bucks have stopped growing their canines, the roots having closed.

A possible explanation for the bucks inside the reserve having longer tusks is that those fittest for fighting hold the best territories, and territories within the reserve are better than those outside. Some first-year bucks that move out to the farmland may return to the reserve when they mature, either to fill in gaps in the territorial system or to oust weaker, older males. Those defeated old males then retire to the landscape beyond the reserve. Thus, loss of young males on the farmland (e.g. by shooting) may subsequently contribute to reduction of fitness or population declines inside the reserve.

2.5   Ageing deer

Considerable experience is required before muntjac can be confidently aged in the field, especially for people like me who do not handle deer regularly. Chapman and Harris (1996) and Chapman (2008) provided some useful definitions of age and sex classes:

- Fawns are up to 2 months old and are usually spotted (Figure 3.6a).
- Juveniles are 2–5 months old, are of either sex and up to three-quarters grown.
- Immature females are 5–8 months old and are three-quarters grown to full size.
- Adult females are older than 8 months.
- Subadult males are 5–26 months and vary from developing pedicles (Figure 15.1) to having a hard first set of antlers. Face patterns are developed by about nine months.
- Males become adult between 12 and 26 months when they cast their first set of antlers.

Dead muntjac can be aged by examination of tooth eruption (D.I. Chapman et al. 1985) and wear on the mandibular molars (Chapman et al. 2005). The latter authors acknowledged that variability and unreliability of the method increased with age, but considered that this would affect ageing of relatively few deer in wild populations. Norma Chapman (2007) combined and summarised the previous articles and the Deer Initiative (2008) provided a guide on the Internet on how to age muntjac.

Lack of antlers and facial markings makes water deer harder to age than muntjac, but the fact that they are seasonal breeders more than compensates for that. The researchers at Whipsnade Zoo had their own age classes. Stadler (1991) described fawns as being up to 9 months, yearlings as 10–21 months and adults as older deer; whereas Zhang (1996) had fawns up to 6 months, yearlings between 6 and 12 months and adults older than 12 months. I prefer to refer to fawns, as for muntjac above, as young deer up to about two months and still with a spotted coat (Figure 3.12a). Young are born from late May until early July. By late summer, after they have lost their spots, their coats are softer and shaggier than adults (Figure 3.12b). By the time of their first rut in December, they are about 6 months old and are still lighter in build than adults. By this time, they are sexually mature, and I usually refer to them as 'first-year deer'. By the time they reach their first spring, when they are about 10 months old, I cannot distinguish young females from older ones. Bucks at this time can still be accurately aged in the field if their small tusks are seen (Figure 2.2c). Beyond their first year, I would call them all adults.

In Branféré Zoo in France, Dubost et al. (2008) were able to derive a method for ageing dead water deer by studying eruption and wear of their teeth – similar to what has been done for muntjac. The authors remarked that tooth eruption and replacement occurred earlier than in species of lower or comparable weight. They categorised live deer as newborn, juvenile, yearling (those in their second year) and adult. There was little difference in body weight and breeding performance between yearlings and adults. Eruption of posterior teeth up to 15 months of age has been illustrated in detail for Korean water deer by Seo et al. (2017).

In the early 1980s, Tony Mitchell-Jones and I examined teeth in a sample of 14 water deer found dead in Cambridgeshire. He recorded eruption and relative amount of wear on the premolars and molars, while I prepared the first molar in the mandible for examination of incremental layers of cementum. Each molar was transversely cut between the cusps, the cut surface being polished and viewed by reflected light. This technique had been widely used for ageing mammalian species (Grue and Jensen 1979) and has since been confirmed to reveal age reasonably accurately in muntjac (Revington 1996), although this author stated that discerning layers was more difficult in the teeth of the small species of deer. In the molars of adult water deer, where an adequate thickness of cementum was present, alternating white and translucent bands could be seen. In three individuals the pattern of banding was unclear, while in others the band could be seen across the width of the tooth. There was a positive relationship between amount of tooth wear and the number of cementum layers. The water deer were not of known age, but studies on other species suggested that the bands were likely to be deposited annually. On the basis of this assumption, age at death to within a month or so could be calculated for 11 of the 14 deer – all 11 were from Woodwalton Fen and their ages are given in Section 2.10.

2.6   Signs

The following information is mainly abstracted from Lawrence and Brown (1967), Cooke (1995), Chapman and Harris (1996), Cooke and Farrell (1998), Chapman (2008), and Hewison and Staines (2008). The most obvious signs to look for are the black or dark brown dung pellets and footprints (slots) made with the two cleaves of the hoof, but it can often be difficult to distinguish between signs of muntjac, water deer and roe because of overlap in size for these two types of sign (Table 2.6).

When deer walk, the print of a hind foot is superimposed on the print of the forefoot (Figure 2.3a). I was puzzled how deer achieved this until a camera trap video showed a water deer placing its hind foot precisely under its corresponding forefoot as the latter was leaving the ground. A clear track can look as if it has been made by a biped rather than a quadruped, and this is probably the origin of 'devil's footprints' that have been reported from time to time. As prints in snow thaw and refreeze, they can look as if they have been made by a larger creature! The stride lengths given in Table 2.6 are distances between the left and right feet when the animals are walking.

Deer often share paths (also known as racks) with other animals, such as badgers. Muntjac use paths through grassy areas that may be 15 cm in width, but well-used paths through sparse vegetation can have worn centres up to 40 cm wide and be in

Table 2.6   Information on slots and faecal pellets for muntjac, water deer and roe.

Feature	Muntjac	Water deer	Roe
Slots	Small, pointed, often with a longer outer cleave that curves around inner one (Figure 2.3a). Typically 3 cm × 2 cm, but up to 4 cm long on soft ground.	Cleaves fairly similar in size with straight inner edges (Figure 2.3b). Typically 4 × 3 cm, but 4–5 cm × 3–4 cm in soft mud or snow.	Similar to water deer but slightly larger. Size up to 5 × 4 cm.
Sign of dew claws in snow or soft mud	No	Yes	Yes
Stride length	30 cm	35 cm	35–40 cm
Faecal pellets	Variable shape, but at least one end usually tapered (Figure 2.4a). Typically up to 13 mm × 11 mm. Sometimes adhere together. Latrines occur at high densities, often next to trees.	Pointed at one end and rounded at the other (Figure 2.4b). Typically up to 15 mm × 10 mm. Used to mark territory in winter. Latrines can occur.	Variable shape. Typically up to 12 mm × 11 mm, but up to 20 mm long if cylindrical. Latrines can occur.

use for many years (Figure 2.5a). Traditional paths occur because these deer have glands in their feet, so deposit scent as they walk. Other deer follow these scent trails and paths persist. A muntjac seen walking slowly along a rack through a wood with its head down is more likely to be following a scent trail than feeding. During the seven years that I have been deploying camera traps, the positions of some paths have remained virtually unchanged. This is especially true for those made by muntjac because in wet habitat favoured by water deer, unused paths can become

Figure 2.3   Slots: (a) a muntjac track on the beach at Holkham, Norfolk, where the walking deer has put its hind feet precisely on top of the prints made by its forefeet; (b) a slot of water deer in peat soil.

Figure 2.4   Dung: (a) a large group of muntjac pellets;
(b) a small group of pellets from a water deer.

Figure 2.5   Deer paths: (a) a well-used muntjac path (rack) through a muddy wood;
(b) water deer will often create a narrow path along a ride connecting areas of cover and
feeding – here they have taken advantage of a frozen, snow-covered ditch. See also Figure 7.1.

overgrown very rapidly. Muntjac paths in a wood usually go across rides from one
patch of cover to the next, whereas water deer paths at my study population in
Woodwalton Fen often trail down a ride, showing where deer regularly commute
between cover and better feeding – such trails may be only 10–15 cm wide and go on
for hundreds of metres. Their paths made through reeds are about 20–25 cm wide
and may display tunnels through collapsing vegetation that are roughly 50 cm high.
Muntjac also make rather smaller tunnels through vegetation such as dense dead
bracken *Pteridium aquilinum*.

Couches are places where deer have rested and flattened the vegetation, usually
on dry ground. In the case of an adult muntjac, these may be about 40 cm × 30 cm.

Couches made by water deer are rather larger – the average of a sample that I measured in fen vegetation was 60 cm × 40 cm.

Damage to vegetation gives rise to a number of signs which can reveal what species has made them. Browse lines result where there has been much feeding on foliage as high as the deer can readily reach. For muntjac and water deer, this is about 100 cm while they are standing on all four feet (Figure 2.6). Muntjac can reach to at least 125 cm by standing on their hind legs, but I have never seen this behaviour in water deer. An amusing video of a buck water deer showed him straining to reach sallow leaves that were just out of reach; eventually he gave up, shook the rain out of his coat and walked off. Had that been a muntjac, it may have stood on its hind legs. Although this behaviour is not often witnessed, I have recorded it more frequently on camera traps. Roe deer can reach to about 125 cm while standing on four feet. Muntjac will also place the forefeet on logs to gain height and even climb on to fallen trees to browse on side shoots or ivy that had been growing up the tree before it fell.

Brown hares *Lepus europaeus* and particularly rabbits *Oryctolagus cuniculus* cause browse lines but these are lower, being about 50 cm for rabbits and slightly higher for hares. Deer bite tooth against a gum pad and leave ragged stalks or parts of leaves still attached to stalks (Figures 4.1 and 14.1), while rabbits and hares bite tooth against tooth and are much neater feeders, leaving clean oblique bites on stems. But beware – ride management in woodlands can leave both ragged and clean 'bites' and may extend far higher than deer can reach!

Muntjac have another trick of biting through, or partially through, a woody stem to bring down the palatable leaves at its tip to a height that is convenient for browsing (Figure 2.7a). Such damage can be seen on coppice regrowth (Figure 2.7b) or on easily broken species such as privet *Ligustrum vulgare* or elder *Sambucus nigra*. Stems that are attacked are usually in the height range 70–200 cm; breakage height

Figure 2.6   A muntjac browse line on ivy *Hedera helix* growing on an oak (contrast with Figure 9.3a).

Figure 2.7   Breakage of woody stems: (a) a doe muntjac breaking a woody stem
by biting it in order to be able to eat leaves at its tip (camera trap photograph);
(b) breakage damage on coppice regrowth of field maple *Acer campestre*.

is between 40 and 100 cm and thickness where the stem is targeted is 2–10 mm
(Cooke and Farrell 1995). Muntjac and roe will also walk over flexible stems such
as suckers to defoliate tips that are normally out of reach, although this behaviour
usually does not leave clear evidence of what has occurred. I have not seen either
type of behaviour displayed by water deer. Indeed, buck water deer with long tusks
would probably be unable to be able to position stems in their mouths so they
could chew through them.

An additional aspect of muntjac behaviour that can lead to damage of woody
stems is territorial fraying. Bark is scraped away primarily with the incisors in the
lower jaw at a height that is usually between 20 and 40 cm, but can be as high as
60 cm (Section 3.1.1; Figures 3.3a and b). Coppice stems and young willow, *Salix*
species, are particularly favoured, with diameter of the latter being as thick as
5 cm. This form of damage should not be confused with bark gnawing by rabbits,
in which the bites with lower and upper incisors can be clearly seen, or with the
fraying and antler-cleaning activities of roe and larger deer which often extend
much higher up a sapling. Water deer do not seem to indulge in fraying behaviour
resulting in any obvious signs. Neither do they leave scrapes that are likely to be
noticed, whereas both muntjac and roe scrape the ground sufficiently vigorously
for this form of territorial marking to be very visible. Muntjac scrapes can be up
to 1 m across and take the form of shallow dishes in the ground, often with hoof-
prints and dung visible on their surface (Section 3.1.1).

Grazing on flora can be obvious if it is searched for, but too often a visitor
to a site with deer is concentrating on what is flowering and fails to notice that
something is missing. With flowers such as bluebells *Hyacinthoides non-scripta* or
orchids, the whole flower head will be consumed by muntjac leaving a vertical
stalk until it rots away. However, many other creatures are attracted by these treats,
including rabbits, hares, small mammals, birds and invertebrates, especially slugs
and snails. Most of these animals take parts of flowers or leaves, but feeding by
rabbits and hares can be difficult to separate from that of muntjac. And where they
occur, the potential impact of water deer should not be forgotten. In wetland sites,
their biomass can be considerable and their feeding activity is usually impossible to
distinguish from that of muntjac.

2.7   Senses, communication and social behaviour

Sight, smell and hearing are important to these species of deer, warning themselves and conspecifics of impending danger, helping to orientate themselves within their ranges, deter or avoid rivals, distinguish friends and family, and attract mates. Water deer tend to occur more frequently in open areas than muntjac and may be more dependent on sight to detect danger. Stadler (1991) described the staring and scanning behaviour of water deer at Whipsnade Zoo. In the first type of behaviour, the head is rapidly raised into an upright position and the deer stares attentively in one direction for a variable amount of time with ears pricked and body still (Figure 2.1b) – although the head might be moved up and down or from side to side. The deer might resume its previous activity after a few seconds or could remain in that position for minutes. Staring is a posture I know well as it is often adopted by an individual I am observing through binoculars; I am hoping that it will relax and show me its profile so I can be sure whether it has tusks, but often I give up with aching arms before it does. Scanning behaviour involves a more relaxed check on the general environment to ensure that all is well. Stefan Stadler found that staring was generally much more common than scanning, but in a less disturbed situation the reverse may apply. It is often said that deer are alerted more by movement than by shapes, but it is also true that something new and conspicuous in their home range may disturb them. For instance, a new post in the middle of a grass field may be avoided for some time.

Deer have an acute sense of smell and water deer can sometimes move away when 200 m downwind. The ability of muntjac to detect danger is similarly well tuned. During 1993/4, I undertook much counting of muntjac in woodland plots in Monks Wood. I was walking along rides focusing totally on seeing muntjac. Because of how I was recording I could calculate the percentage of muntjac seen (Cooke 2006). Of those that will have been within 30 m of a ride, I was only detecting 25%. Most were evidently detecting me before I noticed them and were hiding or moving away unseen. In another small-scale study in the wood, I radio-tracked eight muntjac – these had been fitted with radio collars by Ian Wyllie of the ITE (as in Figure 1.3). Each time they were tracked, I attempted to see them. Initially, I expected regular success, but on average I only saw them on 21% of occasions with a range of 0–43% for individuals. In addition to illustrating how difficult deer can be to see, even when you know where they are, it also illustrates their individual nature. Kathie Claydon has recounted to me similar variations in behaviour displayed by individually tagged muntjac in the major study undertaken in the King's Forest in Suffolk (Blakeley et al. 1997).

Scent marking was described above in relation to establishing and maintaining deer paths and is covered in depth in the next chapter when dealing with territories and breeding. One has only to look at the face of a muntjac with its massive preorbital glands and large forehead glands to appreciate the importance of smell in their lives. A muntjac will often lick its preorbital glands (Figure 1.1b), which may act to dissipate scent (Chapman and Harris 1996), reinforce recognition of its own scent (Chapman 2008) or be one of a series of alarm gestures (Smith-Jones 2004).

Turning to vocalisations, their most commonly heard sound is the bark. Muntjac make a fox-like noise, less gruff than the bark of a roe. Deer that have been alarmed

Figure 2.8   A water deer in Woodwalton Fen barking at an intruder (GP).

may bark, as will females in oestrus. Barking is also used to communicate presence and may be repeated many times. Oliver Dansie (1983) reported muntjac barking at intervals of four or five seconds for 45 minutes or longer, that is, barking more than 500 times. The bark of a water deer is more of a growl, scream or grumble. I hear it most often when they are alarmed by my presence. If the animal runs into cover, the bark may be repeated many times over several minutes. Sometimes the deer will stand its ground and bark at an intruder (Figure 2.8), as will a muntjac. Other, unseen water deer will bark from cover, but it is impossible to know why. On the open fields at Whipsnade, the response of conspecifics to barking varied from indifference to immediate flight, but the most usual behaviour was staring to determine the level of any threat (Stadler 1991).

Incidence of barking by water deer has been routinely recorded since Lynne Farrell and I began our study at Woodwalton Fen in 1976. During the first three years when we visited regularly each month, there was a marked seasonal change in the percentage of barking recorded for deer that were seen (Figure 2.9). Incidence of barking was highest in the months June–August when the vegetation was at its

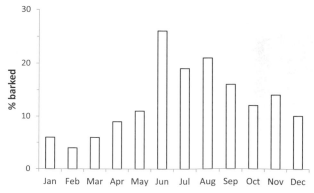

Figure 2.9   Water deer at Woodwalton Fen: the percentage of deer seen that barked during January–December, 1976–1979 (adapted from Cooke and Farrell 1981).

most dense and lowest in the winter when the vegetation had died down. Stadler (1991) reported that his impression at Whipsnade was that barking was more prevalent in the summer and could more easily be induced at night. Some of the barking may be associated with maternal behaviour as females will be tending young fawns in summer. But some could be due to the denser habitat meaning that sound is a better way of communicating danger rather than using visual signals. At Woodwalton Fen, individual water deer were more likely to bark in winters when the population was low.

Incidence of barking by muntjac was studied in Monks Wood over a 12-month period, May 1993 to April 1994 (Cooke 1998b). The seasonal trend was similar to that displayed by water deer with more barking during May–October. As the muntjac breeds throughout the year this trend in barking was more likely to be due to environmental conditions than to tending fawns.

Both species make a range of other sounds, including loud screaming when caught or trapped, while courting animals will squeak. Similar noises are made by submissive water deer and by a mother to her fawn, while very young fawns can emit a soft scream (Stadler 1991). Muntjac females utter a submissive squeak during courtship, and a fawn that has been caught squeals and bleats to attract its mother (Chapman and Harris 1996).

The deer have other ways of expressing their feelings. Muntjac that have just noticed a human may stop and stare, much as described above for water deer, but they may occasionally stamp their feet and even more rarely click their teeth. Both of these actions seem to register annoyance. In captive muntjac, bucks were found to click mainly as a threat to another buck (Buckingham and Buckingham 1985); how they did it was not clear, but the 'back part of the mouth' seemed to move. In Monks Wood during July 1993 to April 1994, I recorded the behaviour of muntjac that had seen me on 889 occasions: deer stamped on five occasions, but no clicking was recorded (Cooke 1998b). Muntjac stood and watched me walk past on only 32 occasions. Male water deer also make a noise that is described as clicking, whickering or chittering, and this can be quite commonly heard during the midwinter rut when a buck is chasing a rival. First-winter males and females have been heard making this sound and it is probably made with the molars (Stadler 1991).

Figure 2.10   Escape behaviour: (a) a muntjac jumping and flagging its tail in alarm (ML); (b) bounding gait sometimes adopted by water deer (MB).

Muntjac have the well-known habit of 'flagging' when alarmed; they move off and lift their tails to show the white underside (Figure 2.10a). I studied this behaviour in Monks Wood in 1993/4 (Cooke 1998b). Most deer that ran away flagged. However, none flagged out of those that did not move – the likelihood of tail flagging was associated with the speed of the escape. This behaviour is presumably to warn other muntjac of danger but also to inform a predator that it has been seen. When water deer perceive danger they will sometimes crouch and try to blend into the background even if there is only limited cover. On other occasions they will run or less often bound away. In their bounding gait, they fling up their hind legs in what may be their equivalent of flagging or pronking (Figure 2.10b). Bounding seems to occur more frequently in open landscapes.

Information on group sizes is given in Table 2.7. Results from China are for particular seasons whereas information from Whipsnade Zoo and Woodwalton Fen was gathered year-round. Definitions of what constituted a group varied between the studies, from animals within 10 m of one another in the high-density population at Whipsnade (Stadler 1991) to animals within 50 m at Woodwalton Fen (Cooke and Farrell 1981). At Woodwalton Fen, many animals are living a solitary existence when seen and sizeable groups form only rarely. Even when groups of several individuals are seen, they are often more likely to be chance encounters of deer coming together on favourable feeding areas. Should such groups be disturbed, the deer will probably disperse in different directions. Dubost et al. (2011b) studied social interactions in the captive water deer at Branféré Zoo in France and found: females spent less time with other deer prior to giving birth; females tended to shun the company of other females at the height of the rut; territorial males were more often with females during the rut than were non-territorial males; and there was no durable bond between mothers and their young.

Stephan Harding (1982; 1986) regarded muntjac as 'for the most part' solitary throughout the year, while to Norma Chapman (2008) they were 'basically' solitary.

Table 2.7 Group sizes of Chinese water deer in different situations. Values are percentages of the total number of groups. Chinese data are from Sheng (1992b), Whipsnade data from Stadler (1991) and Woodwalton Fen data from Cooke and Farrell (1981). See also Section 10.2.

Group size	China		Whipsnade Zoo	Woodwalton Fen
	Early summer	Midwinter		
1	84.6	55.0	34.3	77.8
2	13.0	31.9	31.8	17.7
3	1.6	9.2	16.4	3.5
4	0.8	2.7	8.4	0.8
5	0	0.5	4.6	0.2
6	0	0.5	2.2	0.1
7	0	0.2	0.6	0
More than 7	0	0	1.8	0
Number of groups	123	1,080	16,811	1,715

Table 2.8   Group sizes of muntjac. Values are percentages of the total number of groups. Data for Monks Wood are year-round, 1993/4, and data for Woodwalton Fen are for the winter of 2009/10.

Group size	Monks Wood	Woodwalton Fen
1	90.7	89.6
2	8.6	9.0
3	0.6	1.4
4	0.1	0
Number of groups	870	212

Oliver Dansie (1981) evidently did not wish to qualify the word solitary, which he defined as 'living habitually alone', and on that basis he decided that muntjac could not be described as solitary. He spent hundreds of hours watching for muntjac from a high seat in Watery Grove Nature Reserve in Hertfordshire: about 80% of sightings were of single animals, 16% were of two, and 3% were of small groups of up to five. However, my observations of muntjac seen during surveillance visits to reserves in Cambridgeshire indicate an animal more likely to be seen alone (Table 2.8), possibly because of the comparatively superficial nature of my observations while moving slowly at dusk. I have selected two situations when muntjac sightings were especially numerous; the relative frequencies of group sizes were very similar. One strand of evidence cited by Oliver Dansie for muntjac being relatively sociable was that ear-tags used to mark individuals were often chewed – suggesting periods of mutual grooming. And I agree that, when two muntjac are seen together, there are more likely to be bonds between them than when two water deer are together. Two muntjac together are usually an adult buck and doe, two does or mother and her young one (Chapman and Harris 1996). Two adult females together are likely to be mother and daughter, as they often have overlapping ranges.

In both species, mothers groom their fawns (Chapman and Harris 1996; Zhang 1996; 1998), but it is only in the muntjac where there is mutual grooming by an adult pair. Such grooming is more frequent and intensive when they are moulting into summer coat. In water deer, any physical contact between adult bucks and does is extremely limited outside the rut.

2.8   Habitats

Muntjac prefer habitat that is dense and varied at ground level (Chapman and Harris 1996; Chapman 2008). They are only able to run fast in short bursts so are less often encountered far from cover during daylight hours. Habitats in this country with the densest populations include deciduous woodland with a diverse under-storey and young plantations not yet thinned. They seem to habituate readily to the presence of humans and colonise well-vegetated and overgrown habitats in towns and cities such as alongside railway lines or in cemeteries and mature gardens. This behaviour has meant their distribution on a 10 km basis appears continuous over a significant tract of England, including major cities such as Birmingham and

Figure 2.11   Muntjac habitat: (a) Putuo Island, China (NC); (b) Hapen Reserve, Taiwan (NC).

central London (Figure 5.3). In contrast, water deer living in this country prefer the countryside, generally living in wetlands, on arable and grass farmland, and also in woodland and scrub (Cooke and Farrell 1998). Their densest unmanaged populations occur in the Norfolk Broads and on Woodwalton Fen in East Anglia in sites where the habitat closely resembles where they occur or once occurred in China. Discussion on habitat use in this country is developed more fully in Chapter 4. The remainder of this section concentrates primarily on habitat preferences in their native ranges.

Reeves' muntjac are adaptable and live in dense shrub-lands and forests in temperate and subtropical zones (Timmins and Chan 2016). The species was reported as living in hilly areas with dense cover in China, but was rarely found in high forest or grassland (Sheng 1992a; Figure 2.11a). Snow cover placed a northern limit on its range with mortality occurring if snow was deeper than 10 cm and lasted for more than ten days. In Taiwan, McCullough et al. (2000) reported that muntjac occurred in most of the broadleaved forests in the mountains of Taiwan in a zone above intensive agriculture but below the winter frost line. They studied a population in virgin forest near Small Ghost Lake. Regional climate tended to be tropical with cooler winters and hot wet summers, but in the study area summer temperatures were ameliorated by frequent fog. Signs indicated muntjac lived in most of the available microhabitats and concentrated their activities in dense tangles of vegetation – such as bamboo, shrubs and herbs – growing in gaps in the forest. Chiang (2007) reported that, in the largest remaining lowland primary forest on the island, muntjac were most frequent in habitat with a diverse under-storey on gentle slopes at low altitudes (Figure 2.11b). The photographs from China and Taiwan are strikingly dissimilar from muntjac habitat in this country in their luxuriant remoteness and extent.

When Robert Swinhoe first described water deer in 1870, he included a brief account of their natural history and his experiences of them. He reported them from large riverine islands above Chinkiang (Zhenjiang) on the Yangtze. They lived among the tall reeds, but moved to the mainland when the reeds were cut for thatching in the spring – and departed for cover in the hills. When the vegetation had regrown in the autumn, deer returned to the islands with their young and overwintered there. These populations disappeared many years ago and water deer are now found in only four areas of China: the Yancheng coastal area in Jiangsu, about 250 km to the north of Shanghai; the Zhoushan Islands of Zhejiang in the mouth of the Yangtze River; around Poyang Lake in Jiangxi, about 500 km to the south-west of Shanghai; and in the eastern parts of Anhui Province. As can be seen in Figure 1.2, there are huge distances between these remaining populations. When species' ranges contract in this way, the habitats in which they live now are not necessarily where they lived in the past, but at least these are traditional areas for them and it is worth examining what is known about their current habitat.

The Yancheng Nature Reserve extends to 4,500 km² and is, therefore, on a totally different scale from reserves in this country. Several spatially separated groups were found along a 200 km stretch of coast occupying mainly areas of short grass with taller grass for cover (Xu et al. 1996). In a later study, deer tended to select dense areas with tall vegetation that were fairly close to water but well away from human disturbance (Zhang et al. 2006; Figure 2.12). At Dafeng, there is a famous State Nature Reserve for Père David's deer, the milu *Elaphurus davidianus*. Zhang and Zhang (2002) reported water deer took to wetland in the Dafeng Reserve during daytime to avoid human disturbance or remained hidden in tall grassland; they then fed in wheat or rape fields at night.

On the Zhoushan Archipelago, Sheng and Lu (1984) undertook a preliminary study in order to understand the basic natural history of water deer. They reported that a feature of the species' survival was an ability to swim several kilometres between islands, especially to escape from heavy hunting pressure. Their ideal habitat was described as localities where cogon grass *Imperata cylindrica* was abundant and shrubs were sparse. They were also evidently a problem by feeding on farmers' fields. Wang and Sheng (1990) researched habitat selection on the islands and found the deer avoided various types of forest, but did occur in 'shrub-grass' and 'high grass'. Guo and Zhang (2002) surveyed 25 islands in the group for signs of water deer and reported that the major positive factors were large size of island and proximity to the main island (Zhoushan Island) or the mainland of China. The principal negative factor was poaching.

When Sun and Sheng (1990) studied the populations at Poyang Lake in the late 1980s, water deer occurred in three main areas. The lake lies to the south of the Yangtze River and is the largest lake in China during the rainy season. At that time the deer were forced by flooding to move into the surrounding hills where food was

Figure 2.12   Water deer habitat, Yancheng Nature Reserve, China (NC).

less abundant and had lower nutritive value. As a consequence, deer raided farm fields and many were killed by farmers. In winter, deer could return to lake-shore grassland, but much was cut by the farmers, leaving the deer struggling for food once again.

The population in Anhui Province seems the least numerous of the four listed. A preliminary survey by Chen (2012) discovered relatively sparse evidence of presence.

Less is known about water deer on the Korean peninsula. In South Korea, they were said to occur in a variety of habitats by Won and Smith (1999), preferring forested areas and early successional vegetation and being found in the lowlands of montane regions and in riparian areas. They were also found in agricultural areas, which brought them into conflict with farmers (Kim et al. 2011; Jung et al. 2016; Chun 2018). Jung et al. (2016) developed a habitat suitability index which demonstrated the deer avoided roads, preferring quiet areas with wetland and forest. They were said to be an 'edge species', using forest cover during the day but emerging to feed in open areas, including farm fields, during the night. Water deer avoided areas of high human population density, particularly the Seoul conurbation. Habitat selection in North Korea seems not to have been assessed in part because frequent introductions confound knowledge of natural distribution (Harris and Duckworth 2015).

Certain features crop up in several of these accounts of water deer habitat in their native range, such as avoidance of areas disturbed by man unless they are driven to raiding farm fields. Cover in which to shelter and hide was provided in a range of habitats from coastal landscapes to forests. Deer were accustomed to making seasonal or diurnal movements for reasons of avoiding environmental or anthropogenic dangers or moving between food and cover.

The name water deer suggests a liking for wet places and, while this is true, they are not often seen in water. Videos and photos exist of them swimming, both in their native range and in this country (e.g. Chapman 1995; Anon. 2013). They are evidently good swimmers, taking to water in the Yangtze (Swinhoe 1870) and the sea (Sheng and Lu 1984). A boatman on Horsey Mere in the Norfolk Broads told me in 2017 that he had twice seen water deer swimming across Meadow Dyke, a navigable waterway connecting the Mere to Heigham Sound. These deer were moving between areas of prime habitat. I have only once seen a water deer swimming and that was in the canalised river running along the eastern edge of Woodwalton Fen; in this situation, there is little reason for a water deer to leave the sanctuary of the reserve on a regular basis for the farmland on the other side. Simon Leatherdale, who was Forest Officer in charge of Sandlings Forest District on the Suffolk coast, recounted to me an unusual incident witnessed in December 2013. A storm surge had caused sea water to encroach far into Dunwich Forest. He saw three water deer in the inundation, including an individual that was apparently browsing on flooded gorse *Ulex europaeus*. When he approached, it swam away through water that was about 1.5 m in depth.

Muntjac will also swim if there is a need (Anon. 2013), but photographic evidence of such behaviour seems very rare (Figure 2.13). Norma Chapman told me that, when she visited the Zhoushan Islands in China, she learnt that muntjac do not swim between the islands, unlike the water deer. As a consequence, muntjac have been wiped out by hunting on several islands in the archipelago. There is some

Figure 2.13   A mother muntjac and her fawn cross a deep
fenland ditch in Norfolk (camera trap photograph).

disagreement over the degree to which muntjac drink. Oliver Dansie (1983) stated that dew is insufficient and they drink 'deeply and regularly', whereas Charles Smith-Jones (2004) noted that they drank only occasionally. The frequency of slots in the mud around woodland ponds can demonstrate that such a feature may be well visited, but many woods with muntjac do not have standing water.

2.9   Food and feeding

Hofmann (1985) and Putman (1988) discussed the food of deer in relation to the structure of their digestive systems. All deer are ruminants, but some species are better at it than others – and neither the muntjac nor the water deer is very proficient. They are small deer with relatively simple guts and are poorer at processing plant fibre than a species such as the fallow deer. They rely more on selecting highly nutritious morsels rather than grazing mouthfuls of grass. Water deer can be found in landscapes dominated by grasses, but close examination shows they are selecting the tips of blades not cropping the grasses close to the ground. Food ingested by deer goes first to the rumen where it is stored until the animal rests and ruminates. Then boluses of food are returned to the mouth where they are chewed more finely and passed down the digestive system.

In this book, I use the word browsing to describe feeding on the leaves, shoots, nuts and fruits of woody trees and shrubs including species such as bramble *Rubus fruticosus* and ivy. Grazing is used for feeding on non-woody species such as herbs, grasses, sedges, rushes, ferns and fungi.

Harris and Forde (1986) monitored the diet of muntjac in the King's Forest throughout the year. The forest consisted of older plantings of coniferous species with broadleaved shelters and an understorey of bramble, raspberry *Rubus idaeus*, grasses and ferns (Figure 4.2a). Bramble and raspberry were most important, accounting for 30–40% of the diet in each month. Amounts of other food items varied seasonally. In the winter and spring, ivy and ferns were important, with increasing amounts of grass being eaten. The main time for grass consumption was around March and April as species were growing rapidly (Figure 2.14a). In summer, the foliage and stems of broadleaved trees and shrubs became important. And in autumn, muntjac took advantage of fungi and treats such as acorns, beech mast or sweet chestnuts. Had this been a deciduous wood, then there would have been a summer peak in the consumption of forbs (broadleaved herbs) and more noticeable intake of coppice regrowth and tree seedlings. In this study, there was little raiding of crop fields but some young barley plants were consumed in winter and small amounts of grain were taken in August. Diversity of food items was highest in July and lowest in December. Muntjac were found to be predominantly browsers and the authors noted the similarity between the diets of muntjac and roe deer in forests in southern England. Where the two existed together, there were subtle differences, with roe eating more conifer leaves and ferns in winter and ranging more widely, while muntjac ate more grass (Chapman et al. 1993; Chapman and Harris 1996).

Muntjac cause few problems on crop fields; they may raid the edges of fields of beans or oilseed rape adjacent to a wood (Cooke and Farrell 2001b; Section 11.2), but their bite size is too small to tackle carrots, potatoes or sugar beet (Chapman and Harris 1996). Smith-Jones (2004) noted that when crops are tall, muntjac may take advantage of the cover and food, and live in the fields, with maize being particularly favoured. Once discovered, food is also readily taken from pheasant feeders. Charles Smith-Jones has informed me that he has heard many reports of muntjac taking the eggs of ground-nesting birds – see Dolman (2011) for examples of nestling predation. It would be unwise to assume they are totally vegetarian

Radio tracking in the King's Forest demonstrated that muntjac typically have five active periods of about 3–4 hours interspersed with periods of 1–2 hours when they rest (Forde 1989; Chapman et al. 1993). During the active periods they will be mainly foraging; ruminating will be done when the deer are resting. In Taiwan, muntjac were found to associate with Formosan macaques *Macaca cyclopis* to take advantage of fruit and foliage dropped by the monkeys (Chiang 2007). Unlike the completely diurnal macaques, muntjac were more active at dusk, dawn and during the night, possibly to avoid diurnal predators.

Figure 2.14   Grazing at Woodwalton Fen: (a) muntjac eat more grass during the flush of growth in the spring (KL); (b) a mother water deer (centre) and her two young grazing together in November (GP).

On the basis of what they eat and their anatomy, muntjac are concentrate selectors (Chapman 2008). At the other extreme of the ruminant spectrum are domestic cattle, which are non-selective roughage feeders. Between these extremes are 'intermediate' feeders that are more adaptable; and water deer were assigned, on this basis, to a position between concentrate selectors and intermediate feeders (Hofmann 1985; Putman 1988).

Guo and Zhang (2005) investigated feeding of water deer living on the Zhoushan Archipelago by observing feeding signs in the wild, examining faecal pellets and undertaking feeding trials. This was an area with 39% of forest cover. In the first part of the work, identification of plants that had been grazed or browsed revealed that 60% were forbs, 30% were woody species, 8% were ferns and 2% were grasses. The deer focused on the tender tips of plants. Evidence was also found that deer were visiting paddy fields and crops of sweet potato and peas to feed. Faecal analysis showed that forbs and woody species were the main foods, with more of the latter being taken in winter. Legumes and ericaceous species were among those most commonly taken. These authors also referred to studies undertaken by Yongbei Wu at Yancheng Nature Reserve, where 87% of the diet was forbs with some grasses and 13% comprised woody species. Zhang and Zhang (2002) reported that water deer based in the reserve at Dafeng visited wheat and rape fields at night to feed. Beside Poyang Lake, the diet was primarily terrestrial and aquatic plant species (Sheng 1992b). There is little specific information from Korea on what water deer eat, despite the fact that they damage agricultural crops in some areas (Jung et al. 2016). They were eating leaves of water lilies in an Internet film of them swimming in the Demilitarised Zone.

Walking around Woodwalton Fen and surrounding areas at dusk, most water deer that are out in the open appear to be delicately grazing grasses or sedges (Figure 2.14b), but this tells only part of the story. The deer evidently enjoy a broader diet when in cover away from the rides and open fields. Lynne Farrell and Tony Mitchell-Jones examined the rumen contents from six deer from Woodwalton Fen, and single deer from Holme Fen and Monks Wood found freshly dead during the winter and spring 1977–1980. Rumen contents were expressed as percentages of drained wet weight. At Woodwalton Fen, browse species made up 33% with the main components being bramble leaves 15%, buckthorn *Rhamnus cathartica* leaves 6%, buckthorn berries 3%, ivy leaves 3% and sallow (*Salix* species) leaves 3%. Graze species were 63% with grasses and sedges making up 48%. Less than 5% of fragments remained unidentified. Rumen contents of the other two deer were rather different with the individual from Holme Fen having eaten 79% bramble and 19% ferns, while the Monks Wood animal had consumed only meadowsweet *Filipendula ulmaria*. This small sample reflected the broad range of plant species that might be taken and showed how the ratio of the main dietary components could change from deer to deer and probably between days and seasons. Deer feeding on crop fields just outside the Woodwalton Fen reserve occasionally caused minor, but observable, damage to carrot fields (Cooke and Farrell 1987) and to growing oilseed rape (Figure 11.3). Often, however, their primary focus was on weeds growing around the field edges. When arable fields were converted to grassland, they gathered to graze on the new flush of grass, especially during February–April (Cooke 2009a; Figure 11.2).

Endi Zhang (1996; 2000a) reported that water deer at Whipsnade Zoo had two periods of feeding during the day, one in the morning and one in late afternoon. Feeding bouts, which were interspersed by periods of ruminating, were on average longer in males (22 minutes) than in females (16 minutes). Water deer in the main study area at the zoo faced a restricted diet as regards availability of woody species. Zhang (1996) examined rumen contents from 11 deer in April and June 1994. The sample contained 66% of grasses, 22% of forbs, 4% of woody species with unidentified fragments amounting to less than 8%. Hofmann et al. (1988) examined rumen papilla morphology in 58 water deer culled at Whipsnade and considered that 'seasonal rumen development was poorer than expected perhaps indicating that this population was undernourished'. This could be the explanation for the small size of water deer at the zoo. Although some of the studies into the composition of food eaten by water deer have shown high levels of grass intake, it seems as if this may not be an ideal diet. The conclusion seems to be that they eat different diets depending on what is available, but select the most easily digestible parts of each food plant.

2.10 Mortality

In China there was intense hunting of muntjac for skins with about 650,000 being killed annually out of a total population of over 2,000,000 (Sheng 1992a). This, not surprisingly, resulted in populations dominated by young deer. In a sample of 363 hunted in 1984/5, 88% were aged up to three years, with the remainder being aged up to nine years. Muntjac made up 32% of the diet of leopards *Panthera pardus*, but the decrease in numbers of this big cat meant relatively few muntjac were being killed by them. Hunting of muntjac by man has since been made illegal, but it remains one of the most heavily hunted larger mammals in the world (Timmins and Chan 2016).

The situation with hunting of water deer in China was somewhat similar. Population dynamics of deer on Zhoushan Island were studied in the winter of 1982/3 in order to understand better how the deer could be exploited (Sheng and Lu 1984). Demography was even more skewed than in the muntjac, with 48% of culled deer being in their first year, 36% in the second year and the remaining 16% being aged up to five years. Hunting pressure was judged to be 'rather high' with 70% of yearlings failing to survive to adulthood, and a ban on hunting in midwinter was suggested. The species has been protected for some time in China (Zhang 1996) but poaching continues, both of adults and fawns. In Yancheng Reserve, there has been considerable mortality from both poaching and tidal inundation. For instance, Zhang (1994) reported that more than 100 were drowned when a typhoon struck the coast in September 1990. Xu and Lu (1996) modelled the population to assess the likelihood of extinction from these twin threats. If both threats continued, extinction was predicted within a few decades – further protection was needed. Hunting for use as food and in traditional medicine has caused population declines on the Korean peninsula in the past (Won and Smith 1999). Currently in South Korea, as much as one-third of its considerable water deer population may die annually from hunting and as a result of road accidents (Chun 2018).

Generally, muntjac deer in this country are healthy and free from disease (Chapman 2008). They carry lice and ticks, including the sheep tick *Ixodes ricinus*, the vector for Lyme disease – but parasite burdens are not usually high (Figure 2.15). There is no native wild predator capable of killing an adult muntjac; the principal causes of mortality are stalking and road traffic. In 1995/6, the national cull total was estimated to be 11,000 (Macdonald et al. 2000), which probably represented more than 20% of the total population. The current total will be much higher. An undated report from the Deer Initiative, *Deer on Our Roads*, states that there are 34,000–60,000 road collisions with deer per annum and 25% involve muntjac, which equates to 8,500–15,000 deer. Not all deer will die as a result of the collision, but the Deer Initiative considered that traffic accidents were the second most common cause of death of muntjac in this country after stalking. Hard winters, such as occurred in early 1947 and 1962/3, with persistent snow and/or cold conditions have also caused significant mortality among our muntjac populations (Chapman et al. 1994a). However, such extreme winters have not been experienced for more than 50 years. Starving deer in the large population at Monks Wood suffered unusually high mortality in 1991 and 1994 during fairly typical winter weather (Section 8.2.2). Red foxes *Vulpes vulpes* are probably the most important predators of fawns, but numbers are widely controlled, including in areas where muntjac populations are also controlled. Similarly, muntjac of all ages may be killed by domestic dogs.

Demographic data for muntjac from southern England indicated 56% mortality by one year of age, 69% by two years, 75% by three years, 81% by four years and 88% by five years (Chapman 2008). These data point to high mortality in the first year, but annual survival rates averaging 73% for the following four years. For female muntjac killed on roads through the Breckland Forests, average age was three years and 20% were found to be at least five years old (Chapman and Harris 1998). One female is known to have lived for 13 years in that area and another lived for more than 20 years in captivity.

As with muntjac, water deer are frequently killed on our roads and by stalkers. Less than 1% of traffic collisions with deer involved water deer (Langbein 2011), suggesting that the total killed per annum was in the low hundreds. The number shot is unknown, but will be increasing as the species spreads, particularly into more open habitats. Very hard winter weather can cause considerable mortality.

Figure 2.15   A muntjac buck enjoying the attentions of a magpie
Pica pica, which appears to be ridding it of parasites (KL).

Whitehead (1950) recounted the problems at Woburn Abbey during the early months of 1947, when it was feared at one time that all the water deer might die. The deer refused to eat hay and most failed to recover when found in an emaciated condition and put in a yard with chopped roots. Bucks were especially sensitive, possibly because of their earlier exertions during the rut. Ted Ellis, the Broadland naturalist, told me that water deer emerged from cover during hard weather in 1979 in a starving condition. Nevertheless, they seemed to manage rather better than muntjac during cold conditions in 2009/10, one of the worst winters in recent years.

There was also a major mortality incident at Whipsnade Zoo during December 1933 and January 1934 (Middleton 1937). The population had originated from introductions between 1929 and 1930 and increased to about 200 by late 1933. The number of deaths was estimated at 140 and a lorry was needed to collect the carcases. Deer had been seen moving slowly and erratically and some appeared to have paralysed hindquarters. Two dead deer were examined at London Zoo, where cause of death was diagnosed as enteritis. Several were examined at Whipsnade and found to contain large numbers of nematodes. Generally, park deer carry greater burdens of parasites than wild deer (Chaplin 1977). Water deer are prone to capture myopathy in which stress can lead to muscle damage and death (Cooke and Farrell 1998). Myopathy can be caused by other factors apart from capture, and dead deer have been found at Woodwalton Fen that appeared healthy apart from being dead!

In the early study at Woodwalton Fen, the main periods of mortality coincided with water deer being forced to move on to the farmland because of severe weather causing periods of snow cover or flooding in the reserve (Cooke and Farrell 1981). Over four years, 28 dead deer were recorded with 64% being found during the months of January–March. Of the 16 deer found dead during the hard winter of 1978/9, four were adult females, seven were adult bucks, two were first-year bucks and three carcases were incomplete. Thus, there was no evidence of young deer being especially vulnerable in bad weather. Of 11 deer from Woodwalton Fen, whose ages were estimated from dental cementum layers, two were in their first year, one was second year, five were in their third year, one was in its fourth year and two were in their fifth year. Potential longevity was given by Dubost et al. (2011a) as 11 years.

Loss of adult deer was computed for each calendar year in the study at Woodwalton Fen. Loss represented adult deer leaving as well as dying, and if one adult replaced another, this would register as no net change. Average loss for the first 40 years was 21%, varying between 1% and 42%. Annual death rate in the intensively studied population at Branféré Zoo for deer of one year or older was 20%, and some deer reached eight years of age (Dubost et al. 2008). However, losses up to one year of age in this population amounted to about 70%. Of deer that died between one and eight years old, 25% had signs of advanced periodontal diseases. Over the years at Woodwalton Fen, several mature bucks had splayed front legs which hampered their ability to walk or run, and at least one individual knelt in order to graze.

It will have become clear from these statistics that water deer do not generally enjoy long, healthy lives even if they have the security of a zoo or nature reserve. Muntjac seem to have the potential to live longer, but often this is not realised. Both species are unusual as the males grow tusks that they use for fighting and, in the case of muntjac, for defending themselves against predators. Water deer are vulnerable to attacks by dogs. At Woburn they have been caught by dogs without a chase, seeming to become mesmerised by their killers (Whitehead 1950). Unlike other deer, they do not seem to fight back or struggle if caught. Muntjac may also be chased and attacked by dogs, but bucks in particular can come off better in such a confrontation. Because of their small size, both species require readily digestible food, which if unavailable can apparently lead to poor growth or die-offs.

CHAPTER 3

Breeding

3.1 Muntjac

British naturalists, including those in armchairs, are familiar with the noisy and conspicuous rutting gatherings of larger deer species such as red or fallow deer. The ruts of muntjac and water deer are very different. These are relatively solitary animals and they do not have spectacular events where males fight to retain and mate with harems of females as they come into season. Muntjac in particular tend to remain dispersed rather than clump together when males are rutting; and a muntjac population ruts at all times of year depending on the reproductive cycles of the individual females. The word 'rut' is defined as the mating season of ruminant animals, and deer in particular – but it is debatable whether it can be applied to muntjac. Nevertheless, it is used here to describe breeding strategy and processes up to and including mating.

3.1.1 *Rutting and territorial behaviour*
Most of the information on breeding in muntjac comes from wild and captive populations in Europe rather than from China, where it has been largely neglected by researchers.

As stated above, muntjac are unusual in breeding throughout the year. No tendency to favour any particular season has been observed either in China (Sheng 1992a) or in Britain (Chapman and Harris 1996). Bucks are fertile throughout the year, even when they have just cast their antlers or are in velvet (Chapman and Harris 1991). After a gestation period of about seven months, females give birth and come into oestrus a day or two later (Dansie 1983; Chapman and Harris 1996). If they do not conceive again on that occasion, they enter oestrus every two weeks until they do. So females can be pregnant for nearly all of their adult lives. Sexual maturity occurs at five to seven months in China (Sheng 1992a); and at about seven months for does and nine months for bucks in this country, with time to attain maturity depending more on rate of growth than actual age (Chapman and Harris 1996). Threshold weights seem to be about 10 kg for does and 12 kg for bucks.

A home range is the area of land occupied by a deer during its normal activities over a period of time. A territory is that part of the home range that is defended. Bucks tend towards a solitary independent existence. They are terri-torial throughout the year although they can be reasonably tolerant of one another, such as when sharing communal feeding areas (Smith-Jones 2004). I once saw four bucks in the mid-1990s all behaving amicably in Monks Wood in an area of about a quarter of a hectare. Average range size for bucks around that time in Monks Wood was 16 ha and was similar in Rockingham Forest (Wyllie et al. 1998), while in the mainly coniferous King's Forest, average home ranges for two-month periods were 20–28 ha (Chapman et al. 1993). Females tended to have smaller home ranges with a greater degree of overlap, range size being 6 ha in Monks Wood, 13 ha in Rockingham Forest and 11–15 ha in the King's Forest. In contrast, average home range in Taiwan was much larger at 108 ha (McCullough et al. 2000).

Home ranges of does in this country can be remarkably stable both seasonally and from year to year (Chapman and Harris 1998). There is no pair bond (Chapman and Harris 1996), but there is some evidence that bucks may favour the oldest doe that has given birth to the most fawns, where they have access to more than one doe (Dubost 1971). Adult bucks will aim to mate with those females whose ranges overlap with theirs, the system being polygynous. And conversely a doe may mate with more than one buck, although there is a lower chance of this happening since her range is likely to be smaller. Information on sex ratio is equivocal. In the King's Forest, the ratio of females:males was 1.7:1 (Claydon et al. 1986), while cull samples of adults were 1.1:1 in Monks Wood and 0.75:1 in Holme Fen; but males may walk further and be more conspicuous (Chapman et al. 1993), making them more likely to be shot. Sex ratio is an aspect that warrants further study.

It is to a buck's advantage to establish and maintain as large a territory as possible. Territories are held on average for 3–4 years (Chapman and Harris 1996), during which time the resident bucks continually mark them. The most common type of marking observed in a study with captive deer in China involved use of the facial glands to apply scent to the ground or low vegetation, combined with urination and/or defecation (Lai and Sheng 1993). Some observers (e.g. Harris and Duff 1970; Clark 1981) have suggested that dung on its own may serve as a territorial marker. The type of marking that I see most frequently with camera traps is when a buck dips his head and touches the frontal glands in his forehead on the ground, often without breaking stride (Figure 3.1). Sometimes this is repeated in exactly the same place over a period of days or weeks as he patrols his domain. It is a very brief subtle movement and is likely to be missed when watching muntjac in the field.

Although vigorous scraping with the forefeet was less common in the Chinese study, it can be more frequent where there are higher densities of deer (Dansie 1983). This is an obvious sign of marking in my local sites, with the vegetation and soil being raked over a patch up to a metre across. Often droppings are visible on the scraped area and presumably urine may also have been deposited. At Woodwalton Fen, such scrapes are frequently at the edge of rides adjacent to trees and shrubs such as alder *Alnus glutinosa*, birch *Betula* species, bog myrtle *Myrica gale* and hawthorn *Crataegus laevigata*. I have seen muntjac standing on their hind legs next to bushes but evidently not feeding. The first one was in 1995; an adult buck reared up on his hind legs on a scrape and stretched up into an adjacent myrtle bush. I thought that they may be anointing vegetation as high as possible with scent from their facial glands to demonstrate their size to rivals.

In 2015, Michael and Jum Demidecki positioned a camera beside a scrape next to an ivy-covered elder in their study site at Wilstone Reservoir in Hertfordshire (Demidecki and Demidecki 2016). In August of that year, the camera recorded a buck with long antlers going through a sequence three times of standing on his hind legs, then landing and scraping the ground vigorously with his fore feet. At least one of his preorbital glands was wide open, and the whole performance took almost a minute. Later, in February 2016, in a remarkable period around dusk of about 80 minutes, three different bucks were videoed standing on their hind legs at the same scrape. Females were also filmed sniffing the scrape and the ivy, but not standing on their hind legs.

Michael Demidecki alerted me to what the camera had recorded and I found a suitable scrape beside a mature birch in the mixed woodland at Woodwalton Fen. During 81 days of observation from early March until late May 2016, my camera recorded a total of 22 events when two separate bucks stood on their hind legs (Figure 3.2) and scraped the ground. These events occurred at night as well as in daylight, and one buck displayed this behaviour after shedding his antlers. Several videos showed, with reasonable clarity, the preorbital glands being used to mark small side shoots on the tree and possibly the bark. Odour profiles from the preorbital glands of both muntjac and water deer have been shown to be different for individual deer (Lawson et al. 2000). This activity could be aimed at avoiding territorial disputes and communicating social status (Demidecki and Demidecki 2016). Muntjac deer have previously been reported marking twigs with their preorbital glands (Dubost 1971) and frontal glands (Sheng 1992a), but not standing on their hind legs to do so. Although muntjac have glands in their hind feet (Dubost 1971; Chapman 2008), they do not occur in the front feet, so the act of scraping does not impart any scent. However, when on their hind legs, they have to keep moving like a stilt walker to maintain their balance and this could leave scent in the scrape, as could defecating and urinating. In addition, frequent scraping with the forefeet should help to dissipate any odour. Water deer sniffed both the scrape and the tree on rare occasions.

Figure 3.1   A muntjac buck scent-marking ground vegetation (camera trap photograph).

Figure 3.2   A muntjac buck vigorously marking a birch tree at Woodwalton Fen while standing in a scrape on his hind legs (camera trap photograph).

Figure 3.3   Fraying willow stems: (a) frayed stems in Monks Wood;
(b) a buck muntjac rubbing the base of his antlers and his frontal
glands on the frayed area (camera trap photograph).

Muntjac fray stems of coppice regrowth, saplings or shrubs to mark their territories (Figure 3.3a). *Salix* species are specially favoured and can be frayed up to a diameter of 5 cm (Cooke 1995). The process of fraying has been described in different ways in the literature. For instance, Michael Clark (1981) was able to observe and sketch fraying by captive deer. He observed the lower incisors being used to leave characteristic twisted ends of bark above the scraped patch and the frontal glands applying scent. Some other authors have, however, stated that the canines and preorbital glands are involved. Gérard Dubost (1971) studied this behaviour in semi-captive Reeves' and Indian muntjac in France and described use of the frontal glands and rubbing with the pedicles. Norma Chapman and I investigated fraying using a captive pair of muntjac and a battery of camera traps set to take either videos or photographs. Stems of willow were cut and stuck in the ground to resemble coppice regrowth. The buck began fraying within minutes of us leaving the paddock and continued through the hours of daylight and into the night. His attack was sufficiently robust to pull several stems out of the ground. He used his incisors to fray the bark and vigorously rubbed with the base of his antlers and his forehead (Figure 3.3b). There was no indication of the tusks, preorbital glands or pedicles being employed. He also occasionally inspected or smelled his handiwork, as did the doe. Fraying appears to have both visual and olfactory functions.

Bucks will also bark repeatedly to advertise or defend their territory against others (Downing 2014). People who are able to imitate the bark may be able to persuade a deer to give away its position or even entice it into the open (Smith-Jones 2004).

3.1.2   *Fighting and mating*
Having stable territories is likely to reduce physical contact between bucks and thereby minimise the likelihood of fights occurring (Harris and Duff 1970) – but fights do happen. One situation that can lead to a fight is if a female's range overlaps those of two bucks, and one of these animals follows the female into the other's territory. Barrette (1977b) and Chapman and Harris (1996) have provided a graphic account of fighting and territorial defence. Bucks demonstrate a variety of aggressive behaviour, including clicking the cheek teeth, scraping the ground, and

parallel walking. The fight consists of pushing and twisting using the pedicles and antlers. They try to throw one another off balance so the rival can be slashed with a tusk. Bucks are protected around the neck by thicker skin than is found on other parts of the body.

There are several videos on the Internet of bucks fighting and one is left with the impression that the protagonists are very reluctant to withdraw. Michael Clark's account includes a sketch of two bucks pushing and shoving with heads lowered and a comment about two bucks attempting to fight despite being on opposite sides of a fence. In Monks Wood, adjacent faeces-laden scrapes could in the past be readily found on either side of the deer fences where rivals marked their territories and possibly occasionally tried to fight. Kevin Robson photographed one buck standing over a squeaking and submissive rival at Wicken Fen in Cambridgeshire, but it was unclear what had led up to this situation (Figure 3.4).

Fighting can lead to broken tusks. In a sample of bucks 3–5 years old, 51% had broken one or both tusks (Chapman 1997a). Bucks with broken tusks or antlers in velvet are likely to be at a serious disadvantage if engaging with a fully equipped rival, and both disabilities can lead to loss of territory (Soper 1969; Chapman and Harris 1996). Fights often result in injuries. For instance, a broken tusk tip was found in the leg of a shot buck (Downing 2014), and a piece of antler tine turned up embedded in the head of another (Chapman 1996). Damaged antlers, as on the buck in Figure 2.14a, will also be a disadvantage in a fight. Three of 80 bucks examined by Oliver Dansie (1983) had scars on their necks suggestive of tusk damage, and significant ear tears were found by Norma Chapman (1996) to be roughly ten times more frequent in adult males as in adult females.

Michael Clark (1981) recounted how his 'tame' captive buck killed a water deer buck that shared his pen. The muntjac had become excited by a muntjac doe in oestrus. Tellingly, the water deer made no attempt to retaliate, not recognising his killer as a rival. There are many stories of muntjac attacking or defending themselves against dogs, often to the detriment of the dog. Again, there are a number of media stories blaming water deer for injuring dogs, but it is more likely

Figure 3.4   A dominant buck muntjac standing over a submissive rival (KR).

that muntjac were responsible in most, if not all, of these instances. In some cases, such stories emanate from parts of the country where muntjac are common, but water deer are not. And as Graham Downing (2014) has explained, care is needed when butchering a dead buck muntjac as many stalkers have 'received post-mortem retribution from the tusks'.

Bucks follow the scent of does in oestrus and mating follows a period of pursuit (Dubost 1971; Chapman 2008; Figure 3.5). However, most sightings of muntjac are of single animals, and sightings of an adult pair can be relatively rare. The best winter for muntjac sightings at Woodwalton Fen was 2009/10, when there were more than 200 encounters with one or more muntjac (Table 2.8). Of these, only about 4% were of an adult pair together. Michael Clark's book (1981) contains two drawings of a buck closely following a doe with her tail raised in typical fashion. In the first, he is trotting after her and in the second, just prior to mating, he is licking and pressing her. On several occasions, when working with Norma Chapman's captive pair, the buck has been recorded by my camera traps chasing the doe around their paddock during the day and the night. During the final stages of courtship, pursuit can result in well-marked tracks around bushes or dense clumps of vegetation (Soper 1969; Dubost 1971). Muntjac chases through woodland are usually heard before the deer are seen clearly, and when the participants are identified they are often found to be a doe and a buck. In May 2013, my camera traps recorded an amusing sequence along a deer path in Monks Wood. First, early one morning, a doe skipped down the path with a buck about five seconds behind. After five hours, the doe again ran down the path, with the buck following her scent trail about two minutes later. The final video was five days later and showed the buck pelting after the doe and only two seconds behind. There was no way of being certain that these were the same two individuals as they were always running away from the camera. But in view of the typical spatial arrangement of muntjac, I tended to assume that they were – and I have often wondered about the outcome.

As a female comes into oestrus very soon after giving birth, she is likely to be pursued by a buck while trying to care for her newborn fawn. And during oestrus,

Figure 3.5   A doe muntjac with her tail raised is pursued by an attentive buck (KL).

she may be mated a number of times. The female remains still during copulation with her head lowered and her tail raised, while the male rests his chin on her withers and licks her (Dubost 1971). Duration of copulation has been stated to be 3–5 seconds (Clark 1981) or 1–2 minutes (Sheng 1992a). Norma Chapman has suggested to me that this apparent disparity might be explained by each attempt at mounting lasting only a few seconds with the entire sequence of multiple couplings lasting a minute or two. The doe may be pregnant again within a few days of giving birth.

3.1.3   Gestation, birth and beyond

The egg to be fertilised can arise from either ovary, but implantation almost always occurs in the right horn of the uterus (Chapman and Dansie 1969; Dansie 1983). Embryonic development occurs as the mother lactates to feed her current fawn. In a flourishing population, most females will be pregnant at any one time. For instance, in a sample of 27 does in China, 89% were pregnant (Sheng 1992a). A study of uteri from 80 pregnant wild deer in England found 42 foetuses were male and 38 were female, a ratio of 1.1:1 (Chapman 2008).

In her groundbreaking study on wild muntjac in a Hertfordshire garden, Eileen Soper (1969) estimated gestation times for five fawns as 28–37 weeks. Raymond Chaplin (1977) reported gestation as taking about 206 days. More recent authors have settled for 210 days or seven months, including Helin Sheng (1992a) for muntjac in China. Norma Chapman (1993) recorded the average time between successive births in her captive population as 243 days for a total of 56 observations; the range was 210–522 days, but only three instances exceeded 280 days. Average productivity is likely to be 1.6 fawns per doe per year (Chapman et al. 1997a).

As a spot to give birth, a female muntjac will seek out an area of dense vegetation within her range but away from her usual resting places (Dubost 1971). Females can be aggressive to other deer prior to giving birth (Soper 1969). Multiple births are very rare in muntjac and in the past there has been much discussion about whether they ever happened. Graham Downing (2014) remarked that there were sufficient reports of twin fawns from reliable observers to show that this happened on rare occasions. Twins have been born in captivity in China (Sheng 1992a).

The longest-lived doe in the captive population of Norma Chapman (1993) survived to an age of 14 years, and gave birth to 22 fawns, of which 18 lived for at least one year. Perinatal losses among her deer amounted to 13 fawns out of 69 (19%), of which five also resulted in the death of the mother. Losses of fawns in captivity occur especially when does are giving birth for the first time (Dansie 1983). In an investigation by Oliver Dansie (1977) into the deaths of 168 wild deer, four mothers were found to have died during fawning. Sex ratio at birth was given by Norma Chapman as 55% males and 45% females for a total of 69 fawns, a ratio of 1.2:1. Average weight of fawns during their first day of life was 1.2 kg for a sample of 30 with a range of 0.9–1.5 kg (Chapman 2008). Weight gain in the first 6–8 weeks of life amounts to about 0.5 kg per week (Dansie 1977).

The only detailed description of the mother–fawn relationship appears to be by Gérard Dubost (1971) on captive muntjac. This particular account may be based on Indian muntjac rather than Reeves' muntjac, but the author stated that no outstanding differences existed between the behaviour of the two species. The fawn can walk after a couple of hours, but initially stays fairly immobile. Its lightly spotted coat helps it blend into the forest background and the fact that its glands do not function lessens risk of predation (Figure 3.6a). The doe returns and the fawn runs to meet her. They communicate by softly calling (Sheng 1992a). After suckling, the doe licks away its faeces and urine, again to reduce predation risk. A fawn begins to accompany its mother after several days, and she can be very courageous in defence of the fawn (Clark 1981). By two months of age, the fawn is weaned and its spots have faded (Harris and Duff 1970; Smith-Jones 2004). There is much grooming, including of the ears (Figure 3.6b). Occasional suckling may be seen up to 17 weeks (Chapman 1993).

In addition to losses associated with being born, predators such as foxes, dogs, carrion crows *Corvus corone* and stoats *Mustela erminea* pose a real risk to very small fawns (Harris and Duff 1970). In the King's Forest, about 50% of fawns were lost by the age of two months, mainly because of fox predation (Chapman and Harris 1996). Such predation is likely to be highest in winter when there is less cover and fawns are more likely to be discovered (Clark 1981; Smith-Jones 2004). However, fawns survive snow and low temperatures well in captivity. During a major die-off of muntjac in Monks Wood in 1994, it was the newly weaned and/or independent young ones aged 20–40 weeks that were especially vulnerable to starvation (Cooke et al. 1996; Figure 20.2).

Their life is, though, not totally dominated or consumed by hardship and fear. Several authors have described play behaviour in fawns (Soper 1969; Dubost 1971; Clark 1981), which can occur at any time of day or night (I have occasionally recorded a gambolling fawn on a night-time video). The fawn runs, jumps, bucks and generally capers – often around objects such as trees or other deer. Older deer, including adults, will also sometimes indulge in play; Norma Chapman has described to me a captive male which bucked and tossed logs.

She and Stephen Harris (1996) summarised dispersal from their study area in the King's Forest. When deer numbers in the forest were low, young females generally stayed close to their natal ranges and might continue to associate occasionally with their mothers. Most dispersal was by young males, often about one year old, which tended to move elsewhere within the forest. As numbers built up, however, more young females dispersed and longer movements to new areas occurred. Younger animals that moved away sometimes returned to the better habitat offered by the forest after they had matured.

Figure 3.6   (a) A muntjac fawn with its spotted coat (MC).
(b) A mother muntjac grooms her young one in cover (MB).

3.2   Water deer

The best place in this country to watch the midwinter rut was, and may still be, Whipsnade Zoo. A public footpath runs outside the perimeter fence along the eastern side of the zoo. In the recent past it was possible to see large concentrations of water deer in grass fields, some of which were not open to the paying public inside the zoo. I last watched the Whipsnade rut in December 2008 when I estimated there were about 100 deer in two fields. At the time there was constant courtship going on between males and females, many males chasing rivals and much vocalisation. Total number of water deer counted at the annual census has, however, fallen by an order of magnitude since that time (Marc Baldwin, personal communication). The best area to see them inside the zoo is in the drive-through paddock known as 'Passage through Asia'.

3.2.1   *Rutting and territorial behaviour*

The rut of water deer is to a human observer a secretive affair. Kenneth Whitehead (1950) said of the situation at Woburn Abbey, 'One would hardly realise that the rut was on.' And that was in open parkland – in a reed bed, there is little chance of seeing rutting behaviour. The rut is, though, seasonal. Rutting occurred in December and January at Zhoushan Islands in China (Sheng and Lu 1984). Lixing Sun and Nianhua Dai (1995) found that the population at Poyang Lake in China rutted slightly earlier, with males beginning courting in late October and mating being seen from late November until mid-December. Roy Harris and K.R. Duff (1970), who studied Bedfordshire park deer, reported that the rut started in early November and continued until early December. The population of water deer at Whipsnade Zoo was observed by Stefan Stadler (1991) and Endi Zhang (1996) who found that males began courting in early November with mating occurring mainly in December. Gérard Dubost and his colleagues (2011b) reported that bucks in a captive population in Branféré Zoological Park in France began territorial behaviour in late October; this peaked and mating occurred mid-November to late December. In the population at Woodwalton Fen, signs indicated that December was the main rutting month with a lessening of activity in January (Cooke and Farrell 2001a). So there seems to be variation in timing, at least some of which may be caused by differences in climate or by weather conditions in certain winters.

At Woodwalton Fen, whickering chases between bucks may occur as early as October, but it is usually late November before the bucks start to become more noticeable when setting out their territories. They do this by marking both the ground and vegetation, and this behaviour has been studied in detail at Wuppertal Zoological Park in Germany (Feer 1982), at Whipsnade (Stadler 1991) and at Poyang Lake (Sun et al. 1994). Marking the ground is the more common type of behaviour. The ground may be first inspected, sniffed and scraped with the forefeet, before faecal pellets or a few drops of urine are added, the deer being in a slightly hunched posture with its tail lifted horizontally. The whole process takes only a few seconds. Bucks indulging in such marking often walk around with their tail sticking out. They are faced with having to economise on use of both faecal pellets and urine, especially as they do not feed so much during the rut. In the Chinese study, it was shown that they reduced the number of pellets per defecation in the rut so as to increase the number of places they could mark (Sun et al. 1994). At Woodwalton Fen, the number of fresh faecal depositions noted during our surveillance walks reached a peak in January (Cooke and Farrell 2001a). At Whipsnade, it peaked during November–January and a reduced frequency of faecal marking was seen through the rest of the year, but marking the ground was rare at Poyang Lake outside the rutting season. The presence of small glands in the forefeet of water deer was observed in an early study (Pocock 1923), so scraping with forefeet probably has some olfactory purpose in addition to that of the deposited urine and faeces (a thickened glandular rim around the anus was also reported by Pocock).

Exactly how they mark vegetation is not clear. In the three studies listed above, bucks were described rubbing their foreheads 1–15 times against twigs, plant stalks, tufts of grass or pieces of wood. Forehead-rubbing was sometimes accompanied by sniffing, nibbling, gnawing or scraping the object with the incisors. This behaviour was less commonly seen than marking the ground. Forehead glands have not been reported in water deer, but Norma Chapman has commented to me that in the absence of frontal glands they may have unusually large sweat or sebaceous glands as are used by roe bucks for territorial marking (Danilkin 1995). I have a single camera trap video of such behaviour from Woodwalton Fen and that appears to show a thin sallow stem being brushed by the preorbital gland as the forehead rubs against it. Helin Sheng (1992b) referred to bucks in China scent-marking twigs or grasses with their preorbital glands. Yu et al. (2013) observed captive bucks 'forehead scrape-marking', which was described as using the infra-orbital gland (also known as the lachrymal gland) to mark fences, branches or turf.

Harris and Duff (1970) described a different type of behaviour when a buck hooks his tusk around a thin plant or very small tree stem, and moves his head up and down, rubbing the stem. The edge of the tusk marks the stem and the stem may pick up scent from the preorbital gland. In a small, closely observed breeding population in a paddock with woodland and grassland in West Midlands Safari Park, bucks marked saplings up to about 12 mm in diameter in exactly the same fashion (Lawrence 1982).

These two different descriptions of how vegetation is marked present a puzzle. Both types of behaviour involve vigorous head movements and use of the teeth, but the second refers to the tusks and the first does not. It is odd that several sets of authors reported one type of behaviour or the other, but not both – although both have been filmed on rare occasions at Woodwalton Fen. Further explanation is needed.

Sex ratios have been reported in wild and captive populations, but are variable. Sheng (1992b) and Xu et al. (1996) reported ratios of females:males of 1:0.93 and 1:0.77 for two Chinese populations. At Whipsnade Zoo, the ratio varied from 1:0.81 (Middleton 1937) to 1:1.46 (Zhang 1996), while at Branféré Zoo it was 1:0.83 (Dubost et al. 2008). At Woodwalton Fen, based on sightings of individuals in the spring when developing tusks become apparent on young males, the ratio was 1:0.92 for sightings in April and May, 1977–1979, but 1:1.05 for sightings April–June (Cooke and Farrell 1981; 2000). The average of these samples was 1:0.97.

In the water deer, females do not hold territories during the rut, but there is some evidence of them doing so prior to giving birth (Cooke and Farrell 1998). The proportion of the annual home range that is defended by a male can vary considerably depending in part on whether seasonal movement occurs. And the size of a home range can depend on whether the male is territorial. Thus at Whipsnade Zoo, the average annual home range of adult males was about 12 ha, but if they were territory holders then average range was 2 ha with a seasonal range of 1 ha (Stadler 1991). Yearlings and older females had annual home ranges of about 25 ha. In contrast, average seasonal home ranges of water deer released into Shanghai's Nanhui East Shoal Wildlife Sanctuary varied between 100 and 300 ha (He et al. 2016).

The main focus of a buck in winter is on mating, but the principal concern of a doe is on feeding. And the females, at least in some situations, are the deer that determine where territories are held. The situation at Poyang Lake was described in detail by Sun and Sheng (1990), Xiao and Sheng (1990) and Sun and Xiao (1995). This is the largest freshwater lake in China with an area in the wet season in summer of nearly 3,000 km². In the dry season in winter, it can reduce to less than 400 km², exposing large tracts of grassland. About 1,500 water deer were believed to occur at that time within the region of the lake. Studies on the water deer were based in and around Poyang Lake Nature Reserve where about 500 were counted. They were forced into the surrounding hills in summer but returned to the grasslands to feed and breed. Average home ranges in the hills in the summer floods were 18 and 30 ha depending on methodology, whereas home ranges on the grassland were 37 and 46 ha. Within one study area of roughly 1 km², deer were displaced by human activity from one grassland area, with the females leaving first. They moved to another area of grass on the edge of the lake's beach, with males following after a few days. The females rested on the quiet beach during the middle of the day, returning to feed on the grass in late afternoon and staying during the night. Males set up small territories of about 0.5 ha where the females fed. There were no territories on an area which the females avoided because their food plants grew less well. Bucks rarely left their territories. They did, however, move elsewhere if they failed to establish a territory or held a territory that females vacated. There was a positive correlation between female density and territorial stability indicating that males wanted to be where there were the most females. This was interpreted as female density playing a pivotal role in whether and where a buck might set up territory.

At Whipsnade, territory-holding bucks were described as defending an area that contained resources valuable to females (Stadler 1991). In other words, they were not defending females – and the latter were not necessarily present all of the time. It was only when females were in or close to oestrus that males attempted to keep them within their territories by herding them. At Whipsnade, female territories may be more clumped during the rut – and males with territories in such localities should benefit.

At Woodwalton Fen, water deer are found throughout the reserve at virtually all times of year. Any doe is interested in finding a quiet area with cover and good feeding. Thus a buck based in such an area will almost certainly have at least one doe in his territory during the rut. The reserve is gridded by dykes (ditches) and rides into about 50 roughly square compartments, each of about 4 ha. Many of these compartments have dykes on three or four sides and can serve as a complete territory for a buck. However, some bucks may defend another one or two neighbouring compartments, especially if the winter population is low. Elsewhere on the reserve, blocks as large as four compartments may lack any obvious internal barriers, and these sometimes have areas of fen vegetation that have been mown by the time of the winter rut. These mown areas can provide grazing for a number of bucks and does, and so also function as rutting stands. Here territories are likely to be smaller than the minimum size of 4 ha elsewhere in the reserve.

Little is known about territorial stability. Stadler (1991) stated that territories at Whipsnade Zoo were usually maintained 'year-round'. Although he provided no data on stability from year to year, he did recount observations on one particular tagged buck known as M630. In February 1987, when M630 was 8–9 months old, he was not a territory holder and 'floated about' showing typical ranging behaviour for bucks of that age. Then during the rut of 1987/8 he fought and displaced a more mature buck, and maintained that territory through the rut of 1988/9, after which time the study ended.

3.2.2   Fighting and mating

A buck lacking a territory can try to displace a rival or could perhaps establish itself between territories and gradually increase its patch (Stadler 1991). The second method is probably less viable in those areas of Woodwalton Fen where adjacent territories are separated by dykes. When a buck spots another in his territory, he is likely to walk swiftly towards it with a stiff gait, with his head up and perhaps with his tail stuck out. This is usually sufficient for the intruder to move rapidly away, probably being chased by the whickering territory holder. Chases are common and can be seen at any time of year, but peak around the time of the rut. If an intruding buck does not back down, the two may indulge in parallel walking where they size one another up by striding along 10–20 m apart. If neither backs down, then a fight will result.

Fights between water deer bucks are very different affairs from contests between our larger species of deer, in which the contestants lock antlers and attempt to push or twist their opponent. Water deer bucks dance in front of and around one another, and try to land strikes with their tusks or front legs. Although there may be more dancing and acrobatics than actual aggression, loss of a tusk (Figure 4.6) can put a buck at a dangerous disadvantage. Stefan Stadler's doctoral thesis (1991) contains two photographs of fights at Whipsnade, involving an attempted front leg strike and two bucks crouching and in contact. At a semi-natural site such as Woodwalton Fen, fights are rarely seen because they are brief, usually lasting less than a minute, and often occur in deep cover or at night. Local wildlife photographer, Gary Dean, was in the right place at the right time on 19 December 2009 to take a sequence of photographs of a fight at the height of the rut (Dean and Cooke 2015). The reserve was covered at the time by a blanket of snow about 10 cm deep: season and conditions were perfect that day. The fight was on a ride fringed by tall reeds. It lasted only 12 seconds, during which time Gary took ten photographs. One buck feinted to go clockwise, but then cavorted in the other direction around his more static opponent. They both attempted to stab one another. Immediately after the last photograph, the deer separated with one moving into the reeds and the other walking away down the ride. I know of no other fight on the reserve that has been seen, let alone photographed. Had the deer been just off the ride, then they would have been invisible in the tall dense reeds.

In this country, there is a greater chance of witnessing a fight in more open habitat such as arable fields or grazing marshes. Mike McKenzie took the amazing photographs in Figure 3.7 at Claxton close to the south bank of the River Yare in Norfolk in February 2013. The whole affair lasted about a minute. One buck seemed to stumble into another's territory and was initially chased off. The first one, however, turned and a fight began with them spinning round trying to land blows with their tusks. The territory holder saw off the intruder. There is also a video on the Internet of several altercations fought out in the Demilitarised Zone between North and South Korea.

As with other deer species, water deer do not always escape from these fights without serious injury, and some fights end in death. On one occasion, Stadler (1991) found a dead buck at Whipsnade that had been stabbed through the heart by a tusk. He also recounted one particularly vicious fight that left the protagonists dazed, bloody and limping. Bob Lawrence (1982) described an extreme case at West Midlands Safari Park of a vanquished buck lying on his back squealing like a pig, while his rival stood over him. Bucks carry scars and other signs of fighting, the

Figure 3.7   Photographs depicting a fight and subsequent chase between rival water deer bucks: (a) they circle attempting to stab one another with their tusks (MM); (b) the territory holder chases off his rival (MM).

Figure 3.8   A tuft of hair lost in a fight between rival male water deer.

most frequent and obvious damage being to their ears. Such injuries range from small nicks to large tears with ears almost being torn in half (e.g. Figure 19.3c).

Analysis of surveillance information from Woodwalton Fen during the three winters 2010/11–2012/13 revealed that 20% of deer seen with long tusks had ear damage (Cooke 2013a). During the same period, just 1% of females and young bucks without visible tusks had ear damage. This supports the contention that virtually all ear damage is caused by older bucks fighting rather than via snagging ears on thorns or barbed wire. It also suggests that first-year bucks rarely attempt to fight. The incidence of ear damage was higher in winters when deer density was greater, which had probably led to more intensive fighting. Checking back through records to 1990, I found 78 instances of damage to the left ear, 41 to the right and 7 to both ears. The damage to left ears was greater than could be explained by chance implying that deer behaviour was leading to a greater probability of injury to that side. A left-sided injury is more likely to occur if, when bucks confront one another, they tend to circle and dance in an anticlockwise fashion, as in Figure 3.7a. Both of Stefan Stadler's photographs showed deer moving in this way, as did the majority of Gary Dean's photographs and the videos of fighting deer in Korea.

When bucks fight, they frequently lose hair. Tufts on the ground give away the location of fights or chases where the retreating buck has not escaped scot-free (Figure 3.8). The coarse hair is hollow and about 40–55 mm in length. It is whitish over most of its length, giving way to a dark brown band and finally being pale brown, buff or ginger at its tip. At Woodwalton Fen, where hair might be found in perhaps 10–20 locations per winter, the frequency of finding newly lost hair generally peaks in December (Cooke and Farrell 2001a). Clumps of hair can remain obvious in short grass for months afterwards. When the winter coat is shed in the spring, it is usually lost as individual hairs which are much less obvious and unlikely to be confused with hair lost when fighting. Hair tufts at Woodwalton Fen are often found near to where two compartments are joined by a bridge or ride, suggesting an intruding buck has taken on a defending deer just inside the latter's territory. In larger mown areas where several males are active, finding hair on the ground is a relatively rare event, indicating bucks quickly establish a hierarchy and so minimise fighting. But in such situations, chases are commonplace.

While males are engaged in holding territories, females contentedly carry on feeding – but their peace is punctuated with increasing frequency as the peak of the rut approaches. When advancing on prospective mates, bucks lower and stretch out their necks and rotate their heads to produce an almost comical slapping movement of their ears (Figure 3.9a). They often utter a whistling noise when engaging with a doe; Stadler (1991) recorded whistling at Whipsnade as early as September, two months before the start of the rut. Prior to entering oestrus, a doe will often run away from the advances of a buck (Figure 3.9b). Bucks will monitor the oestrous state of a doe by lifting their upper lip to test the air, a process known as 'flehmen' (Figure 3.9c).

Eventually, when a doe comes into oestrus, she will usually accept a male's advances and mating is likely to result. The couple will stay together for a number of hours and mating may take place several times. Zhang (1996) recorded six instances of successful mating at Whipsnade; they all occurred in December and each lasted only 4–9 seconds. In the population at Branféré Zoo in France, only 11 matings were seen from mid-November until mid-December in 317 hours of observation, with duration ranging 1–17 seconds (Dubost et al. 2011b). Sun and Dai (1995) reported 15 instances at Poyang Lake involving seven different pairs from late November until

Figure 3.9   Water deer courting: (a) a buck approaches a doe rotating his head and slapping his ears (MB); (b) an unreceptive doe being pursued by a buck (MB); (c) a buck attempting to detect the scent of a doe in oestrus (MB).

mid-December, each lasting 4–13 seconds. Using a camera trap, I managed to video mating twice at Woodwalton Fen on the night of 23/4 December 2014 (Section 3.2.3); this probably involved the same pair and mating times were four seconds and at least three seconds (the ten-second video stopped while they were still mating). A pair filmed by Stephen Plummer (personal communication) in Bedfordshire on New Year's Day 2012 mated for three and ten seconds.

Sheng and Lu (1984) stated that rutting behaviour occurred in December and January on Zhoushan Island off the coast of China. Stadler (1991) did not specify in which months he observed mating at Whipsnade, restricting his comments to rutting happened 'about December'. Gérard Dubost and his co-workers (2008) calculated from dates of birth and mean gestation period that dates of mating continued into early January in their captive population in France. Furthermore, they found during the three-year study that numbers of matings were highest during the coldest weeks of the rut. In the exceptionally mild winter of 2015/16, mating at Woodwalton Fen almost certainly occurred in January although this could not be confirmed. Earlier analysis of data at the reserve showed that the frequency of rutting signs in January was second only to that in December (Cooke and Farrell 2001a).

Another study on the deer at Branféré in France involved linking analysis of sex hormone levels in faeces to behaviour (Mauget et al. 2007). Female water deer had oestrus episodes during the early stages of the rut but were eventually sexually receptive for only a few hours. The brief window for successful mating could explain why the males constantly monitor females from several weeks before the event. Ideally, a male must sniff out the onset of oestrus in each female in his territory. High frequency of associated behaviours displayed by the bucks was concomitant with the highest faecal hormone levels. The authors speculated that such a mating system may be very prone to failure at low deer densities. In a similar study on captive deer in China, Yu et al. (2013) found that several types of rutting behaviour by bucks were most frequent when faecal testosterone levels were high.

Chinese water deer reach puberty in this country at 4–7 months (Chaplin 1977); in China, time of puberty was given as 5–6 months for males and about 8 months for females (Sheng and Lu 1984). However, younger bucks are unlikely to be as successful as mature individuals, especially if overall density is high. Both Stefan Stadler and Endi Zhang observed first-year males courting older females at Whipsnade but without any sign of successful mating. However, the mating videoed by Stephen Plummer on an arable field in Bedfordshire in 2012 appeared to show a young buck successfully mounting an older doe. Zhang (1996) found 5 out of 15 regularly observed young females showed signs of late pregnancy. First-winter bucks in the less crowded environment of Woodwalton Fen have apparently occasionally held compartments, but their success at mating remains unknown. Stefan Stadler (1991) commented that only a small proportion of yearling males held territories at Whipsnade.

3.2.3   Camera trap studies at Woodwalton Fen
At Woodwalton Fen, camera traps were used during the three winters, 2013/14–2015/16 primarily to obtain better information on when the rut occurred and to determine the daily activity pattern. Because some forms of aggressive or sexual

interaction had not been seen despite decades of surveillance, it was assumed that much activity must happen at night. A camera was set up overlooking a mown area of 1–2 ha where deer were regularly seen at dusk; cameras were changed at roughly weekly intervals so I was aware of events soon after they occurred. The first and third winters were exceptionally mild, and both ruts were drawn out affairs extending into January and with no clearly defined peak. In contrast, although 2014/15 could not be classed as a cold winter, it was more typical with a cold spell in the second week of December; deer activity peaked 19–24 December with relatively little happening in January (Figure 3.10). Here the camera was used as a sampling device to record when deer were active in a small but representative part of the mown area.

Daily activity pattern was similar in all three winters. The pattern for 2014/15 is shown in Figure 3.11. Sunrise occurred at roughly 08.00 with sunset varying from just before 16.00 until about 16.30. Only 10% of activity occurred between sunrise and sunset. Deer were most active 16.00–18.00. The best chance of seeing rutting behaviour would be in the two-hour period around sunset.

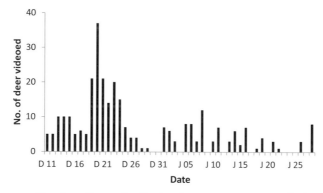

Figure 3.10   Number of records of water deer on a camera trap during the rut in Woodwalton Fen, 11 December 2014 to 28 January 2015. The camera was set to take ten-second videos at intervals of one minute when triggered.

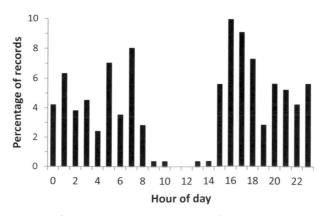

Figure 3.11   Daily activity pattern of water deer during the rut recorded by a camera trap at Woodwalton Fen, December 2014 to January 2015.

During the day, deer tended to keep hidden in the surrounding reed beds and carr. In the wetter winter of 2014/15, when the mown area was partially flooded, several bucks spent time lying up in the islands of saw-sedge *Cladium mariscus* that had been retained in the mown area. These were marginally above the shallow floodwater and allowed the bucks to remain close to the action. Surprisingly, this behaviour was not repeated in the drier winters.

A subsidiary aim of the midwinter studies was to see if informative videos could be obtained. The cameras failed to record a fight although they picked up many instances of bucks chasing rivals. In January 2015, I searched the mown area for lost hair at the end of the rut but only found it in one corner, well away from the camera position. So there may have been just a single head-on fight despite the intensive level of rutting activity. The two instances of mating recorded at 22.55 on 23 December 2014 and at 01.21 on the 24th were, as far as I am aware, the first time mating had ever been videoed at night. Other interesting information on behaviour included the reactions of a grazing doe at 05.03 on 22 December 2014. She was evidently in oestrus, because as she was approached by a subdominant buck she went to him and touched noses. However, they were disturbed by the dominant male, identifiable by his badly torn ear, and the first buck ran away. This video suggested that the doe was interested in any buck, rather than just the dominant one. That evening, what was probably the same doe was videoed about to mate with the dominant buck. In the two videos of mating captured on the night of 23/4 December, the buck involved was not the dominant animal.

In water deer, females are largely unimpressed by any male until they come into oestrus – then it seems they may mate with the first one they encounter. While this is likely to be the territory holder, who is tending them closely, this is not necessarily the case. On the crowded pastures at Whipsnade, Stadler (1991) monitored three rutting seasons and found that out of the 18 mating pairings he observed, the territory-holding buck was involved in at least 14. In the small concentration of water deer at Poyang Lake studied by Sun and Dai (1995), nine males had opportunities to mate: four of them were seen to do so and three of these mated with two females. The mating system is described as polygynous.

3.2.4 *Gestation, birth and beyond*

The water deer is unusual in having litters of young. Raymond Chaplin (1977) knew of one example of seven foetuses in the uterus and three instances of Woburn deer having six foetuses. He found that as the number of ova per ovulation increased so the implantation rate decreased; for instance, implantation rate was 100% for a single ovum but on average only 85% for three. Average number of foetuses was 2–3 in both Britain (Chaplin 1977) and China (Sheng 1992b). Pregnancy rate was found to increase in China from February to March, with first-year does lagging behind older animals. Length of gestation period estimated by various authors averaged out at about 170 days. During the final stages of pregnancy, a mother's weight may double (Chaplin 1977), causing Stefan Stadler (1991) to speculate that her ability to escape from a predator must be diminished at that time. Heavily pregnant females negotiating flooded deer paths at Woodwalton Fen were noticeably cautious and unsteady in my camera trap videos.

The consensus is that May and June are the main months for births (Cooke and Farrell 1998). First-year females gave birth later than older animals at Branféré Zoological Park (Dubost et al. 2008). At Whipsnade, females paced around just prior to giving birth, appearing both uncomfortable and nervous (Stadler 1991); there was no obvious selection of birthing site, fawns being found in open grass fields, patches of nettles or thistles and clusters of conifers. Fawns found at Woodwalton Fen were in open fen fields with vegetation of variable height. On a farm near Aylesbury, fawns were found in the centres of hayfields rather than around the edges (Marc Baldwin, personal communication).

Water deer generally produce litters of two or three, although this observation is largely based on number of foetuses rather than number of fawns (Chaplin 1977; Sheng and Lu 1984; Zhang 1996). Litter sizes of up to six have been recorded in China but this was exceptionally rare. In Bedfordshire, four was the largest litter reported by Chaplin (1977), but Childerley (2014) referred to five or more fawns being 'not uncommon'. At Branféré Park, the percentage of females giving birth increased with age (Dubost et al. 2008): young deer 31%, yearlings 47% and older deer 75%. Average litter size also showed a tendency to increase with age: young deer 1.6, yearlings 1.8 and older deer 2.1. During the course of the study at Branféré, deer numbers increased inside the 12 ha enclosure, densities being roughly 5 per ha in 2002/3, 6 per ha in 2003/4 and 7 per ha in 2004/5. This was associated with a decrease in average litter size from 2.3 in 2003 to 1.9 in 2004 and to 1.7 in 2005, an apparent density effect. Litter size was quite low even in 2003, so high density might already have been having an effect at the start of the study.

Weights at birth are in the range 0.6–1.1 kg (Chaplin 1977; Cooke and Farrell 1998), and initial weight gain is approximately 0.1 kg per day. Sex ratios of young fawns did not differ significantly from parity either in the wild in China or in captivity in England or France (Zhang 1996; Dubost et al. 2008). A description of the early life of fawns has been distilled from these authors and from Dubost et al. (2011b). Fawns are groomed as soon as they are born which cleans the coat and may impart a distinctive scent. A dead fawn elicits no response from the mother. Fawns can stand after about one hour and may move as far as 100 m during their first day, so a litter can become dispersed (Figure 3.12a). After a few hours, they hide and spend much of their early life concealed, especially during the first month. The doe usually returns four or

Figure 3.12   Young water deer: (a) a fawn with spotted coat (MB);
(b) an individual aged 2–3 months, short-faced and shaggy-coated (KL).

five times per day to feed and groom them. After their mother's visits, they select a different hiding place. Vegetation is sampled from a few days of age. Duration of suckling bouts decrease from about three weeks and young are weaned by about three months. They sometimes associate with other fawns which are not siblings.

Water deer are prolific breeders and one might imagine that populations would have a high proportion of young deer. However, this has not been found apart from where a population has been heavily shot, such as on the Zhoushan Islands in the 1980s (Sheng and Lu 1984). At Woodwalton Fen, where typically around 20% of the winter population is comprised of first-year deer, this represents roughly 0.5 young per adult female compared with a birth rate that may be in the region of two fawns per female (Cooke and Farrell 2000). This simple calculation suggests that on average three-quarters of fawns may die or disperse before their first winter.

High fawn mortality has been reported in captive populations. At West Midlands Safari Park, 40% died in the first four weeks, mainly from various birthing complications and from exposure (Lawrence 1982). At Whipsnade, 24 out of 28 fawns tagged in 1993 died during the summer, mainly from fox predation (Zhang 1996). At Branféré, 48% of fawns died during their first month and further mortality occurred after weaning at 2–4 months of age (Dubost et al. 2008). The early mortality was mainly due to birthing problems or death of mothers, and also to the high density in the park leading to fawns attaching themselves to the wrong females and to desertion by mothers. The later mortality in August was during hot summer weather that apparently led to hyperthermia. Raymond Chaplin (1977) had previously noted that 25% of fawns died during their first three days in his farmland study area at Woburn; malpresentation was an important reason for mortality, but also hot weather caused overheating and dehydration in the open habitat. High summer temperatures can apparently be beneficial if fawns are in dense vegetation since, at Woodwalton Fen, high levels of winter recruitment often followed hot summers (Cooke 2009a). Amounts of summer rain did not seem to affect fawn survival either at Branféré Zoological Park or at Woodwalton Fen. Foxes are likely to be the main predator of fawns at Woodwalton Fen (Section 8.3.5). On average at Branféré, each juvenile female had 0.2 fawns surviving at one month of age, while each yearling had 0.5 fawns and each adult had 0.9 fawns.

In China, an important additional hazard for fawns is poaching. Zhang and Guo (2000) undertook an enquiry into poaching on the Zhoushan Islands in 1999 and 2000. One form of poaching consists of taking unweaned fawns in summer for the colostrum in their stomachs. This traditional folk medicine, called Naikuai, fetches a high price and is used for curing indigestion in children. During the 1970s and 1980s about 150–200 fawns were believed to be taken annually, but in the 1990s poaching of fawns escalated due to demand, which was caused in part by newspapers exaggerating the efficacy of Naikuai. Some local poachers even travelled to Poyang Lake and Yancheng Nature Reserve, and earned enough to make a living from poaching. Endi Zhang calculated that the annual toll of fawns was up to 1,500 per annum, compared with the total deer population on the Zhoushan Islands of about 3,000. Min Chen told me in 2014 that she hoped this practice would soon die out as only the older generation believed in the properties of Naikuai.

At Whipsnade, all sex and age categories were found to indulge in play behaviour, but fawns aged 3–8 weeks were the most playful (Stadler 1991). Five different behaviours were identified: a fast gallop, a capriole that involved jumping and twisting, head shaking, play mounting and a version of a fighting dance. Lynne Farrell and I have recorded the second and third type at Woodwalton Fen by older deer, but summer sightings of fawns in the dense vegetation are rare. Most water deer fawns are fortunate in that they have siblings or others of the same age with which to play.

In late summer and early autumn, well-grown fawns can be seen, often still accompanied by their mothers, inside Woodwalton Fen and outside on the farm fields. These young deer are smaller versions of the adults, but with shorter faces and shaggier coats (Figure 3.12b). How many disperse away from the reserve will depend on the number of fawns that have been produced and the pressure on them to disperse. The shooting records of Martin Guy from the adjacent farmland can help provide information on the proportion of young bucks. In the winter of 2014/15, 16 out of 19 bucks (84%) were judged as first-year animals, whereas in 2015/16 it was 7 out of 14 (50%). In 2016/17, water deer were generally hard to find and very few were shot, pointing to minimal dispersal.

3.3   Comparisons and conclusions

The two species are not closely related and there is no reason to expect that their breeding behaviour would be very similar. In fact they do have many aspects of behaviour in common but they also display some fundamental differences. The greatest difference is that water deer breed seasonally whereas muntjac can breed at any time of year. Other species of muntjac in China also breed aseasonally and have a seven-month cycle (Sheng and Ohtaishi 1993), so this trait was probably established long ago. Apart from the captive studies of Norma Chapman, Gérard Dubost and Michael Clark, relatively little attention has been paid by naturalists, by researchers or by the media to various aspects of rutting behaviour in muntjac, perhaps because it involves only a small proportion of the muntjac population at any one time. This makes it more difficult to see, to study and to film in the wild, especially when it is likely to take place in woodland cover. In contrast, there has been increasing interest in recent years in the rut of wild water deer, despite the difficulties connected with seeing this species in the rut.

Nevertheless, Reeves' muntjac and water deer are small, prolific, non-herding species that live fairly independent lives. Both sexes of both species become sexually mature in their first year. Whether or not they have an opportunity to mate at such a young age depends in part on population density and the degree of competition they face from rivals, this being particularly true for young bucks.

Bucks are territorial throughout the year with muntjac being rather more tolerant than water deer, unless there happens to be an oestrous doe in the vicinity. Bucks of both species indulge in territorial marking, with frequent use being made of dung and urine. The main marking device for muntjac seems to be dabbing the frontal glands on the ground or on low vegetation. Water deer lack frontal glands, but do rub their foreheads on thin stems and other vegetation – a process that still requires some explanation. Both species use their preorbital glands for marking, but only muntjac bucks occasionally stand on their hind legs to do so. And both species paw the ground with their forefeet. This is done more vigorously by muntjac deer, which lack glands in the forefeet; its purpose could be to draw attention to the dung and urine that may be present. Water deer have glands in their forefeet and their pawing does not result in scrapes that are readily discernible to the human eye, but presumably they provide an olfactory signal to other deer. Muntjac also fray and anoint the stems of thin saplings. Water deer bucks should be capable of fraying bark with their lower incisors but none has ever been seen to do so; forehead rubbing on thin stems seems to be their equivalent of fraying. Both species bark, but only the muntjac seems to use barking to advertise or defend a territory. In the water deer, the bark expresses concern and alarm (as it also does with muntjac).

In an undisturbed situation, muntjac males and females may retain more or less the same territories throughout the year and for several years. This means that the same buck and doe may mate regularly over a period of years, despite there being no pair bond. Less is known about territorial stability in water deer bucks or the ranges of females. However, rutting strategies in the water deer embrace both maintaining relatively large dispersed territories and defending much smaller ones where aggregations of animals are attracted to communal feeding areas.

Bucks of both species fight, but in different ways because muntjac deer have antlers and water deer do not. Muntjac bucks push and twist whereas water deer dance – but tusks are the main weapons for both species. There is much more contact in a fight between two muntjac. A fight between water deer is brief and usually contains more dancing than violence. Yet injuries sustained by water deer can be more serious than those suffered by muntjac, probably because of the length of their tusks and the thickness of skin around the neck of the latter. Bucks of both species often display fighting damage, particularly ragged and torn ears – again, on balance these are more obvious in water deer.

Muntjac females in oestrus are especially flirtatious, skipping around with their tails in the air inviting attention. Water deer females, on the other hand, seem to remain more focused on other activities until they are approached by an interested buck. In both species, the territorial buck is likely to mate with the doe, and copulation will probably occur several times during the window of oestrus. A muntjac doe that misses being fertilised will come into oestrus again in a couple of weeks, whereas a water deer doe that misses out will have to wait for another year. One might have expected the latter to try rather harder during each rut!

Water deer court and mate in early or midwinter and most of their activity (at Woodwalton Fen) takes place under cover of darkness. Muntjac mate at all times of year, including when the nights are short. Much, if not most, of their territorial and courting behaviour recorded by my camera traps occurred during daylight, so they seem more diurnal in this respect.

Gestation period is longer in the muntjac and its single fawn tends to be heavier than any born to a water deer. And the water deer takes perhaps a month longer to be weaned. Fawn mortality is high for both species, but especially for the water deer, with losses around birth and predation being important. Despite the water deer's multiple births the muntjac may tend to be more productive in terms of live young reaching their first birthday.

CHAPTER 4

Deer in the field

4.1 Detecting the presence of muntjac and water deer

This chapter is about detecting and seeing deer in the field. The difference between these two activities is that: the former involves simply determining, when their presence is uncertain, whether deer occur in a particular place; whereas the latter relates to looking for deer for reasons as varied as pleasure, monitoring or stalking, and predicting the best places and times to search.

Detecting odd individuals in a locality can be difficult. How to distinguish these two species is discussed earlier, but here signs and behaviour are mentioned that are especially useful for identification at low population densities. Early colonisers will probably be both nocturnal and secretive so sightings might be few and far between. Browsed or part-browsed bramble leaves to a height of about 1 m in woods or hedgerows will probably indicate muntjac deer are present (Figure 4.1), although water deer or even roe cannot be ruled out. Small, curve-cleaved slots in soft mud or snow indicate the presence of muntjac. And once you are familiar with its sharp bark, you will be able to distinguish it from the gruffer bark of roe or the variously described roar, grumble or scream of a water deer. Muntjac invading gardens or similarly well-tended areas may first be detected by roses and foliage mysteriously disappearing. The activity of a single nocturnal wanderer might be immediately obvious to a keen gardener. If you catch a glimpse of a muntjac, it is usually unmistakable when flagging its tail (Figure 2.10a). If it does not beat a hasty retreat, you can always try to make it do so!

Some people are concerned at muntjac colonising an area because of impacts they might cause. Water deer are not generally regarded as a threat, but again it is worth knowing what signs of an early presence to look out for. A problem with water deer is that they are unlikely to be very far from the nearest muntjac or roe, so rendering identification based on droppings or slots more difficult, or even impossible. One reliable means of establishing presence and potential breeding at the same time is finding clumps of their distinctive hair resulting from an altercation between bucks or indicating a nearby carcass that has been scavenged (Figure 3.8). Definite identification of water deer based on a fleeting sighting may be rather more of a problem than it is for muntjac. If you see the long tusks on a buck, there

Figure 4.1   Parts of bramble leaves or whole leaves browsed off can indicate the presence of deer.

will be little doubt about identity, but females and young do not have visible tusks or any distinguishing colour pattern on their coat. The latter 'non-feature' can be useful as muntjac and roe have distinctive rear ends (Figures 2.1a–c). Water deer also have a subtly different appearance from the other species because of their relatively upright and close-together ears, and the lighter tone of their coat. A small deer seen in a suburban garden is much more likely to be a muntjac than a water deer, which are more at home in quiet countryside. You might think that, with just three species of smallish deer living wild in this country, identification should not be a problem. The main reason why it continues to be is that many people do not bother to check on exactly how to distinguish between them.

Now that camera traps are cheap and readily available, they can quickly indicate which species are present. Also known as trail cameras, they have for example been used in Northern Ireland to monitor colonisation by muntjac (Dick 2017) and can reveal the presence of muntjac where none had been suspected (Leadbeater 2011). Cameras can be sited looking along possible deer paths or focused on vegetation that is being eaten. These are small deer and a mistake that is often made is to position the camera too high. I often place mine about 20 cm from the ground, looking straight ahead. If attached higher than 50 cm, they will need to be tipped down at an appropriate angle. Later, I briefly discuss how individual muntjac can be identified (Section 7.2.3). Failure to detect deer by camera traps or other means does not necessarily indicate absence – odd individuals in a relatively large site may be very hard to detect. There is now a range of night viewing devices on the market, including thermal imagers, which might prove useful although, from my limited experience, it can be difficult distinguishing between species.

4.2   Seeing muntjac and water deer

Michael Clark (1981) has been critical of books that describe muntjac as 'secretive' or 'difficult to observe', saying individuals follow a routine and are reasonably predictable in their habits. I would reply that wild individuals are both secretive and predictable, especially at low density. Once they are found and their behaviour is understood, then they may become predictable – but first they must be found!

To have the best chance of seeing deer, you need to know where they occur and when they might be out in the open. 'Where' has two main components: their distribution in the country and, secondly, their preferred habitats. It also has a subsidiary component of precisely where in their preferred habitat they might be. 'When' also has two parts: when during the year and when during the day they are most likely to be seen. The 'when' pair of questions is addressed later in this chapter, including relating their activity during daylight to what they get up to at night.

Distribution of the two species is covered in Chapter 5 and their main habitats have been briefly summarised in Section 2.8. Nevertheless, it is worth exploring their habitat preferences in more detail, especially for the water deer, to try to unravel its apparent liking for both highly artificial agricultural land and semi-natural wetland; as well as to decide whether it is attracted to woodland.

4.2.1   Habitats

Muntjac like woodland, but what features do they find particularly attractive? This question was addressed over a long period of years in the King's Forest, Suffolk, by Norma Chapman and her colleagues, primarily by relating sightings of deer or amounts of dung to habitat and changes in habitat (N.G. Chapman et al. 1985; 1993; 1994a; 1997b; Claydon et al. 1986; Forde 1989; Keeling 1995a; 1995b; Blakeley et al. 1997). Views changed subtly over time as the forest matured and also as a result of the destruction caused by the severe gale in October 1987. Muntjac selected woodland blocks with a high diversity of vegetation in the canopy and at ground level. Variety was more important than density of vegetation. The deer preferred blocks with a range of tree species; those with broadleaved trees that produced nuts were particularly favoured (Figure 4.2a). At ground level they relished forest blocks with a variety of shrub species, bramble and herbs. Prime muntjac habitat had less grass cover than areas frequented by roe and fallow deer. Muntjac avoided places during daytime that were not close to cover, but might visit them at dawn, dusk or during the night when the deer were less obvious to predators. Therefore,

Figure 4.2   Muntjac habitat: (a) Norma Chapman in good habitat in the King's Forest, Suffolk, with sweet chestnut *Castanea sativa*, oak trees *Quercus* species and bramble, spring 2018; (b) even small patches of scruffy cover can provide a base (KR).

cover should ideally be well distributed through their home range and be present throughout the year. Fine-grained diversity appeared more important than the presence of specific types of habitat. The result of the gale was to produce more small-scale diversity in habitat, and home ranges of does decreased and density increased.

Dense habitat with a diversity of vegetation is the key to good muntjac habitat generally in Britain (Chapman 2008), supporting the impressions of earlier authors (e.g. Dansie 1983). And it explains why muntjac occur commonly in such places as mature gardens and overgrown cemeteries, and utilise patches of dense cover in what would otherwise be unattractive open landscapes (Figure 4.2b).

I have counted dung to reveal how muntjac use large woodland blocks in relation to the position of rides along their edges. Transects, 80–90 m in length, were set up in both Monks Wood and Holme Fen perpendicular to rides; counts of dung pellet groups were taken to indicate relative levels of deer activity. In Monks Wood during 1993 and 1994, amounts of dung were related to residual bramble cover and to distance from the ride, indicating more activity further into the woodland (Cooke 2006). At Holme Fen, dung was counted annually in May along three transects perpendicular to rides. In 2005 as stalking was starting, marginally more dung was found towards the centres of the woodland blocks. During 2006–2008 before bramble and ferns had shown much recovery (Figure 16.9a), a marked trend was found for dung to be well away from rides. During this period, there was little cover and deer were keeping back from rides where shooting occurred. By 2009–2014, however, when vegetation had recovered sufficiently to provide good cover (Figure 16.9b), distance from the rides made no difference to amounts of dung. Muntjac were taking advantage of growth of bramble and ferns to be more active close to rides.

Woodwalton Fen provides an example of muntjac and water deer living partially separate existences in the same site. When muntjac colonised the Fen from 1980, they tended to settle in its drier and wooded southern end, and displaced the resident water deer to some extent (Figures 1.7 and 9.1). It took many years before muntjac began to be seen in the extensive reed bed at the north end, and even now sightings of water deer are much more common there, with muntjac tending to be largely confined to fringes of dry sallow carr. And muntjac deer are much less commonly seen out on the open farm fields. An example of the reluctance of muntjac to leave woodland cover comes from the survey of Gill and Morgan (2010). They used a thermal imaging technique to assess deer populations in 15 varied woodlands across lowland Britain: muntjac occurred inside seven sites but surveys of adjacent fields revealed their presence only once. In contrast, fallow deer were recorded inside nine sites and outside seven. Water deer were not mentioned.

In this country, water deer are recorded from a variety of habitats, particularly fens, grazing marshes, grassland, arable, scrub and woodland. How then can we decide what are the preferred habitats? The highest densities in areas that are not specifically managed for deer are in wetlands such as Woodwalton Fen and parts of the Broads. When I took Endi Zhang to Woodwalton Fen in the 1990s, he remarked that parts of the habitat resembled areas in China where they still occurred. More recently, when I am visiting other wetland sites in this country with water deer, I often note their similarity to Woodwalton Fen, as seen for example in Figure 4.3.

Figure 4.3   Two sites with long-established populations of water deer: (a) a view in the centre of Woodwalton Fen NNR, Cambridgeshire; (b) Woodbastwick NNR in the Norfolk Broads. They are more than 100 km apart but have similar plant species and habitat structure.

Even in prime wetlands, it helps if the deer have access to adjacent farmland at times of year when season, flood, snow or ice mean that food and cover are less available inside the main refuge. It is not unusual to see water deer feeding and resting on open ground outside wetlands in late winter and early spring. These are deer that rely for most of the year on what the wetland has to offer. However, over time deer spread further afield and some settle and become resident on farmland, ranging over cropped fields and using whatever cover is available. Crops will provide cover especially in summer up until harvest, but anything from dry ditches to tall grass, patches of scrub and woodland, and game bird shelter-crops will be taken advantage of. Outside Woodwalton Fen, elephant grass, *Miscanthus* species, was grown for a few years and provided an ideal base for much of the year until it was cropped in February or March (Cooke 2010).

On both sides of the River Yare in Norfolk, water deer have dispersed away from wetland areas and grazing marshes. Thus a population is well established in Strumpshaw Fen to the north of the river and has spread via Buckenham and Cantley marshes on to arable land (Tim Strudwick, David Farman, personal communications). When crop fields have been harvested, water deer are not difficult to see and some are resident. Road signs warning of collisions with deer are evidence of their abundance, mobility and lack of road sense. On the south side of the Yare, I first visited to see water deer at Wheatfen Broad in 1997. Since then they have spread along the grazing marshes to the south-east, and beyond to the River Waveney. On the Claxton Estate (Figure 11.5), where they are encouraged by the landowner, some deer from the marshes have dispersed to reside on arable land and in woodland on higher ground further away from the river. Water deer are well able to settle and survive in an agricultural environment, but usually at a lower density than in wetland.

Grazing marshes are an extremely important habitat for them in the East Anglian wetlands from Cley on the north Norfolk coast, to the marshes of central Broadland, to Carlton Marshes in north-east Suffolk, and then down the Suffolk coast to Orford Ness, where a significant population has built up. Based on observations in and beside Woodwalton Fen, water deer tend to avoid fields holding livestock (Section 8.3.5), but grazing marshes are so extensive in the Broads and near the East Anglian coast that the total number of water deer they support must

Figure 4.4   The view from the tower at Hickling Broad – an ideal landscape for water deer.

be considerable. The marshes include such quiet spots as Haddiscoe Island and Heigham Holmes, both of which are surrounded by water, despite being inland sites. These 'islands' are within extensive tracts of land that lack through roads. The area around Heigham Holmes, for example, also contains Hickling Broad (Figure 4.4) and the Horsey Estate, and exceeds 20 km².

Suitable wetlands are in short supply in Bedfordshire where water deer first became well established in this country, and there they have colonised agricultural situations. The best-known is probably the Potsgrove area just outside Woburn Abbey where they seem to be attracted to the mixed landscape (Figure 4.5a). Callum Thomson, a former deer manager for the Woburn Estate, told me the growing crops protected fawns from attack by carrion crows and predatory gulls, which caused many losses of fawns inside the park wall where the grass was relatively short. Bernard Nau (1992 and personal communication) was able to survey water deer in Bedfordshire in the early months of the year by scanning fields from vantage points. Surveying stopped in early April, by which time deer could be missed in the

Figure 4.5   Dry farmland habitat of water deer: (a) general view of the Potsgrove area in Bedfordshire; (b) Sawtry Fen in Cambridgeshire, a flat, open, arable area where they use any available cover, including dry ditches.

fields of growing winter wheat and oilseed rape. He described their habitat to me as predominantly undulating, sparsely wooded arable land. The deer favoured sloping fields where they spent the day resting and grazing, well away from any cover. Since then, water deer have spread considerably to the south-west and can, for instance, be found in similar situations in Buckinghamshire (Figure 11.4).

Locally in Cambridgeshire, a population of water deer on Sawtry Fen is separated from Woodwalton Fen by the London–Edinburgh main railway line, and westward movement of the deer is restricted to some degree by a motorway section of the A1. The farming landscape is dominated by open, flat arable fields with very little cover (Figure 4.5b). The deer do, however, utilise the scrub-covered embankment of the railway line, which must also serve as a north–south corridor (Cooke 2010). In 2009, density was at least 8 per km^2 where the main concentration occurred. By 2016, water deer could be readily found on several square kilometres of arable and new conservation grassland in the north of the Great Fen (Figure 1.10). The landscape there is flat and virtually treeless, but local density was at least 10 per km^2 by 2018. The attraction of both areas to the deer is probably that disturbance levels are low and food is available throughout the year.

Occasionally water deer dispersing through a dry landscape may stumble across a suitable wetland site where they can settle. Michael and Jum Demidecki (2016) have recently studied water deer and muntjac at Wilstone Reservoir, which is one of the reservoirs in the group at Tring in Hertfordshire. Signs of water deer were common through the reed bed on to the muddy shore, whereas evidence of muntjac was found more frequently in mature woodland growing back from the reservoir. Footprints in the partly dry arm of the Grand Union Canal indicated how deer used this corridor to move between Wilstone Reservoir and Tringford Reservoir to the east.

4.2.2   *Water deer in woodland*

Water deer do occur in the mixed woodland at Woodwalton Fen and it is pertinent to ask whether woods can be regarded as one of their preferred habitats. At Woodwalton, they occur throughout the reserve and are found in the woodland in small numbers at all seasons. However, there is evidence that in late winter, when bramble bushes have been defoliated up to the browse line, they may tend to move further north in the reserve where they have easier access to the farm grassland outside the reserve. Good photographs of water deer in wetland habitats are easy to source, but pictures of them in woodland are rare and I have used a camera trap image in Figure 4.6; this rarity probably reflects the fact that photographers are not looking for them in woodland.

In Monks Wood, water deer were resident in small numbers until apparently ousted by muntjac in the 1990s (Section 9.2). Sightings were largely restricted to two open fields within the wood, suggesting they were attracted to the fields rather than the woodland, although doubtless they used nearby woodland as cover (Cooke 2006). In contrast, muntjac sightings were dispersed much more uniformly throughout the wood. In the birch woodland in the centre block at Holme Fen, water deer were still comparatively abundant during winters in the early 1990s; they were mainly seen close to the southern edge where they had easy access to farm fields outside the reserve. At that time, muntjac deer were starting to build up in

Figure 4.6   A buck water deer emerges from a thicket of blackthorn *Prunus spinosa*
in the mixed woodland at Woodwalton Fen. He has lost a tusk, presumably while
fighting. This area is prime habitat for muntjac (camera trap photograph).

numbers and their distribution was similar; by 2004/5, they were fairly generally
and densely distributed and the water deer had gone (Figure 9.2).

In Bedfordshire, the water deer surveyed by Bernard Nau (1992) did not use
woods at all. Although Bedfordshire is a poorly wooded county, there are woods
studded along the Greensand Ridge, and I have come across water deer in or near
several woods in the county. From East Anglia, there are records from, for instance,
Thetford Forest and from Bacton Wood near North Walsham. There are areas where
good numbers of deer use woodland for shelter before emerging to feed on arable
or grazing marsh, such as on the Benacre Estate in Suffolk, at Buckenham Marshes
beside the Yare and at Wheatacre Marshes on the Waveney. Graham Downing has
told me that water deer regularly use woodland on the Sotterley Estate in Suffolk,
with increasing numbers being seen in recent years. On the other hand, Rick
Southwood considered solid scrub and carr to be unattractive to them on the Bure
Marshes (personal communication).

My impressions are that large tracts of woodland do not comprise optimal
habitat, but water deer will readily use woodland for food and cover as part of
their home range. Min Chen has confirmed to me that the situation is similar
in China. However, over most of their range in this country, water deer would be
hard pressed to find a wood that did not already have competing muntjac. While
water deer are able to survive satisfactorily in the absence of woodland, it would be
unwise to dismiss them as animals that avoid this habitat.

4.2.3   *When to see deer*

Deer may occur in an area throughout the year but are more likely to be seen at
certain times of year and day. Factors such as weather conditions will also affect
whether deer might be seen on a particular visit. Thus, high winds or heavy rain
can cause them to seek shelter, whereas prolonged cold may lead them to forage
more widely and more often.

Lynne Farrell and I began investigating the water deer at Woodwalton Fen NNR in June 1976. Our initial intention was to determine whether we could see sufficient deer to make a study viable. We knew that other species of deer tended to have periods of greatest activity around dawn and dusk, so we began by visiting in the evening – and saw reasonable numbers of deer, suggesting that a dusk study would indeed be viable. For the first 36 months, we concentrated on finding out basic information about the deer, including how sunset time and season influenced numbers seen. Nearly four times as many sightings were logged in the half-hour after sunset as for 30-minute periods at least two hours earlier (Figure 4.7). This pattern was consistent throughout the year apart from relatively more deer being seen in daytime in midwinter. Difficulty in seeing deer in the developing gloom was probably the reason why numbers seen declined in the period 30–59 minutes after sunset.

There was a seasonal change in monthly averages of number seen per hour (Figure 4.8), depending on the habits of the deer and the height and structure of the vegetation. On average, April was the best month to see large numbers of deer grazing in the evening on rides or open fields. After this, deer became progressively

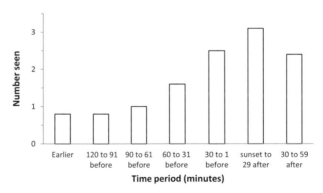

Figure 4.7   Water deer at Woodwalton Fen: average number seen per 30-minute period in relation to sunset, 1977–1979 (adapted from Cooke and Farrell 1981).

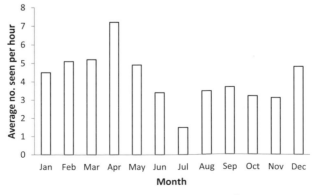

Figure 4.8   Water deer at Woodwalton Fen: average number seen per hour at dusk in each month of the year, 1976–1979 (adapted from Cooke and Farrell 1981).

less conspicuous until July, with females being more secretive prior to and after giving birth and with the growth of lush vegetation. Common reed *Phragmites australis* and other vegetation in the mixed fen compartments grew to a height of several metres and rides became more overgrown. By midsummer, few deer could be seen, despite the fact that the population had been swollen by fawns born in late spring and early summer. Numbers recorded per hour increased again in August and September, when young deer were observed accompanying their mothers. During the autumn, numbers seen remained fairly constant until the main rutting month of December when deer, especially long-tusked bucks, were more apparent. Good numbers were recorded during the first three months of the year as vegetation cover was sparser, and deer often fed on the neighbouring farm fields where they were comparatively easy to see. This general seasonal pattern has persisted through to today, although there have been some modifications related to changes in management of the landscape.

At that time, I was kept informed of the muntjac's colonisation of Monks Wood NNR by the warden, Jeremy Woodward. Monks Wood is only about 5 km from Woodwalton Fen, and muntjac began colonising the wood in about 1970 – and from there some apparently moved to Woodwalton Fen. From 1977 until 1985, Jeremy Woodward spent most of his time working in the wood, and mapped all of his sightings of muntjac. He copied his maps to me and I tallied up the numbers. The warden's sightings showed highest numbers from December to May with a peak in April. During subsequent dusk surveillance at Monks Wood, Woodwalton Fen and Holme Fen, a pattern has emerged of numbers seen per hour increasing steadily from October through to April. There was a similarity between the monthly patterns for sightings of muntjac at Monks Wood and those of water deer at Woodwalton Fen, with minimum numbers being seen in July and maximum numbers in April.

This pattern was likely to be due to a combination of varying amounts of cover and seasonal changes in behaviour for muntjac, as it was for water deer. Muntjac deer breed all year round so their populations should show relatively little seasonal variation in reproductive behaviour. The peak in sightings that occurred in April may be linked to diet. Harris and Forde (1986) studied the faecal contents of muntjac in the King's Forest in Suffolk during 1983–1985. They found that grasses comprised only about 3% of plant fragments in faeces collected in October and November, but this rose steadily to a peak of about 37% in April and May: muntjac were turning to grasses with the flush of growth in early spring (Figure 2.14a). In our local reserves, the spring peak of sightings is largely made up of deer grazing on rides and open fields. Feeding in such situations means individuals are more conspicuous and more likely to be recorded. Muntjac are relatively inconspicuous in the first part of winter unless hard weather forces them to forage more widely. By May, many grasses are seeding and are less attractive as food, as well as being tall enough to conceal these small deer from view.

Two books that cover the stalking of muntjac (Smith-Jones 2004; Downing 2014) refer to the species' field activities through the year. This is of relevance to when to see muntjac because the species does not have a close season, so stalkers would be expected to avoid seasons when deer were difficult to find and focus on times of year when they were conspicuous. Often high seats can be used as vantage points, but I was most interested to see what they wrote about stalking on foot. Both authors found June–August difficult because vegetation was high. In autumn, Charles Smith-Jones spoke of muntjac taking advantage of 'nuts, berries, windfall fruit and fungi' – in years when there has been a good acorn crop, I have seen them virtually camping out under oak trees. As regards winter, he found that the woodland blocks had much less cover, but deer were attracted to the remaining patches of food, such as bramble thickets, where they could be difficult to see. Graham Downing's busiest period of muntjac stalking was in February and March, although high seats were still used. Charles Smith-Jones noted that, in spring, muntjac were more inclined to travel in their search for new shoots and buds. Up until May, he found success 'moving quietly about the woods on foot'. These observations are generally consistent with trends in encountering muntjac in my local reserves.

During the period from May 1993 until April 1994, I routinely walked a fixed route of 8 km around Monks Wood recording muntjac – four dusk walks and four midday walks were completed each month. One aspect of behaviour that was studied was when muntjac were seen relative to sunset throughout the year. Greatest numbers were recorded during the half-hour periods immediately before and immediately after sunset. The difference between these periods and earlier ones was not as striking as for water deer at Woodwalton Fen (Figure 4.7), so muntjac were slightly less crepuscular. The pattern for muntjac was consistent throughout the year, as it was for water deer. Muntjac at Woodwalton Fen have traditionally been seen on surveillance walks rather earlier than water deer, so reinforcing the view that they are often less crepuscular.

Diurnal and seasonal activity may vary in other habitats. Water deer on farmland are of course far easier to see when there are no tall crops in the fields. Richard Lawrence (2009) noted that most Natural History Society records on farmland in Bedfordshire during 2008 were from the winter months; meanwhile in Norfolk, most records in 2007 were from December (Leech 2008). Such trends may reflect when naturalists looked for deer as well as when deer are out in the open. In arable areas where they are numerous, they can often be encountered out in daytime, sometimes in sizeable loose groups; in the middle of the day, most deer may be resting in full view. In a wetland with dense vegetation, water deer probably need to be active to have any chance of observing them, but the same does not apply to animals living in open habitats. Precise data on diurnal behaviour of wild deer in such habitats do, however, seem unavailable at present.

4.2.4   *Activity patterns throughout the day and night*

So far I have concentrated primarily on activity around dusk and have only briefly mentioned that deer also tend to be active at dawn. Gérard Dubost (1971) reported that semi-captive muntjac in a French park were mainly active during early morning and evening. In the summer, there was a long period of inactivity between 12.00 and 18.00, whereas in the winter this period moved forward a couple of hours with some activity between 12.00 and 14.00. Similarly, Zhang (2000b) and Ma et al. (2013) observed captive water deer at Whipsnade in England and in Huaxia Park, Shanghai respectively, where activity peaks were again found at dawn and dusk.

In January 1994, I was interested to determine how counts of muntjac varied during daylight hours, so undertook ten counts along the southern edge of Monks Wood spaced out at hourly intervals from just before sunrise until just after sunset.

The highest count (15) was the latest in the day and the second highest (12) was the earliest. The lowest counts (2) were at midday and just after. This gave me some reasonable information about daytime activity, but it meant I was on my feet for a long time and it told me nothing about night-time. It needs ingenuity or technology, such as radio tracking or camera traps, to help determine activity patterns for the whole 24-hour period. With observers increasingly watching deer in darkness using night viewing devices, it is relevant to report these complete activity patterns.

Richard Yahner (1980) studied captive muntjac held in pens at a zoological park in Virginia. They were observed from a central tower at half-hour intervals during the day and night, with floodlights being used to illuminate the paddocks. In summer and autumn, activity levels were highest around sunset and lowest during daytime, whereas variations in activity were less marked in winter.

Activity of radio-tagged muntjac in the King's Forest was monitored by using signal amplitude to decide whether an animal was moving (Forde 1989; Chapman et al. 1993). In this way it was possible to work out the percentage of the time that muntjac were active for each of the 24 hours of the day for two-monthly periods throughout the year. Steady fluctuations in signal amplitude indicated a deer was travelling or foraging; infrequent fluctuations in amplitude indicated slight movements of the head during rumination or vigilance behaviour and were catego-rised as inactivity. Peaks of activity were found to occur at dawn and dusk, with lower levels of activity during the day and even lower levels at night. Similar diurnal activity patterns were found at Rushbeds Wood in Buckinghamshire (Harding 1986). On average through a 24-hour period in the King's Forest, an individual might be active for about 70% of the time, with a typical day being composed of five active periods of about three or more hours each, punctuated by inactive periods of 1–2 hours. It was perhaps a surprise to find slightly higher activity during the day than at night because deer stalking and dog walking occurred at the time of the study, the latter making the deer particularly wary (Blakeley et al. 1997).

McCullough et al. (2000) radio-tracked muntjac in Taiwan using mercury tip switches which changed the pulse rate depending on the position of the animal's head. Both sexes had peaks of activity in the morning and the evening in most months, with more activity during the day than at night. Chiang (2007) studied muntjac in Taiwan using camera traps, and found them to be especially active at dawn and dusk. Unlike in McCullough's investigation, there was little activity in the day – behaviour which Chiang suggested might be to avoid diurnal predators. Min Chen has informed me that use of radio collars confirmed that the main activity periods of water deer in China also tended to occur at dawn and dusk.

I have used camera traps to provide information on when deer are active during the day and particularly the night. Information on when muntjac were active in front of cameras in situations with and without stalking is given later (Table 12.1). Cameras provide data on activity in relatively small areas which might amount to less than 0.1% of an animal's home range, and results therefore need to be inter-preted with care. Thus a camera on a well-defined path in cover may tend to feature deer walking to and from favoured feeding areas at dawn and dusk; whereas a camera in an open, disturbed area may reveal little activity during daytime, but

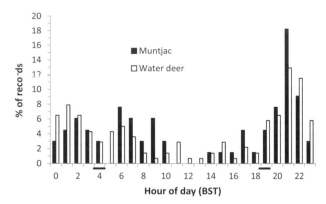

Figure 4.9   Daily activity patterns of muntjac and water deer recorded by camera traps in freshly cut sallow coppice at Woodwalton Fen, May–July 2011. The black bars under the time axis indicate times of sunrise and sunset (British Summer Time).

much more at night. An example of activity patterns for both species in the same habitat is shown in Figure 4.9. Several blocks of sallow coppice had been cut at Woodwalton Fen during the winter of 2010/11. Coppiced areas with cameras were initially flattened and open but became more secluded as reed and other vegetation grew the following summer. The diurnal patterns for the two species were similar with little activity from late morning until pronounced peaks occurred at and just after sunset. Both species were particularly active during the night – results repeated in many other camera trap studies (Figure 4.10).

When I started working with camera traps in 2010, information on the activity of water deer at night was fragmentary (Cooke 2012a). For instance, Endi Zhang's study at Whipsnade Zoo in March and April 1993 indicated feeding peaks from 06.00 to 10.00 and from 17.00 to 21.00 (Zhang 1996), but no other details were given. In the previous chapter, a graph of rutting activity by water deer on an open area at Woodwalton Fen indicated little activity during the middle of the day with much more activity around dusk, continuing until dawn (Figure 3.11). Amounts of cover, however, can affect diurnal behaviour considerably. Cameras in the mixed

Figure 4.10   Night-time activity: (a) a muntjac doe foraging; (b) a water deer moving out in the evening on a flooded deer path through sallow carr at Woodwalton Fen to feed on adjacent farm grassland (camera trap photographs).

woodland in the winter in 2010/11 showed water deer to be more uniformly active during the day and the night with most records around dawn (Cooke 2012a).

In this way, camera traps can help to understand diurnal behaviour in different situations and assist in maximising opportunities of seeing the deer or certain aspects of behaviour. They have proved very useful in showing how deer use the adjacent conservation grass fields to the west of Woodwalton Fen in the late winter and early spring. I had known for some years that the fields were much grazed in the late afternoon and evenings (e.g. Figure 10.1), but I was reluctant to put cameras out on the fields for fear of them being damaged by livestock, taken by people or vandalised. So I concealed cameras along deer paths in the strip of wet sallow carr bordering the reserve. Most of the movement in front of them occurred around dawn and dusk (Cooke 2012a). From the position of the cameras I could tell whether the deer were moving towards the fields or back into the reserve (Figure 4.10b). So, I could work out whether deer tended to stay out for a short time or all night – tendency to remain on the fields all night varied from year to year. Cameras also solved another question: numbers of water deer that I could count on the farm fields around dusk decreased from April to May, but was this due to a real drop in numbers or were the deer simply more difficult to see in the tall seeding grass? Cameras on the deer paths revealed that deer traffic did decrease in May as the grass seeded and became less attractive for grazing and as the females prepared to give birth in the reserve.

4.3   Conclusions

Detecting these species has become much easier since camera traps became widely available. Even without a camera trap, searching a new site for signs, particularly slots, droppings and browsed vegetation, can alert an observer to the recent presence of small deer. A visit a few weeks later should reveal whether the presence was transitory or more permanent. Distinguishing slots of the two species is probably the most reliable way of resolving the evidence in the absence of sightings of the deer or their hair. Their barks are also distinctive but some experience may be necessary to tell them apart. Bear in mind that it is possible for both species to occur at the same site.

The nature of the habitat is a clue to what deer may be present. In woodland in lowland England, muntjac are much more likely to be encountered than water deer. Indeed, anyone wishing to see a muntjac should have little difficulty in doing so over much of England. Their adaptation to living in and around human habitation means that muntjac may be very much closer to you than you realise. It is, however, less easy in my area to see muntjac since culling has reduced numbers and affected the behaviour of survivors.

Water deer can often be encountered within a few kilometres of wetland or agricultural areas that are known to hold concentrations. But visiting wetland sites themselves does of course provide the best opportunity of all. Walking for an hour at dusk at Woodwalton Fen during September–May will virtually guarantee sightings of water deer. Their lifestyle is dictated by the seasons: in early autumn, young deer are more in evidence; rutting behaviour may be seen during November–January; and aggregations feeding on the grass or fen fields occur during February–early May. Norfolk is the county that boasts most water deer and many accessible wetland reserves present opportunities to see them. These sites do not necessarily have a high density of deer, but they do have a high density of knowledgeable observers. Good farmland areas for them include the Buckenham area of Broadland, Potsgrove and Beckerings Park in Bedfordshire and parts of the Aylesbury Ring in Buckinghamshire. Fields in the northern region of the Great Fen in Cambridgeshire currently hold concentrations of the species, but how long this situation will continue remains to be seen. Conditions on wetland reserves can be regarded as reasonably permanent, but a farmland landscape can quickly change – and so can its attractiveness to deer.

I have recommended looking for deer at dusk. Deer can be seen at any time of day, but chances are maximised by looking for them at dusk and dawn. If you own a night viewing device such as an image intensifier or a thermal imaging camera, it is possible to go out at night and detect even more deer. These devices are extremely good for maximising counts of deer. However, the views they give are not as clear as with deer seen in daylight. I prefer to venture out at dusk rather than dawn, in part because my records indicate more deer activity then. The other advantage in going out around dusk is that darkness literally draws a veil over proceedings and means you have to stop. With a dawn visit, there is a temptation to go on for too long.

John Thornley (2016) talked about having a sixth sense as regards encountering fallow deer. This related to feeling that deer were close, followed shortly by deer appearing. I often have this feeling, which for me is brought on by weather and light conditions, by deer signs and by the immediate habitat. After decades of looking for deer, I suspect I am subconsciously aware of the range of conditions often associated with them appearing out of the undergrowth or around a corner in a ride. But I have to admit that I am often disappointed! What does undoubtedly help is building up in the mind an image or images of the shapes, colours and textures of deer and bits of deer that immediately trigger a reaction when glimpsed through trees, bushes or grass. Knowing where and how to look are important.

The Buttolo roe call is a device used to attract roe, which can also be used for muntjac (Collini 2004; Smith-Jones 2004; Downing 2014). Several of our local stalkers use such a call intermittently for muntjac. I have never used one in part because I have no wish to affect their operations by overusing calling, but also because I am usually involved in surveillance and need to keep my ability to detect deer as constant as possible. Now I am in my mid-70s, I wonder how my abilities compare with those of 20 or 40 years ago. Ageing ornithologists are becoming aware that their abilities are diminishing to hear bird songs and calls, so meaning their surveys are now more likely to under-record or completely miss certain species that are present (Appleton 2017). I like to think that my ability to detect deer shapes may be better than ever, so helping to compensate for some deterioration in both sight and hearing. An alternative view on this issue is that experience may increase detection probability if our faculties remain intact.

Occasionally, I feel fraudulent writing about muntjac and water deer when I have never visited China, Taiwan or Korea to look for them. However, people who have looked for these species in their native ranges have experienced the habitat, but often found wild deer to be elusive. England may be the best place in the world to see these species. Min Chen, who is one of the Chinese experts on water deer, was astonished at how many she saw at Woodwalton Fen when I took her there in 2014.

CHAPTER 5

Colonisation at national level

5.1 History of mapping deer distribution

Both species have been colonising England for many years and, in the case of the muntjac, it has now been sporadically reported in Wales, Scotland, Northern Ireland and the Republic. This chapter and Chapters 6 and 8 deal with colonisation at three different scales. This one examines how and why these two alien species have spread regionally and nationally at a coarse level. It is of relevance to the rate and extent of any further spread and, at least in the case of the muntjac, to the cause of even more widespread impacts. It also focuses on attempts made to record and plot colonisation, and begins with a brief history of mapping.

Whitehead (1964) made the first attempt at mapping the spread of muntjac and water deer with stippled areas, numbers and dots indicating the distribution of wild and captive deer. Later maps were usually based on presence reported in 10 km squares on the National Grid. In 1965, the Mammal Society launched a National Distribution Scheme to gather and map data for British mammals, and provisional distribution maps were published (Corbet 1971).

The BDS published maps for our deer species recorded during 1967–1972 (M. Clark 1974). This exercise was not repeated for another 30 years, but records continued to be collected on an ad hoc basis. In the meantime, the national Biological Records Centre (BRC) had been receiving deer data directly from recorders, which they added to the earlier information collected by the Mammal Society and the BDS to compile two sets of maps (Arnold 1984; 1993). When Chapman et al. (1994a) reviewed past, current and future muntjac status, they made a determined attempt to improve the 1993 map for muntjac by making television, radio and magazine appeals, and by circulating questionnaires to a wide range of bodies including the BDS and Mammal Society.

BRC data up until 1993 were incorporated into reactive maps in the NBN Gateway. Other major data sets contributing to these maps have come from the British Trust for Ornithology (the BTO's network of recorders also report on non-avian species), the Mammal Society, county biodiversity information services, the Recording Invasive Species Counts Project of the GB Non-Native Species Secretariat (for the muntjac only) and the Living with Mammals Survey of the People's Trust for Endangered Species (PTES). The NBN Gateway maps are continually updated and can be interrogated in order, for example, to follow colonisation over time.

The Tracking Mammals Partnership was launched in 2003 with the aim of bringing together organisations with an interest in UK mammals so as to improve the quality, quantity and dissemination of information (Battersby and the Tracking Mammals Partnership 2005). The Partnership identified the BDS as the organisation undertaking deer surveys and the aim was to repeat the survey every five years. In addition to the initial work in 1972, the BDS has so far monitored the recent expansions of our national deer populations by collating records up to 2002 and by organising surveys in 2007, 2011 and 2016 (Ward 2005; Ward et al. 2008; Hailstone 2013; Smith-Jones 2017). These seem to be the only recent distribution maps published in paper form on a 10 km square basis. The fourth edition of the 'Mammal Handbook' (Harris and Yalden 2008) relied on shaded maps, as did the earlier editions.

5.2   Spread of muntjac nationally

5.2.1   *Early colonisation*

Details of the early years of muntjac colonisation were eventually painstakingly constructed by the researches of Norma Chapman and her colleagues (Chapman et al. 1994a) – and this account draws on that review. Reeves' muntjac of unknown origin were introduced in 1894 by the 11th Duke of Bedford to his estate at Woburn Abbey in Bedfordshire (Figure 5.1). Releases from the estate occurred from 1901 onwards until perhaps the 1950s. Their early history is confused because Indian muntjac were also kept and sometimes released. Dating from the first part of the twentieth century, specimens of Reeves' muntjac were moved and held captive in other places, from which they might have escaped. For instance, a buck and a doe were donated to Whipsnade Park, Bedfordshire in 1928; this was before the zoo opened to the public in 1931.

Figure 5.1   Woburn Abbey and part of the parkland in more recent times.

Figure 5.2   Maulden Wood in Bedfordshire has held a population of muntjac for 80 years.

Establishment and colonisation of Bedfordshire was slow. Chapman et al. (1994a) suggested a general rate for natural dispersal of about 1 km per year. An appreciably faster rate was thought to indicate movement by people. Evidence from tagging work in Suffolk indicated that dispersing muntjac only rarely move more than 5 km from their natal area (Chapman and Harris 1996).

'Barking deer' were reported around Woburn Sands in Buckinghamshire (about 3 km from Woburn) during the First World War. Then in the early 1920s, there were sightings at Silsoe in Bedfordshire (about 12 km from Woburn) and at Ashridge on the Bedfordshire/Hertfordshire border (15–20 km south of Woburn). There had been uncertainty about whether these were Reeves' or Indian muntjac. Reeves' muntjac were also kept by Lord Walter Rothschild, apparently until at least the 1930s, in his estate at Tring in Hertfordshire, close to the border with Buckinghamshire (Chapman et al. 1994a), and escapes from the estate may have contributed to some of the early sightings. These authors suspected from reports from further afield in the 1930s and 1940s that there was introduction into parts of Northamptonshire and Warwickshire in the 1930s.

By 1938, a population of muntjac was well established in Maulden Wood, Bedfordshire (Anderson and Cham 1988; Figure 5.2), little more than 10 km from Woburn. Palmer (1947) stated that, apart from possible escapes from Woburn Park, there were 'no wild or feral ungulates' present in Bedfordshire. At that time, muntjac were rare but Key (1959) noted they were recorded from several places in the 1950s.

Even by the 1960s, naturalists were unsure whether they were observing Reeves' muntjac or hybrids between this species and Indian muntjac. Skull measurements and chromosome studies finally demonstrated that the only species still present in Britain was Reeves' muntjac (Chapman and Chapman 1982; Chapman et al.

1983; Chapman and Harris 1996). The two species have been known to hybridise in captivity but their progeny were infertile, so had they hybridised in England in the first part of the twentieth century, the young would not have been able to breed.

Between 1947 and 1952, there were several deliberate releases of Reeves' muntjac from Woburn to attempt to establish free living populations elsewhere in England (Chapman et al. 1994a). The 12th Duke considered that an optimum group for release comprised four does and five bucks. Deer were taken by lorry at night and released in at least five areas: Bix Bottom and near Bicester in Oxfordshire, Elveden on the Norfolk/Suffolk border, near Corby in Northamptonshire and an unknown location in Kent. These releases accounted for a number of the early records of muntjac away from Bedfordshire. Later there were many other deliberate releases or accidental escapes which helped to augment existing populations and extend distribution to new regions. Some releases will have been of accidentally trapped and/or injured animals that had been rescued and perhaps treated before being released into what was considered to be suitable countryside.

One factor that checked the spread of muntjac was cold, hard winter weather. Muntjac were not particularly numerous by the late 1940s but there was significant mortality in 1946/7 in Hazelborough Forest in Northamptonshire where a population had apparently developed from one of the suspected releases in the 1930s (Chapman et al. 1994a). These authors reported that the long, cold winter of 1962/3 resulted in significant mortality in Northamptonshire, Bedfordshire and Hertfordshire. No winter since has matched those two for harsh conditions, and this is probably one reason why muntjac have been able to spread and increase so much after escaping or being released. Whitehead (1964) detailed reports from 16 English counties, including several where they were well distributed. This information will have largely predated the cold winter of 1962/3.

5.2.2   Mapping the spread

By the 1960s, muntjac were generally distributed through Bedfordshire, Hertfordshire, Buckinghamshire and Northamptonshire into Warwickshire, as well as having small, scattered populations elsewhere in southern England (Whitehead 1964; Chapman et al. 1994a). The spread of muntjac from 1960 to 1985 was mapped by Anderson and Cham (1988) using BRC data; the number of 10 km squares recorded increased steadily from 12 to 280 during this period.

Published distribution maps of colonising muntjac will always lag behind reality, but it seems that some may lag further behind than others. The number of squares recorded will be affected by the amount of effort and care that went into the survey as well as actual distribution. Large numbers of recorders may be enlisted, but deer can be difficult to identify and there is something to be said for tending to rely on specialist recorders rather than canvasing records from all sections of the community and taking all information at face value. Nevertheless, with a small well-known species, such as the muntjac, a broad request is probably acceptable. It would be less so for the water deer which is still more poorly known. All of these surveys seemed to rely entirely on sight records of animals. Muntjac can be secretive and difficult to see, and this is especially true with colonising individuals or low-density populations. In such situations, looking also for their distinctive slots, deer paths or browsed vegetation may considerably improve the likelihood of demonstrating presence. There may, however, be reluctance on the part of survey organisers to accept such evidence.

It is instructive to compare the maps of Arnold (1993) and Chapman et al. (1994a): the former BRC data set amounted to 421 occupied 10 km squares, whereas the latter survey amassed records for 751 squares. They were published one year apart, but the latter had roughly 80% more records. Were these valid records that the other data set missed or were they suspect records or possibly sightings of isolated muntjac that subsequently moved on or died? The map of Chapman et al. (1994a) showed a general expansion around the main core range depicted in the map of Arnold (1993) and, assuming that both natural spread and movement by humans were (and still are) occurring continuously, this is the type of expansion to be expected. Chapman et al. (1994a) recorded presence in areas as widely spread as Yorkshire in the north, Hampshire in the south, Norfolk and Lincolnshire in the east and parts of Wales in the west. Results from subsequent BDS surveys reflected such expansion, and muntjac became established in some of these areas through to the present day (Smith-Jones 2017).

The BDS map of 2002 (Ward 2005) had a similar number of occupied squares to the map of Arnold (1993), but differed in indicating more complete presence in East Anglia and Lincolnshire but less cover in western England, including the Midlands. Despite its later date, it had considerably fewer occupied squares than the map of Chapman et al. (1994a) in the west and south of England, perhaps because of few keen Society members in those areas. On the other hand, it had more records in Lincolnshire and northern Cambridgeshire – trends which continued to be displayed by the later BDS maps of 2007 (Ward et al. 2008) and 2011 (Hailstone 2013). This evidence of the colonisation of Lincolnshire is consistent with the increase reported by Manning (2006); and, as is described elsewhere in the following chapter, the fens of north-east Cambridgeshire were colonised later than the rest of the county and still hold relatively low densities. The 2007 BDS map was similar to the map of Chapman et al. (1994a) in its depiction of the western and southern areas reached by muntjac.

By 2011, the BDS map showed a main area with few unoccupied squares extending from the east across the Midlands and from the south coast north as far as Lincolnshire. There was still expansion at the fringes, most notably in north-east England. Records in central and northern Scotland that appeared on the 2007 map were omitted without any explanation from the 2011 map (but not from the analysis). The first records for Northern Ireland did, however, appear on the map for 2011.

For the 2016 survey, maps indicated which old records had not been confirmed that year. The map for muntjac (Figure 5.3) demonstrated that many old records scattered across the south-west of England, southern Wales and the northern England had not been confirmed. Some of these apparent losses might have involved translocations of individuals that failed to prosper or had been eradicated by shooting (Smith-Jones 2017). On the other hand, there was evidence of expansion around the edges of the main range in southern England and the Welsh Borders; and there was a particularly remarkable mass of new records to the north-west of the previous main range. The map also showed a few new records in Scotland, where the species was thought unlikely to persist, and some consolidation of its

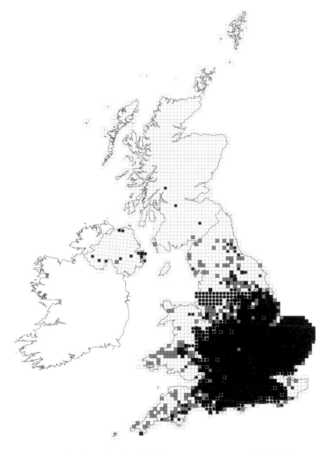

Figure 5.3   Muntjac distribution in the BDS survey 2016 (Smith-Jones 2017):
records confirmed only in 2016 (black dots), earlier records reconfirmed in
2016 (black squares), earlier records unconfirmed in 2016 (grey squares).

presence in Northern Ireland. The overall impression was of some increase in
the area of established colonisation but with limited permanence of presence out
beyond its fringes.

Average *annual* rates of expansion calculated for the periods between BDS
surveys were 8.2% for 1972–2002 (Ward 2005) and 11.6% for 2002–2007 (Ward et al.
2008). Losses during the first period were taken into account, but it was not clear
whether this was the case for the second period. *Overall* expansion was 10.3% for
2007–2011 (Hailstone 2013); this calculation assumed that grid squares recorded in
2007 still held muntjac, even if no deer were recorded. In the report on the 2016
survey (Smith-Jones 2017), the *overall* increase between all squares reported up to
2007 and those reported in 2016 was 8.7%. It is difficult to compare these figures
directly but they may suggest a decreased rate of spread since 2007. A reduction in
rate of expansion might have been anticipated since 2007, in part because the coast
had been reached in the east and much of the south.

Chapman et al. (1994a) compared their map of muntjac distribution on a 10 km
square basis with land classes and found that muntjac tended to select a variety

of arable land classes and avoid marginal upland ones. Land classes could also be used to predict muntjac distribution using models. The models suggested that the greatest increase post-1994 was likely to be in the south-east, and the 2011 and 2016 maps indicated further colonisation in south London and the extreme south-east of England. Conversely, the models predicted more limited further colonisation of the west of Britain, including Wales, and these predictions are supported by the most recent BDS survey.

5.3   Spread of water deer nationally

5.3.1   *Early colonisation*

Just as with the muntjac, the founder stock of the water deer that occur in this country were brought to Woburn Abbey by the 11th Duke of Bedford. The first deer, a doe, was imported in 1896 (Chapman 1995), two years after Woburn's first Reeves' muntjac. By 1913, 19 had been imported and they bred well resulting in a total count of 126. Their source of origin was unrecorded, but recent genetic research by Richard Fautley (2013) has indicated them to be from the Chinese mainland.

In 1929 and 1930, a total of 32 water deer were taken approximately 15 km from Woburn to Whipsnade Zoo. The first time that releases or escapes from Woburn Abbey were mooted was in 1949 when the 12th Duke of Bedford referred to the park gates being occasionally left open during the Second World War, enabling some water deer to escape. The first report of the species away from Woburn was in Buckinghamshire in either 1940 (Lever 2009) or 1945 (Lever 1977). If the latter date is correct, then it is possible that Woburn was the origin of that sighting as the park is less than 3 km from the Buckinghamshire border. However, both Fitter (1959) and Whitehead (1964) suspected that some deer escaped from Whipsnade. As one edge of Whipsnade abuts the county boundary with Buckinghamshire, a date of 1940 for the first feral animal suggests it came from that site rather than Woburn. On the other hand, Nau (1992) doubted there was ever an established population based on Whipsnade escapes, but that does not rule out the occasional animal escaping and dispersing.

During the 1930s, small numbers of water deer were moved from Woburn to captivity near Leek in Staffordshire, to Wormleybury in Hertfordshire and to Cobham in Surrey, but there was no evidence of escapes or releases (Whitehead 1964; Nau 1992; Chapman 1995). In about 1944, Woburn also supplied water deer to Leckford Abbas near Stockbridge and Farleigh House near Basingstoke (Whitehead 1964). Apparently no deer escaped from the first of these Hampshire sites, but they proved adept at penetrating the fences at Farleigh House resulting in a small population on the Hampshire Downs, with one being shot in 1953 about 6 km away. A water deer shot north of Farnham, Surrey in November 1945 (Lever 1977) might also have come from Farleigh Wallop, 20 km to the west.

Introductions of water deer from Woburn to Studley Royal Park, Ripon, Yorkshire, provided further information on the distance these deer may disperse (Whitehead 1964). A total of ten deer were taken to the park in 1950 and 1952. By 1954, because of deaths and some escapes, only three or four remained on the estate and all of those were outside the park. In 1952, one deer was found injured about 14 km away near Harrogate.

In about 1950, two pairs of water deer were sent to Walcot Park near Bishop's Castle in Shropshire (Whitehead 1964). Despite the death of one buck, they bred well and had built up to a 'considerable number' when the owner sold Walcot Park in 1956. The owner removed about 20 deer to his new residence at Hope Court near Ludlow 25–30 km away, from which some escaped (Chapman 1995). Many of the remaining deer at Walcot Park were shot but some escaped into the surrounding countryside, and by 1963 it was estimated that these numbered about 20.

Thus the distribution map constructed by Whitehead (1964) showed small numbers living in the wild in Bedfordshire (and Buckinghamshire) around Woburn and Whipsnade, as well as near the translocation sites in Hampshire, Yorkshire and Shropshire. Kenneth Whitehead described the situation in Bedfordshire as the deer being established in some woods within about 8 km of Woburn Abbey, but mainly to the east towards Amptill and Flitwick, and south where the range had been extended by deer from Whipsnade. Previously, Whitehead (1950) had mentioned that he had seen a few in Northamptonshire and a friend had flushed one in Oxfordshire, but no exact locations or dates were given. In his 1964 account, Whitehead reported another deer near Bicester, Oxfordshire, in 1961, as well as a possible sighting at Weston Turville in Buckinghamshire in the same year. Bernard Nau (1992) considered that the Northamptonshire and Oxford sightings could stem from Woburn, and that the Buckinghamshire deer may have come from Whipsnade. However, Weston Turville is 20–25 km from both Woburn and Whipsnade, so either site could have been the source. The Northamptonshire border is a similar distance from Woburn. Although such a distance might be bridgeable by an individual deer, it was clearly not an isolated record and a deliberate introduction from Woburn in around 1950 might be suspected, as was documented near Woodwalton Fen in Cambridgeshire (Chapman 1995).

Water deer from Woburn were released near Woodwalton Fen sometime between 1947 and 1952. Although deer were seen in the reserve during the 1960s, no one initially realised that they were water deer. Water deer were also recorded in the Broads during the late 1960s, and information has been abstracted from the annual mammal reports of the Norfolk and Norwich Naturalists' Society. The first two water deer were seen in 1968 near Hickling after escaping from a collection at Stalham less than 5 km away; one of these was subsequently killed on the Stalham bypass. There were other road casualties between Outwell and Nordelph in 1969 and near Terrington St Clement in 1970 – both of these were on fenland roads in west Norfolk. Questions were asked in the report for 1970: how many were there, how long had they been there and where had they come from (Goldsmith 1972)? The only suggestion offered was that they had wandered from Woodwalton Fen. The first of these two deer was roughly 35 km north-east of Woodwalton Fen. Although this is a considerable distance to have travelled, the route would have consisted of quiet arable farmland with ditches in which to hide, and it is not inconceivable that these deer originated from Woodwalton Fen, where the population was building up at that time. Information from my local sites indicated a much slower rate of movement (Section 6.4). However, although most colonisation may be slow, that does rule out the possibility that individuals may move considerable distances in relatively short periods of time. They are animals that do not shun open spaces and will run some distance to escape perceived threats.

Figure 5.4   Water deer habitat along the River Ant in Norfolk in the 1980s – an ideal landscape for both living in and moving through.

In 1970, a water deer was also recorded at Cley reed beds on the north coast of Norfolk (Goldsmith 1972). This was another isolated record, being roughly midway between the sightings at Terrington St Clement and the Broads, but it proved to be the first of many sightings at this popular birdwatching haunt. By 1972, seven Broadland parishes were listed with water deer, which were thought to have originated from Horning as well as Stalham. The area colonised was more than 20 km wide and encompassed three river systems, the Bure, Thurne and Ant (Figure 5.4). By 1980, water deer were the most commonly seen species of deer in the Broads and were spreading out along the rivers, including the Yare.

Water deer remained generally scarce in Bedfordshire during the 1970s, despite Whitehead's assessment in 1964 and the fact that 20 were shot on farmland outside Woburn Park in 1972 (Chapman 1995). Indeed, in the reports of Bedfordshire Natural History Society, they were described as rare even in the 1980s, although they were by then starting to increase in number and distribution (Anderson 1973; 1989). In Hertfordshire, a water deer was recorded in 1974 on the Ashridge Estate, where they subsequently became established (Clark 2001). Ashridge is less than 5 km from Whipsnade.

The counties that had become the major centres of water deer in England (Bedfordshire, Cambridgeshire and Norfolk) held established, if not necessarily large, populations by the 1980s. The earlier feral populations living outside distant parks did not, however, fare well, although their exact fate is difficult to unravel. Those near Farleigh Wallop in Hampshire were said to have died out during the severe winter of 1962/3 (Carne 1964), but this author later wondered whether some of the muntjac being reported from Hampshire were in fact water deer (Carne 1981). As regards the two feral colonies initiated in Shropshire in the 1950s, the Bishop's Castle colony survived at least into the 1970s. Peter Carne (1999; 2000) was told by a foxhunting farmer that the deer near Bishop's Castle were sufficiently numerous that it was necessary to 'quieten' them; they then increased again by 1970, but there had been no recent news of them by the end of the millennium. When I enquired in Shropshire about these colonies in 2008 and 2009, no one I asked had any knowledge of them. Carne (2000) noted that the feral population near Ripon did not survive for long.

A possible explanation for the eventual failure of the four colonies is that they lacked a significant protected core population that could provide a safeguard against shooting or other factors which might cause extinction in a small isolated population in dry open habitat. Over the years, there have been other releases and escapes. For instance, vandals freed water deer held at Shinfield in Berkshire for research purposes by Reading University (Chapman 1995). They were reported living in the countryside, but did not appear to survive for long.

5.3.2   Mapping the spread

Numbers of 10 km squares, including outliers, on various maps are shown in chronological order in Table 5.1. The numbers show a steady increase from 1972 until 2016. Up until 2002, the range was expanding in and outside Bedfordshire, in Cambridgeshire and in Norfolk. During the latter part of this period, isolated occupied squares appeared in southern England and on the Suffolk coast, such as at Minsmere (Macklin 1990), and a cluster began forming in south-west Norfolk and west Suffolk, often being associated with the Breckland Forests. Escaped deer from Kilverstone Wildlife Park near Thetford were thought to be the source of some the reports in this cluster (Chapman 1992b; 1995).

The average *annual* increase in national distribution from 1972 until 2002 was calculated at 2.0% (Ward 2005). Between 2002 and 2007, this rate rose by more than an order of magnitude to an astonishing 22.2% per annum (Ward et al. 2008). Although this calculation did not take losses into account and was based on a lower number of records than originally reported for 2002, it seemed that water deer were suddenly taking over parts of eastern England. The BDS map for 2007 showed them to be around the entire north Norfolk coast apart from the Wash and to have connected up with previously isolated squares near the Suffolk coast. They had also

Table 5.1   Cumulative number of 10 km squares
shown on national distribution maps of water deer.

Year	No. of occupied 10 km squares	Source
1972	23[a]	Ward (2005)
1993	37	Arnold (1993)
1998	54	Cooke and Farrell (1998)
2002	62[b]	Ward (2005)
2007	136	Ward et al. (2008)
2011	149	Hailstone (2013)
2016	187[c]	Smith-Jones (2017)

[a] 12, excluding records before 1967 (Clark 1974)
[b] Given as 50 by Ward et al. (2008)
[c] Water deer were reported/confirmed in only 127 squares in 2016

spread inland and their distribution had virtually coalesced with the Breckland cluster. Moreover, there was a new scatter of squares in south and south-western England.

How had they achieved this remarkable expansion after decades of much slower growth? One reason was that in coastal and Broadland Norfolk and Suffolk they had found much suitable habitat through which they could quickly move. And steady colonisation progressed on the peripheries of the other main centres. The coarser scale of recording in the BDS survey also made the presence of water deer appear continuous over much of Norfolk; they were recorded in nearly all of the coastal 10 km squares in Norfolk by 2007. The reality was that they were generally distributed in Broadland, but still tended to be rare elsewhere in the county, as illustrated by Mike Toms and Dave Leech with a tetrad (2 × 2 km) map of distribution in the annual county mammal report for 2004. In addition, this was a time when many field observers were at last appreciating that animals called water deer really existed, and that they might be colonising their own areas. It was a case of naturalists catching up with the species. In the Norfolk mammal report for 2007, Dave Leech (2008) said, 'Public fascination with this introduced deer has led to it becoming the most frequently recorded mammal species in the county.'

Colonisation in Bedfordshire progressed much more slowly than in Norfolk, perhaps because the deer were dealing with drier and more open landscapes. Nau (1992) summarised records of the Bedfordshire Natural History Society, the first of which was in 1969, peaking in 1991 with 24 records. Most sightings came from the south-west of the county close to Woburn and Whipsnade. In spring 1992, Bernard Nau undertook a detailed survey of roughly 200 km² around Woburn Park and estimated 20–50 pairs to be living there. He pointed out that the species' range was 'curiously limited and discontinuous', with the deer having disappeared from areas previously occupied. He considered it likely that local attitudes towards shooting the deer were at least partly responsible. The Natural History Society undertook surveys of the whole county in 1995–1999 (Tack 2000) and 2000–2006 (McCarrick 2007) and assigned the water deer to a grouping of mammals that were 'scarce', with 15% of the county's tetrads occupied in the first period and 16% in the second. Although the animal was well distributed in the west of the county, it was rare in the east. Water deer are known to have occurred in good numbers in both dry and wet habitats in neighbouring areas to the west of Bedfordshire, but more detailed published information seems to be lacking.

The BDS survey in 2011 (Hailstone 2013) suggested that colonisation by water deer slowed suddenly after 2007 with an *overall* expansion rate of 9.6%. As with the rate for 2003–2007, this figure seemed misleading. But this time the calculated rate probably under-estimated reality. Thus in Norfolk, the Naturalists' Society continued to record water deer in new tetrads away from the Broads, particularly along the north coast, whereas the BDS map only showed four new 10 km squares for the county in the whole period. The BDS map showed expansion elsewhere, especially in Buckinghamshire, and there were records from the Lincolnshire coast around Gibraltar Point.

Results for the latest BDS survey for 2016 (Smith-Jones 2017) are shown in Figure 5.5. This differed from earlier maps by showing previously occupied squares where no records were received in the new survey. This provided a welcome new

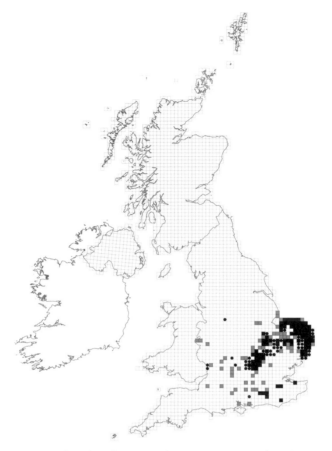

Figure 5.5   Water deer distribution in the BDS survey 2016 (Smith-Jones 2017): records confirmed only in 2016 (black dots), earlier records reconfirmed in 2016 (black squares), earlier records unconfirmed in 2016 (grey squares).

perspective, but what did it tell us? In some cases, where populations appeared to die out many years ago, it provided further evidence of what had been long suspected. But what are we to make of instances where squares that were first recorded in 2011 had no records in 2016, while neighbouring previously unoccupied squares had records in 2016? I believe this was likely to indicate groups of squares which water deer were moving into or through, so the species could not be guaranteed to be present at any particular time. The largest area where this seems to have happened is in East Anglia between the population in western Cambridgeshire and the main continuous mass of records further east in Norfolk and Suffolk. Water deer have been recorded at some time in recent decades in most of the 10 km squares in this area, but they have only become established in a few places. In time, if populations become well settled in suitable sites and provide a source of animals that will be recorded in less favourable areas, this area is likely to acquire a more solid and permanent appearance on distribution maps. The situation in Buckinghamshire is subtly different. Animals appear to be settling more quickly and expanding the core distribution in a (south-)westerly direction, possibly because the densities on farmland are higher there than in East Anglia and so small populations are less likely to die out or move on.

A comparison of the number of occupied squares recorded in 2016 (127) with the cumulative number collected up to 2007 (136) not surprisingly shows an overall decline of 6.6% (Smith-Jones 2017). Comparing cumulative totals, however, reveals *overall* increases of 37.5% between 2007 and 2016 and 25.5% between 2011 and 2016, demonstrating that the species is continuing to reach new locations, even if its presence is currently transitory in many instances.

5.4   Comparisons and conclusions

5.4.1   *An overview*

Figure 5.6 shows the cumulative number of occupied 10 km squares for the two species as recorded by the BDS. It is clear that both species have spread and are continuing to spread – and muntjac are more widely distributed than water deer. These may be obvious statements, but they are worth making in the light of the relatively high rate of expansion that has been reported in the past for water deer, especially during 2002–2007 (Ward et al. 2008). The graph demonstrates that although the percentage rate was higher for water deer, the muntjac had a much wider distribution, and so its expansion was greater in terms of the number of 10 km squares colonised between 2002 and 2007 – the number of squares recorded in 2016 was seven times higher than for water deer. The current distribution of the water deer is roughly comparable to the distribution of muntjac 40 years ago. There is no sign from this information that water deer will one day be as well distributed as the muntjac. Indeed, the gap in the extent of their spread has consistently widened. Nevertheless, it should be remembered that muntjac were probably running wild in this country for roughly 40 years before the first water deer attained its freedom.

Both are alien species and have been added to Schedule 9 of the Wildlife and Countryside Act, 1981. For muntjac, it has been an offence since 1997 to release them or to allow them to escape into the wild; and for water deer, which were added

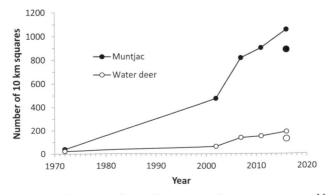

Figure 5.6   Cumulative numbers of occupied 10 km squares reported by the BDS indicating the spread of muntjac (black dots) and water deer (open dots) in the UK. Larger dots show number of squares reported occupied in 2016.

to the Schedule by a later amendment, it has been an offence since 2010. Legislation does not seem to have had any obvious effect on the spread of either species.

If left alone, muntjac have a rate of spread of about 1 km per year (Chapman et al. 1994a). Without human intervention apart from locally at Woburn, they might by now have reached Nottingham in the north, Winchester in the south, Colchester in the east and Gloucester in the west. That deliberate and accidental human assistance has been important is illustrated by the fact that the species has been well established beyond these limits for many years. But even though it was quite well spread by the late 1960s, it was not a common animal except in certain localities in its core area. Since then the increase has been considerable – over much of central, southern and eastern England, it is now a familiar presence. The lack of really severe winter weather has helped. Even in one of the hardest of recent winters, 2009/10, a survey organised for the BDS by Jochen Langbein (Anon. 2010) failed to detect any significant direct mortality.

It is pertinent to compare the 2016 BDS maps for the two species with those on the NBN Gateway in the same year. For the muntjac, both versions had more or less total coverage for central and southern England from the Wash westwards to Wales, apart from patchy cover in Kent and incomplete cover in the south-west where, in the BDS map, many records were not confirmed in 2016. The BDS map had more records continuing northwards up to the Humber, but had fewer current records further north in England and Scotland because many older records failed to be confirmed in 2016. A similar situation occurred in Wales. In Northern Ireland, the BDS map had more records because of those added in the 2016 survey. For water deer, the two maps differed particularly in the area between western Cambridgeshire and Norfolk because the NBN Gateway map displayed all records irrespective of whether deer were observed in 2016. This map also showed a number of outlying records where deer had not been recorded for many years.

Other organisations are now contributing to our knowledge of changes in deer populations. For instance, mammal data collected by contributors to the BTO's Breeding Bird Survey confirmed that muntjac spread from central England to East Anglia and established a high-density population (Newson et al. 2012). So far, water deer have not been sufficiently numerous to enable changes to be monitored. In 2016, they were recorded in only 11 of the BTO's 1 km survey squares compared with 201 which held muntjac (Harris et al. 2017), a ratio of 1:18.

5.4.2 *Habitat and spatial considerations*

Muntjac have been able to colonise much of lowland England because they have found habitat to their liking, and woodland with species-rich ground and shrub layers and varied tree composition has been particularly important. Amounts of woodland were generally not great in counties in lowland England when muntjac were first released in the early twentieth century, but soon after the First World War the Forestry Commission was set up to increase timber production – and muntjac will utilise coniferous forests. Densities may not be as great as in some deciduous woods, but a few conifer forests are huge by English standards and the deer populations can be very large. In recent decades, there has been a greater emphasis on varying forest composition and structural diversity that favours a greater range of wildlife, including muntjac. At the same time, there has been conservation and

Figure 5.7   Water deer are most at home in wetlands, such as Woodwalton Fen (KR).

enhancement of existing broadleaved woodland by the Wildlife Trusts and the Woodland Trust, as well as creation of new woodland such as community forests or plantations initiated under various grant schemes.

A certain amount of dense cover is required into which muntjac can dive if danger threatens. Away from woodland, scrubby railway embankments, shelter belts, fields of biofuel crops and overgrown hedgerows are just a few of the habitats that muntjac colonise in the countryside. In addition, muntjac are not especially shy of humans and now live in villages, suburban areas and even city centres across their main range in England. Charles Smith-Jones (2017) drew attention to the fact that official statistics demonstrated that considerable growth had occurred in the human population of the UK between 2007 and 2016. Any increase is less likely to have affected muntjac than water deer, which have tended to avoid human habitation.

In the vast Yancheng reserve in China, water deer favour undisturbed, tall, damp grassland or reed bed (Zhang et al. 2006; Figure 2.12), and they form their densest populations in such habitat in this country (Figure 5.7). Wetlands have been reclaimed by man for millennia and were relatively rare and fragmented in England in the early part of the twentieth century. More recently, however, conservation organisations have been protecting existing wetlands and creating new ones, especially in order to conserve birds such the bittern *Botaurus stellaris* and marsh harrier *Circus aeruginosus*. This has led to water deer turning up in coastal and inland reserves of the Royal Society for the Protection of Birds (RSPB) at Titchwell in Norfolk, Minsmere and Lakenheath Fen in Suffolk and Ouse Fen in Cambridgeshire (Figure 6.5). Lakenheath Fen, for example, is converted arable land and now holds water deer, muntjac and roe. Moreover, water deer occur in wetland sites of Natural England, the National Trust and county Wildlife Trusts, and especially in the Broads where the National Park is managed by the Broads Authority.

However, having an optimal habitat that is still rare means dense populations of water deer have to date remained limited away from places where they are managed for shooting. In contrast, if colonising muntjac could find a wood with good cover and of reasonable size, a large population might develop. Also the only documented deliberate and successful introduction of water deer into the wild of which I am aware was at Woodwalton Fen. Their susceptibility to capture myopathy may have deterred some attempts at deliberate introduction and prevented others from succeeding (Cooke and Farrell 1998; Smith-Jones 2017). While it is naïve to imagine that there have been no other successful translocations, expansion of the species in this country has probably been largely dependent on their abilities to escape from captivity and to disperse naturally. In this respect, their propensity to live and run in open country may have meant they are much more able than muntjac to disperse quickly over considerable distances. Moving 16 km from Ripon to Harrogate in either a matter of months or a couple of years in the 1950s was an early example. More recently, there have been several other possible instances of longer jumps occurring and it is becoming increasingly difficult to dismiss such long-distance movement out of hand. This is a subject about which we need more information, possibly by catching and satellite-tagging young bucks dispersing away from centres of dense population.

At this stage, it is worth collating what information is available on rate of dispersal of water deer. Most individuals living wild in this country occur in three populations that have developed and expanded in south-west Bedfordshire, western Cambridgeshire and the Broads. On the more recent BDS maps, these populations are starting to coalescence. Rate of dispersal and colonisation of Cambridgeshire is dealt with in detail in the next chapter – dispersal away from Woodwalton Fen to form other populations has occurred at a rate of much less than 1 km per annum. The other situations appear to have started as escapes from captivity and their rates of spread have varied.

The rate of expansion through the drier landscapes of Bedfordshire has been slow. The distribution map for 2000–2006 showed the main range of water deer to be largely restricted to the west of the county, extending only about 20 km to the north of Woburn (McCarrick 2007). The first record of wild water deer in the county documented by the Natural History Society was in 1969 (Anderson 1973), giving a rate of expansion of less than 1 km per annum.

By 1970 in the Norfolk Broads, water deer were only known from the odd sighting and road casualty, but two years later they occupied an area more than 20 km across. By the year 2000, they were 30 or more km away from the centre of the Broads in certain directions, although some of the more distant deer may have owed their presence to other sources. Nevertheless, the overall colonisation rate seemed to be in the region of 1 km per annum. The Broads is, though, a uniquely wet area with corridors of suitable habitat to aid dispersal (Figure 5.4).

CHAPTER 6

Colonisation of a county

6.1 Introduction

The previous chapter looked at national colonisation at a coarse level. This chapter describes at a much finer scale how muntjac and water deer have colonised Cambridgeshire. It charts in as much detail as is available how the two species moved through the landscape. Colonisation of specific sites is covered in Chapter 8.

Anyone living where these deer are currently colonising or might colonise in the future could use information in this chapter to try to work out potential routes, rates and consequences of colonisation and how the situation might be monitored and managed. In this respect, consulting Ordnance Survey maps and other depictions of the countryside, such as Google Earth maps, can prove very useful.

During the colonisation process, there were changes in administrative boundaries. In 1965, a small part of the old county of Cambridgeshire (Vice County 29), the Soke of Peterborough (part of VC 32) and Huntingdonshire (VC 31) were amalgamated to form the new county of Huntingdon and Peterborough (Figure 6.6). In 1974, this county was combined with the remainder of VC 29 to form the new and current county of Cambridgeshire (as in Figure 6.1).

The account begins with a brief consideration of the basic mechanisms involved. This is then followed by a detailed description of when and how the species colonised the county and the chapter ends with comparison of colonisation rates of the two species in the area I know best, the former county of Huntingdon and Peterborough.

6.2 Mechanisms of colonisation

When trying to understand the process of colonisation by deer, such as muntjac, it is helpful to put yourself in their position. What they especially need are food, cover and the occasional opportunity for contact with the opposite sex. Planned introductions of muntjac by the 12th Duke of Bedford in around 1950 led in some instances to successful colonisation (Chapman et al. 1994a), and so such an introduction may be a good place to start. Small numbers of bucks and does were released individually and 'some distance' apart into an area with suitable habitat. The Duke had done his homework, so it is likely that food and cover were sufficient for them to settle down within that general area. Providing they were not too dispersed, their paths would quite literally have sometimes crossed, and they would have been aware of one another. Once a doe came into oestrus, a buck would probably not be far away and mating would occur, so the potential existed for a successful introduction with a population becoming established. If habitat was not initially limiting, there would have been no need for progeny to disperse, but some, especially young bucks, may have done so.

What then happens to emigrants? In a forest situation, a young buck should not have to move very far to escape the attentions of rival adults in order to settle down to a peaceful life with adequate food and cover. In the King's Forest in Suffolk, dispersing juvenile and subadult muntjac utilised less suitable habitat than their parents (N.G. Chapman et al. 1985). The sex aspect might be covered by young of both sexes moving out of their natal areas or by a buck chancing his luck with does on the periphery of the main population. In this way, a forest area could be steadily colonised. In the King's Forest, after the first record in 1963, the species became increasingly common up to the 1980s.

But how will animals behave if they are moving through a mixed landscape with dispersed woodland? Any small fragment of cover may appeal to a migrating muntjac, but how long it stays will depend on what the patch has to offer. In general, if the site is small, lacking variety and if food is scarce or only seasonally available within the vicinity, it will not be long before the deer moves on. If, however, it happens to stumble on a large, mixed wood, its basic daily needs should be catered for and it might settle down for a while. Eventually, though, a lonely buck will start looking for a mate, and his wanderings are likely to take him further and further afield – until, perhaps he leaves the wood behind and ventures once more across open country. If he is a member of the advance guard of muntjac colonising a region, there is no guarantee that he will ever find a mate. But if eventually a buck and a doe do meet, they may mate and reproduce in a new site and the process will begin again. In a big wood, it might be several years before there is significant pressure on young deer to emigrate.

It is interesting to consider whether colonisation occurs at a faster rate in a forest than in mixed countryside. The forest has continuous habitat, so dispersion may occur at a steady rate. In more open country, migration will occur more in fits and starts. And although a buck might move several kilometres through farmland to the next patch of woodland, this will not truly represent colonisation if he cannot find a female before he dies. Sometimes a lone buck is seen at a site many years before the main vanguard arrives, as happened in the nature reserve at Grafham Water (Figure 6.3).

The early stages of colonisation in mixed countryside evidently involve much wandering about. The widespread trails in the snow in southern Cambridgeshire in 1963 (Fordham 1963) were probably the result of this process being accentuated and made more obvious by the persistent snow cover. I recall hoof-prints around a pen containing muntjac in Monks Wood in 1970 and 1971, which may have been evidence of a buck or bucks who, in the very early days of colonisation of the wood, found they had plenty of food and cover, but no female company.

Water deer differ from muntjac in that they seem happier living in a more open arable landscape. Their highest concentrations on farmland tend to be found near dense wetland populations or close to collections from which they or their ancestors may have escaped. Farmland close to the Broads and Woodwalton Fen may host large numbers of water deer, and some areas of Bedfordshire near Woburn Park have sustained semi-managed populations for several decades. Deer in the latter situations are resident on the farmland, but many seen outside wetland sites elsewhere only commute out to feed on arable or grass fields. Water deer have been present on dry farmland in Bedfordshire for many years and sizeable populations have now formed further west.

So long as a colonising water deer has sufficient food and also cover, perhaps in the form of dry ditches or hedgerows, it may settle for a while. But a buck on its own as the midwinter rut approaches is likely to start searching for a doe. Where wet areas exist, deer are likely to gravitate towards them, but if they are absent, then any quiet tract of countryside might suffice if a doe is present. In the previous chapter, it was argued that individual water deer may be more likely than muntjac to undertake quite rapid and long movements through the countryside. This species has, however, shown a reluctance to colonise countryside far from Woodwalton Fen in any numbers. The situation in the Norfolk Broads is different as deer have been able to move along wetland habitat into new terrain (Figure 5.4) and eventually reach coastal marshes. Good habitat is also continuous along parts of the coast of Norfolk and Suffolk, so facilitating 'natural' colonisation.

In the past, people overestimated the ability of muntjac to colonise new areas quickly. But now we appreciate that most of the examples of distant colonisation were due to human intervention, it is important not to make the mistake of underestimating the ability of either species by automatically assuming that all records well away from known populations must be due to movement by people. Norma Chapman has told me that there were examples of tagged muntjac migrating more than 10 km away from the King's Forest in Suffolk.

There are four questions to answer that will help to decide whether any new record is due to natural colonisation or the involvement of people:

1. What are the nearest known sources of wild or captive deer? It is of course possible that muntjac in particular may have colonised more closely than records currently indicate, and such facts should become clearer in time.

2. What is the distance to the nearest record and has there been enough time for the movement to be made? This is the most difficult question to answer, and to a degree depends on the nature of the landscape between the old records and the new one. But most journeys are probably explainable if the computed rate of movement is 1 km per year or less (Section 6.5). However, it also depends on timescale. For instance, I believe water deer can migrate at least 10 km in a single year, but 50 km in five years would be less believable – and 50 km in one year much less so. At present, we have to admit that we simply do not know enough about migration rates and distances attainable for an animal like the water deer.

3. How many deer appear to be at the new site? If there is some difficulty in believing that a single deer could have made the migration under

consideration, then any doubt will be compounded by the observation that several individuals have suddenly turned up.

4. Is it certain there are no deer between the new site and its nearest neighbour with deer? This can be difficult to answer at the time. If it is later revealed that deer were absent from suitable habitat between the two sites, but subsequently colonised close to the new site, then this is further evidence that deer were introduced to the new site and have spread out from there.

6.3   Spread of muntjac in Cambridgeshire

6.3.1   Origins

Muntjac have found their way from their English origins around Woburn in Bedfordshire to occupy much of Cambridgeshire in the space of little more than 100 years. They now occur mainly in western and southern Cambridgeshire because the open fenland in the north and the east of the county is much less suitable for them.

Bedfordshire and Cambridgeshire share a boundary, so it is reasonable to suppose that one route of colonisation involved simply walking in. One of the earliest places in Bedfordshire to be colonised was Maulden Wood (Figure 5.2), where muntjac were well established by 1938 (Anderson and Cham 1988). This wood is on the relatively well-wooded Greensand Ridge and this feature would have facilitated colonisation in a north-easterly direction towards Cambridgeshire – at least until dispersing deer encountered first the A1, then the main London–Edinburgh railway line, which both run north–south through Sandy. The Greensand Ridge continues almost to Gamlingay in Cambridgeshire.

Between 1947 and 1952, there were five known releases of deer from Woburn to attempt to establish free-living populations elsewhere (Chapman et al. 1994a). Two of these are of relevance to colonisation of Cambridgeshire. One was at Elveden on the Norfolk/Suffolk border on the edge of Thetford Forest, to the east of Cambridgeshire. The second was at an unspecified location near Corby in Northamptonshire, on the edge of Rockingham Forest to the west of Cambridgeshire. It is possible that muntjac were already present near the latter site as they had been reported in three forest areas in Northamptonshire in the 1930s, including Rockingham, leading Chapman et al. (1994a) to suggest that there must have been earlier, additional releases in the county.

In the following discussion of colonisation of Cambridgeshire, it is generally assumed that there has been no human intervention unless facts suggested otherwise. It could be argued that a county as poorly wooded as Cambridgeshire would not suggest itself as an ideal place to liberate muntjac, but ...

6.3.2   Establishment

The first mention of muntjac in what is now Cambridgeshire was in old Huntingdonshire at Great Raveley, where Andrew de Nahlik believed them to be present in 1952 (Chapman et al. 1994a). So we start off with a suspect record. Great Raveley is about 45 km from Maulden Wood and 55 km from Woburn. It seems unlikely that there could have been natural colonisation from Woburn or Maulden in that time, especially as there were no records from further south in

Huntingdonshire. Moreover, there were no other records from within 10 km of Great Raveley until Monks Wood began to be colonised in about 1970. It was not until the late 1990s that a significant population built up in Raveley Wood, which is adjacent to the village of Great Raveley and between Monks Wood and the town of Ramsey (Figure 6.1, and see also Figure 1.4). The 1952 report appears to have been of one or more isolated, translocated deer or, alternatively, a recording error. It is conceivable that what had been seen were water deer. The introduction from Woburn that led to the Woodwalton Fen water deer population was sometime between 1947 and 1952 'in the vicinity' of Woodwalton Fen (Chapman 1995). Raveley Wood is only 2 km from Woodwalton Fen and is beside a road – just the place for unloading deer from a lorry. If a few muntjac were included with the water deer destined to found the Woodwalton Fen population, then these evidently soon died out.

The first definite records of muntjac were from Brampton Wood in Huntingdonshire in 1959 (Worden 1960) and Hayley Wood in old Cambridgeshire in 1961 (Symonds 1983). These are 33 and 26 km respectively from Maulden Wood and roughly 40 km from Woburn. So how did muntjac reach these two sites and what happened next?

Colonisation of Hayley Wood by natural dispersal from Woburn is believable (Symonds 1983; Chapman et al. 1994a) as they had as much as 60 years to travel 40 km from Woburn or up to 30 years to travel 26 km from Maulden. Although they had to negotiate the A1 and the main railway line, a chain of woods links Hayley Wood to the sites in Bedfordshire with no gaps greater than 3 km. There are several Cambridgeshire woods of significant size on, beside and beyond the Greensand Ridge all the way to Cambridge. It is one of the most wooded areas in Cambridgeshire and, once established in Hayley, muntjac should have had little problem moving from wood to wood. A problem they probably did encounter was that competing fallow deer may have kept their population generally low – Hayley Wood has never had a dense muntjac population.

The long, cold winter of 1962/3 will have killed muntjac locally and may have affected their expansion. Starving muntjac will forage more frequently and further, and long-lasting snow will reveal their wanderings. Their tracks were 'quite widely' distributed in early 1963 in parts of the south-west of old Cambridgeshire (Fordham 1963). A road casualty was recorded near Fowlmere in 1967 (Chapman 1977). In 1980, muntjac deer were reported from Waresley Wood (Jefferies and Arnold 1981), only 3 km to the north-west of Hayley Wood. Peter Walker has told me that, by 1985, they were well established and breeding in Waresley and Gransden woods. In the early 1980s, muntjac were widely distributed across the south of old Cambridgeshire (Symonds 1983). Ray Symonds' survey map showed clusters of dots in such well-spread places as woods by Madingley and Linton, Fowlmere and the area of Ditton Park and Widgham Woods. They appeared to reach Fulbourn Fen in 1982. In the same survey, muntjac were also reported within Cambridge itself and from Burwell. By the mid-1980s, they had reached Wicken Fen (Carne 2000), about 4 km along the local lode from Burwell or 15 km along the River Cam from central Cambridge.

The first reference to muntjac in Brampton Wood was not just to a chance sighting of a transitory individual. They were said to be resident throughout the

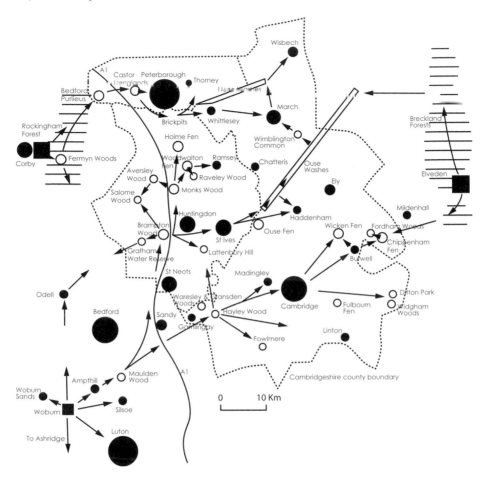

Figure 6.1   A map of Cambridgeshire and relevant surrounding areas with the county boundary and former internal boundaries shown as dotted lines. Towns and cities are shown as solid dots, countryside features are open symbols, forests are horizontally hatched and the A1 road is depicted as a line. Recorded sources of escaped or translocated muntjac are shown as squares. Arrows show possible colonisation routes of muntjac. A few places mentioned in the text do not appear on the map, especially those close to Grafham Water. In this region, the River Great Ouse flows through Bedford, St Neots, Huntingdon and St Ives, before dividing with the main river going on to Ely. The River Nene flows through Peterborough, along the Nene Washes and on to Wisbech.

year (Worden 1960), implying that colonisation began some time before 1959. It is not impossible that they walked all the way, but it seems a less likely explanation than walking to Hayley Wood as they would need to have travelled, for example, 33 km from the Maulden area in a little over 20 years. From the Greensand Ridge to Brampton Wood is a distance of 20–25 km, either by keeping to the west of the A1 and moving through quiet, well-hedged farmland, or by crossing the A1 and moving north along the valley of the Great Ouse. The route to the west of the A1 seems unlikely as muntjac were in Brampton Wood more than ten years before they were recorded in any nearby woods, including those to the south, which would have been expected to be colonised first. In fact muntjac had been in Brampton for more than ten years before they were definitely recorded anywhere else in the old Huntingdonshire. This would seem to rule out the route to the east of the A1 as well as that to the west. A direct introduction to the wood may therefore be the most plausible explanation. While there is no written evidence for this, Brampton Wood is a site that may well have appealed to the 12th Duke when he was selecting release sites during the period 1947–1952 (Chapman et al. 1994a). It is within easy reach of Woburn, is about 1 km down a quiet lane off the A1, and has a number of other woods nearby – and the timing fits.

Whitehead (1964) said that the range of muntjac in Bedfordshire then extended north from Woburn as far as Odell, a distance of 25 km. A band of woods curves round from Odell in a north-easterly direction, and after another 25 km ends at Brampton Wood. Thus, since the 1960s, it is likely that muntjac were colonising this arc of woods from either end. In any event, it appears they reached the cluster of woods around Grafham Water via Brampton Wood (Figures 6.2 and 6.3). The early 1970s saw a clutch of records from within 12 km of Brampton Wood from Grafham Water nature reserve in 1970 and Agden Green in 1973 (Jefferies and Arnold 1977), and from Southoe, Brington and Covington in 1973 (Clark 1974). By 1977, most woods to the west of the A1 and within 6 km of Grafham Water were colonised (Jefferies and Arnold 1978).

Another route of entry into Cambridgeshire appears to have been via Northamptonshire. Initially, muntjac did not prosper in Rockingham Forest, but some evidently survived in the late 1950s since territorial fraying was noted in Fermyn Woods near Corby in 1957 (Chapman et al. 1994a). Possibly the translocation near Corby in about 1950 played some part in this. The Rockingham chain of woods extends northwards, without any gap of greater than 2 km, to Bedford Purlieus about 15 km away. This wood is just in Cambridgeshire, and has held muntjac from the early 1970s at least (Welch 1975), presumably as a result of the Northamptonshire

Figure 6.2   A view of part of Brampton Wood taken by a drone in 2017, showing Grafham Water in the upper left-hand corner. The closest shore of the lake is only 2 km from Brampton Wood, but the nature reserve where the skull of a muntjac buck was found in 1970 is about 6 km from the wood (JH).

Figure 6.3   A photograph of Grafham Water taken in 1972
from the new plantation in the nature reserve.

releases. Another 7 km to the east is Castor Hanglands NNR, to which muntjac
were occasional visitors by the late 1970s (Harris 1981). Castor Hanglands is one
of the largest of a cluster of woods between the A1 and Peterborough that muntjac
will have been able to colonise readily. In this way, muntjac reached woods in the
Peterborough area several years before some woods further south.

6.3.3   Spreading in the south and west

Cambridgeshire has very little woodland and should therefore be intrinsically
unattractive to muntjac. In its favour, however, is that, during the colonisation
process, other species of deer were absent from large areas of the county. One of
the best wooded areas still devoid of populations of fallow deer is in the west of old
Huntingdonshire, and is centred on Monks Wood, Woodwalton Fen and Holme
Fen. Monks Wood was the first of these sites to be colonised, with the process
starting in about 1970 and being completed by 1985 (Cooke 2006). Woodwalton Fen
and Holme Fen are about 5 and 9 km respectively from Monks Wood (Figure 1.4),
and muntjac were first definitely recorded in these reserves in 1980 (Carne 2000;
Cooke and Farrell 2002). There was a record of muntjac from Woodwalton in 1972
and another from Holme in 1973, but both may have been misidentifications of
water deer (Clark 1974). Chapman et al. (1997b) suspected that muntjac colonised
Monks Wood as a result of a deliberate introduction, but I have no evidence to
support this assertion.

   Monks Wood is only about 11 km from Brampton Wood, so it is tempting to
assume that muntjac reached it via Brampton Wood (in 11 years or so). One possible
route would have been to have gone north and colonise via Aversley Wood and other
nearby sites, and then cross the A1 to Monks Wood. However, their population was
still low in Aversley Wood in the 1990s. Monks Wood was perhaps more probably
colonised by the route to the east of the A1 along the valley of the Great Ouse,
branching north up Alconbury Brook as the main river turns east at Huntingdon.
Brampton Wood is only 3 km from the Ouse and was probably the main source
of deer in the river valley. Some may have turned north up the Ouse from the
Greensand Ridge in Bedfordshire, but there seems to be a critical lack of records

along the Ouse in areas such as St Neots to support this suggestion. During the 1970s, Peter Walker and I independently undertook long-term studies on other natural features beside the Ouse at St Neots but failed to notice any signs of muntjac.

The first muntjac reported by Alconbury Brook was in 1970 walking along a dry offshoot of the brook (Brian Davis, personal communication). Migration along the stretch of the Ouse beyond the brook is consistent with deer being recorded further east down the valley at several locations in the St Ives area by 1976 (Jefferies and Arnold 1977). Currently, it is possible to find signs of muntjac in many places beside the Ouse in Cambridgeshire. In 1977, muntjac deer were noted in several small woods beside Ermine Street at Lattenbury Hill (Jefferies and Arnold 1978). This area is quite isolated in open farmland, but is 8–9 km from Brampton Wood (and 5 km from the Ouse), so this population could also have originated from Brampton stock. From Lattenbury Hill, deer could have continued to spread to the east and the south, before meeting deer moving north from old Cambridgeshire.

Monks Wood eventually became the source for other woods in the old Huntingdonshire cluster, including Aversley. It is perhaps surprising that Aversley Wood was not colonised earlier from Rockingham Forest in the west. It is only 20 km from Fermyn Woods and there are even closer woods in the Rockingham belt. Woods to the west of Peterborough that are a similar distance from Fermyn held muntjac in the 1970s. Perhaps, when deer migrating east from the Fermyn area left the forest and arrived at the River Nene, they tended to move along the river to the north or south rather than cross to the open farmland beyond. This farmland now seems to have a thinly spread population of muntjac on both sides of the county boundary, although a few concentrations occur in isolated woodland, such as Salome Wood. There were recording gaps for this area in my 2010 map for the former county of Huntingdon and Peterborough (Cooke 2011), but these appeared to be due to lack of recorders rather than total lack of muntjac. In 2008, I searched six previously unrecorded tetrads to the west of the A1 and had little trouble finding signs of muntjac in all of them. This area may have had a dribble of migrants from several directions during the last part of the twentieth century – Salome Wood is 9 km from Brampton Wood and 5 km from Aversley Wood.

By the 1990s, it was noticeable that muntjac in the county had their densest populations in the bigger, quieter woods with larger blocks of woodland. The smaller woods tended to fill up later. At that time, muntjac were being found more frequently in even less optimal habitat such as brick-pits, shelter belts, game cover crops, railway embankments, hedgerows and gardens. Since 1973, I have lived in the Ramsey area, about 5 km to the east of Woodwalton Fen. I first saw a muntjac in my garden in 1995.

6.3.4   Colonising the east and north

Muntjac occurred in scrubby patches in the expanses of old brick-pits to the south and east of Peterborough during the 1990s. Even by 2004, however, they had seemingly failed to begin colonising the Nene Washes, as indicated by a lack of records on the provisional county map (Bacon 2005). In contrast, the Ouse Washes were marked by a finger of tetrad dots extending into the Fens as far as the boundary with Norfolk. While it is probable that deer moved down the Ouse beyond St Ives and on to the Washes at their south-east extremity, we should not forget the release of animals at Elveden in Norfolk, sometime during 1947–1952 (Chapman et al. 1994a).

Sightings in the Breckland Forests in Norfolk began in 1953 and gradually increased, until by the 1980s, muntjac were seen throughout Thetford and the King's forests and also occurred to the north in Swaffham Forest (Chapman et al. 1994a). The northern part of Thetford Forest is less than 20 km from the top of the Ouse Washes, so it is likely that the Washes have been colonised from both ends.

Suitable but isolated fenland sites, such as Chippenham Fen, may have been colonised from the east. Alastair Burn told me that muntjac built up at Chippenham in the early 1990s; their slots were reasonably abundant when I visited with him in 1998. Mike Taylor, the Site Manager, informed me in 2015 that they have increased in recent years. This reserve is less than 15 km from the southern end of the King's Forest, with a well-wooded corridor between. Norma Chapman lives in the Mildenhall area in Suffolk, between the King's Forest and Chippenham Fen; muntjac were rare there during the 1980s, but more common by the early 1990s. However, Chippenham Fen is only 6 km from Burwell, which muntjac reached in the early 1980s. This illustrates that as time passed, so more isolated spots might be colonised from more than one direction.

The village of Haddenham can seem isolated from good muntjac habitat if you drive through on the A1123, but it is only 5 km from the Ouse Washes to the north-west and 3 km from the Ouse to the south – and muntjac are well known to the villagers. Ely is the classic fen 'island'. I have been told of occasional muntjac in neighbouring villages dating back to the 1980s, but it is only recently that muntjac have become more familiar to inhabitants of the city. The deer that have managed to get there might have arrived from the north-west, the south or the east.

At first glance, large parts of the Fens can, even now, seem devoid of muntjac. This applies particularly to land from the Ouse Washes north to the Lincolnshire border or south as far as Wicken Fen. The provisional county map for muntjac showed few records away from the Ouse Washes for this area of more than 800 km² (Bacon 2005). This is a tract of flatland with virtually no woodland apart from orchards. A new *Cambridgeshire Mammal Atlas* for 2004–2014 has been published (Hows et al. 2016), but displays individual records not occupied tetrads as in the provisional map. This makes comparison of the two maps difficult, but it is clear that muntjac have recently become much more numerous and widespread in this part of the Fens. Where there were just six occupied tetrads up to 2003, there were about 70 records during 2004–2014. Nevertheless, by 2016, muntjac remained at lower density on the Fens and were still rarer towards the north-eastern fringes of the county. In 2008, after I searched six unrecorded tetrads to the west of the A1, I looked for signs in another six on fenland to the east of Ramsey, but found them in only one tetrad.

In 2013 and 2014, I decided to check some of the more promising sites in this poorly wooded, arable fenland to understand the colonisation process better. The Lattersey reserve in Whittlesey is abandoned, scrubbed-over pits. The reserve wardens told me in early 2014 that they had only ever seen one muntjac there, but my search for signs showed a small population was resident. This site is near the eastern end of the pit system in the Soke, so it was not surprising to find evidence of muntjac. Old wooded gravel pits on Wimblington Common stand out on the 1:50,000 Ordnance Survey map as the only green blotch in an otherwise white-dominated landscape. Signs of muntjac were easy to find there in 2013, and the proprietor of the garden centre beside the pits described how muntjac sightings had recently increased after gravel extraction stopped. This site is only 6 km from the Ouse Washes and this must be the likely colonisation route.

A small wood was planted in 1994 on the west side of the town of March. Slots, a well-used path and signs of browsing were evident in 2013, suggesting a small population. This site is about 5 km from Wimblington Common, but may well have been colonised from the west rather than the east. They also occur in the Wildlife Trust's Norwood Road Nature Reserve which is next to the yard of March railway station. Muntjac are rare in Wisbech in the extreme north of the county, but occur in sites such as the Pocket Park at the old General Cemetery and at a caravan park on the town's west edge. A likely route from March would involve walking north along the defunct, scrub-covered railway line, past Whitemoor Prison Nature Reserve (where they were fairly abundant by 2015) and then turning north-east along the River Nene. The caravan park, mentioned above, is only two fields from the river.

We are probably now at the stage when a wandering muntjac might turn up anywhere on the Fens or elsewhere in the county, and, if it finds somewhere that offers cover and food, it will stop for a time. Should it be a pregnant doe and if other migrants find the site, then a small population might result. Visits in 2013–2015 to other far-flung sites revealed the presence of muntjac from Dogsthorpe Star Pit and Eye Green Lake to the north-west of Peterborough, to Beechwoods Reserve and Cherry Hinton Limekilns Pit just to the south of Cambridge and to Fordham Woods close to the border with Suffolk.

6.3.5   An overview of muntjac colonisation

For a colonising muntjac, Cambridgeshire must have seemed an unremarkable county. It is of about average size, not overpopulated with humanity and does not

Table 6.1   A summary of the colonisation and management of muntjac in Cambridgeshire, 1950s–2010s.

Decade	Events
1950s	A toehold was established within the county; muntjac were on its southern, western and eastern fringes as a result of releases and escapes.
1960s	Despite losses in the hard winter of 1962/3, deer spread thinly across the south of the county.
1970s	Gains in the south were consolidated, deer spread up the west side and radiated out in the centre.
1980s	Colonisation gained momentum as significant populations developed in some sites and the first conservation impacts were noticed. Deer management started.
1990s	Prime sites reached maximum densities and muntjac became common in much of the countryside. Deer management increased.
2000s	Deer spread to fenland with colonisation being multidirectional. Any wood of appropriate dimensions was likely to hold muntjac. Management intensified.
2010s	Muntjac now occur throughout the county, turning up in all habitats including urban areas. Populations are largely controlled in woodlands.

have much woodland cover, but it was in the right place as regards the early stages of muntjac colonisation of Britain. What happened here can give pointers to what occurred elsewhere in the past and what might occur in the future in areas where muntjac are still increasing or are yet to become established. In counties on the fringes of the current range, where conditions are less favourable, colonisation may occur more slowly and the final densities may be lower, but the processes should be similar. In Table 6.1 the sequence of events in Cambridgeshire is laid out decade by decade; reference is made to deer management which is discussed in later chapters. It took 30 years from first colonisation until problems were noticed and management began, by which time muntjac were well entrenched.

6.4   Spread of water deer in Cambridgeshire

Woburn was the source of wild water deer in Britain, as it was for muntjac, having been brought to the park from the late 1800s (Chapman 1995). There were no deliberate early releases in the vicinity of Woburn, as was done for muntjac, but some escaped when the gates were left open during the Second World War, and there will also have been later escapes.

How water deer colonised Cambridgeshire is much less complex than the story for muntjac, although the start of the process is not entirely clear. The first mention of them was in 1958 when they were said to have been seen in Brampton Wood (Worden 1959). This was the year before muntjac deer were reported as being resident throughout the year in the same locality (Worden 1960). There were no more reports of water deer in the wood, whereas the muntjac population still survives today, so the report of water deer is probably a case of mistaken identity. However, mistaking muntjac for water deer would have been highly unusual (whereas the reverse was commonplace), and there is another intriguing possibility. I have argued above that muntjac may owe their existence in Brampton Wood to an introduction from Woburn in about 1950. If this is true, perhaps the 12th Duke of Bedford decided to include a few water deer with the muntjac? This could account for a transient presence within the wood in the 1950s.

This is pure speculation, but what is more certain is that Cambridgeshire's current main population did originate from another introduction from Woburn. Norma Chapman (1995) reported that a small number of water deer from Woburn were released in the vicinity of Woodwalton Fen between 1947 and 1952 – that is at the same time as the muntjac introductions to other sites. Signs of small deer were first seen in the NNR at Woodwalton Fen in 1962. The deer were eventually identified by Raymond Chaplin in 1971, by which time the population was well established. They were still rare in the wild in Bedfordshire in the early 1970s (Anderson 1973).

Woodwalton Fen is about 5 km from the other NNRs at Monks Wood and Holme Fen. I saw a live water deer by the road next to Monks Wood in 1977. Signs of small deer had been seen in the wood earlier in the 1970s, but they were probably evidence of muntjac. To reach Monks Wood, water deer had to cross the main railway line, either over the tracks or underneath by tunnel or culvert. At Holme Fen, hair and small deer droppings were noted in the mid-1970s. These were probably signs of water deer; the species was first definitely identified in 1977 (Cooke 1998c). Small populations of water deer resulted at both sites. By 2010, their main concentration could be found through a tract of quiet farmland that included these three reserves and measured roughly 10 km from west to east and 12 km from north to south (Cooke 2010; 2011). Many of these deer were resident on the arable farmland rather than in the reserves. By 2015, mainly because of loss of fields of elephant grass, numbers had declined although the area occupied had increased to 10 × 15 km.

Outlying records, often road casualties, have occurred at distances of up to at least 15 km from the core area, suggesting some individuals dispersed further. Cambridgeshire is a county where, if a water deer can avoid becoming a road casualty or being shot, it could live and move through normally peaceful countryside for as long as it wanted. Elsewhere in the country, there have been instances where escapees have quite quickly migrated more than 10 km, and sometimes apparently much further (Whitehead 1964; Lever 1977; Nau 1992). The motorway stretch of the A1, which runs 5–6 km to the west of Woodwalton Fen, may have inhibited western movement to some degree. In Bedfordshire, the M1 motorway was first known to be crossed by water deer in 1975 (Nau 1992) and pockets of water deer occur beside it today.

In 2006, I was told by Dave Muttock about a population that had existed for about 30 years in and around Ditton Park Wood (Figure 6.1) on the Stetchworth Estate some 20 km east of Cambridge. Stetchworth is about 50 km from Woodwalton Fen and how they got there is a mystery, but introduction is a likely explanation. A single water deer could conceivably have walked from Woodwalton to Ditton Park, but it is stretching the laws of probability to suggest that enough deer to start a population could have done so by 1976.

A water deer was first seen in Wicken Fen in 1982. This appeared to have been a single individual, and probably arrived there unaided by man. Wicken is roughly 35 km from Woodwalton Fen, but less than 20 km from Ditton Park Wood. The habitat at Wicken Fen should be very suitable for water deer, but even now individuals are seen there very rarely, possibly because the resident roe deer and widespread muntjac have inhibited colonisation (Figure 6.4). To the south and

Figure 6.4   Water deer have occurred on the reserve at Wicken Fen, but other species of deer, including muntjac, are much more numerous.

Figure 6.5   A small population of water deer has become established
at Ouse Fen. This population should grow as the reserve develops and
could provide a nucleus of animals for colonising along the river.

south-west of Wicken Fen is a very large area of quiet countryside which is crossed
by the River Cam and several lodes. Much of this is included in the expansion
scheme for Wicken Fen. Water deer already occur in this area, and are likely to
benefit when the Wicken Fen Vision comes to fruition. And water deer also occur
on a farm at Stuntney, about 7 km to the north of Wicken.

Water deer have also been recorded at the RSPB's wetland site, Fowlmere, in the
south of the county. And since 2011, a small population has been reported by the
RSPB as breeding in the reed bed on its Ouse Fen reserve near the village of Over
(Figure 6.5). When finished, this reserve will have 460 ha of reed bed, so there is
potential for a much larger population to develop.

The habitat of Chippenham Fen is similar to that of Woodwalton and Wicken
fens, but water deer have never been reported from there. It is only about 10 km
from Wicken Fen and 12 km from Ditton Park Wood, although the latter journey
would involve negotiating the town of Newmarket and the A14 trunk road (there
have been no reports of casualties along that section of the road). Chippenham Fen
has had significant deer populations for several decades – first of roe, and then of
muntjac – which may have had an inhibitory effect on both the resident roe deer
and on any migrating water deer.

The *Cambridgeshire Mammal Atlas* (Hows et al. 2016), using data collected during
2004–2014, displayed maps of individual records. That for water deer showed 41
records, more or less evenly dispersed through the county. Despite their very patchy
distribution, wandering animals might conceivably be encountered anywhere in
the county. A number of records coincided with roads indicating traffic casualties
or sightings of live animals close to the road. Water deer are still confused with
muntjac and roe deer, and as a consequence are comparatively under-recorded,
although not as much as they once were. The fact that they are now well dispersed
in the county means that they are probably on the cusp of becoming established
in a number of particularly suitable localities and probably also establishing

lower-density populations on farmland. Two of the records in the new atlas were within 3 km of Chippenham and it would not be a surprise if water deer were attracted there now the muntjac population is being controlled. Similarly, parcels of quiet farmland with more cover than most fenland might provide opportunities – the area beside the River Cam between Waterbeach and Upware could be one such area.

6.5   Comparison of colonisation rates in the former county of Huntingdon and Peterborough

In order to be able to examine and compare rates of colonisation, tetrad maps for muntjac and water deer in the former county of Huntingdon and Peterborough up to 2014 are shown in Figure 6.6. These have been constructed using information from Cooke (2011), Arnold and Jefferies (2012; 2013; 2014; 2015) and Flows et al. (2016). The two species seem to have arrived at about the same time, but have very different distributions with muntjac being recorded in 247 tetrads and water deer in only 48. Muntjac distribution remains patchier on fenland to the north-west but infilling will, doubtless, continue. Both species may have initially colonised from introductions of Woburn stock to single sites in around 1950. Ignoring the earlier doubtful reference to Great Raveley, muntjac were first reported in Brampton Wood in 1959; and discounting the report of water deer in Brampton Wood in 1958, their signs were first noticed in Woodwalton Fen in 1962. It is pertinent, therefore, to compare the rates of spread of muntjac from Brampton Wood and of water deer from Woodwalton Fen.

By the early 1970s, muntjac deer were reported from as far afield as Covington (12 km to the west of Brampton Wood), Southoe (5 km to the south) and Monks Wood (11 km to the north). By the late 1970s, they were at St Ives (12 km to the east), and, by 1980, they had reached Holme Fen (18 km to the north). This rate of spread is consistent with the rate of 1 km per annum proposed by Chapman et al. (1994a). The picture in later decades becomes more complicated because of colonisation from other sources and directions but, using data up to 2010 (Cooke 2011), muntjac had been recorded in 182 tetrads in old Huntingdonshire, 67 of which were within 10 km of Brampton Wood and 171 within 20 km. In theory at least, most of old Huntingdonshire could have been first colonised by Brampton stock and, had there been no other source of muntjac, current distribution in this part of Cambridgeshire might not be substantially different.

Water deer from Woodwalton Fen were not definitely recorded at Monks Wood or Holme Fen until the mid/late 1970s. Both sites are about 5 km away, so the rate of colonisation was much less than 1 km per annum. Even allowing for the fact that significant pressure to disperse from Woodwalton did not occur before the population was well established in the late 1960s (Cooke and Farrell 1998), the rate of dispersal was still less than 1 km per annum. By 2014, the core area of the population only measured about 10 km by 15 km, more than 40 years after the population was well established. There were outliers elsewhere in old Huntingdonshire, but no other known populations. By 2014, there were 31 recorded tetrads in the core area, with a further 17 more distant tetrads. The total of 48 was just 19% of the number of tetrads occupied by muntjac by the same time.

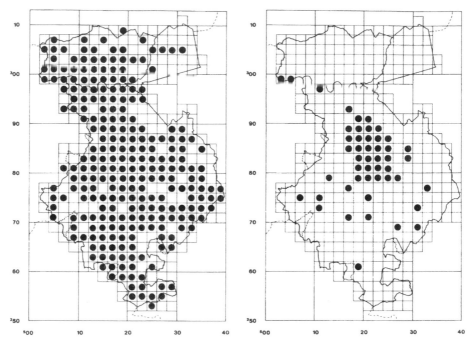

Figure 6.6   The recorded distribution of muntjac (left) and water deer (right) on a tetrad basis in the former county of Huntingdon and Peterborough up to 2014 (updated from Cooke 2011). The base map is divided into old Huntingdonshire in the south, the Soke of Peterborough in the north and a small part of VC 29 in the north-east.

So even though water deer had the benefit of a major wetland reserve as a sanctuary, whereas muntjac deer were faced with one of the least wooded areas of Britain, the latter easily out-colonised the former. The reserve at Woodwalton Fen has been a source of emigrants in most years, but water deer have struggled to establish themselves over a significant area of farmland. Elsewhere, however, such as in Bedfordshire and Buckinghamshire, their colonisation of farmland has been more successful. The reasons for this disparity are not clear but are likely to be connected to differences in habitat suitability, dispersal opportunity and the attitudes of landowners and deer managers.

One has only to look at national distribution maps of the two species to conclude that the muntjac has been the more successful coloniser (Figures 5.3 and 5.5). However, on a national scale muntjac deer were released first – and the relative number and locations of introductions of the two species are largely unknown. The advantage of analysing the situation in Huntingdonshire is that the two seem to have started at single local sites at around the same time, although in recent years introductions outside Huntingdonshire are likely to be contributing. If the different rates of colonisation are applicable to the country has a whole, this has relevance for surveillance and management more generally. Greater vigilance and control will be needed to prevent muntjac becoming established in new areas, regions and countries, should this be considered necessary.

CHAPTER 7

Methods for determining population size and changes at sites

7.1   Introduction

How can deer numbers be estimated or monitored over a period of years? This is a question that has taxed scientists and deer managers for a long time. It is easier to follow changes in population size than to estimate number of individuals because certain simple indices of abundance can be used. And it is easier to monitor changes qualitatively rather than quantitatively because an index may not be linearly related to population size.

When I began studying deer in the 1970s, I had been counting populations of waterfowl and amphibians for national and local monitoring programmes for a number of years, so it seemed logical to count deer that I saw. I was aware that on any single visit to a site I would see only a small proportion of the deer population. However, this seemed a reasonable way of monitoring trends providing that I counted in the same way for a fixed time or related numbers seen to effort so as to derive an index. Muntjac and water deer have relatively small and often individual home ranges in which they may reside throughout the year (and perhaps for several years). In this respect they differ from the larger herding species, such as red and fallow deer, which move around in groups over wider areas. So following trends in numbers of muntjac or water deer by counting is probably more informative than trying to do the same for the larger species.

This chapter outlines how counting has been used primarily to follow changes in population size and to monitor measures of recruitment and survival. In order to avoid arriving at misleading conclusions it is necessary to acknowledge and, if possible, take into account the limitations of this simple method. To do so involves understanding how deer behaviour and other factors influence observations. These considerations are also of fundamental importance to anyone who wants to see deer or study them, and so are examined below in some detail. Before that, however, several other methods are discussed.

7.2   Field methods

I had some knowledge of ecological techniques for estimating numbers of animals, based on catching and marking individuals. I had done this with frog tadpoles, but they were of course considerably easier to catch in meaningful numbers than deer dispersed through woodland or fen. Nevertheless, I took advantage of this and other techniques if opportunities arose. As time passed, new techniques were developed, and a standard text on estimating deer population size was published (Mayle et al. 1999). This book detailed 21 methods for estimating populations of the British deer species and gave examples of how these had been used. However, none of the examples mentioned water deer and only one referred to muntjac. This example was an estimation of numbers of red, roe and muntjac in blocks in Thetford Forest using a high-resolution thermal imager. Thermal imaging, combined with distance sampling, was developed as a method for estimating deer populations during the 1990s (Gill et al. 1997), long after I had started counting deer. This is now the method of choice for professionals estimating deer densities in woodlands. Thermal imaging equipment is, however, expensive and unlikely to prove as effective in low, dense vegetation such as reed beds. A BBC film crew had only limited success when it brought a thermal imaging camera to Woodwalton Fen in December 2015 to film water deer during the rut (and deploying a drone was also ineffective). I have looked at water deer at night through small monocular thermal imaging devices that are now available at a more reasonable price. Using one would allow me to record additional deer on evening surveillance walks, but after so long recording without one, I did not wish to change conditions that would render before and after comparisons difficult or impossible. In about 1980, I borrowed a military-grade image intensifier, but failed to see significantly more water deer at Woodwalton Fen than I did with good 10x50 binoculars, in part because I found it difficult to handle and use.

7.2.1   *Counting dung*
Several other methods depend on counting dung pellet groups (e.g. Putman 1984; Mayle et al. 1999; Ellwood 2000). One way is by counting dung that has accumulated in plots of known size – this is termed the 'standing crop'. Estimation of deer density requires knowledge of defecation rate (number of pellet groups produced per day) and decay rate of dung. An alternative method is to make clearance counts, where dung is removed from plots and then counted on a second occasion, the interval being less than the decay time of the dung. Here, a less detailed knowledge of decay rate is needed, but a figure for defecation rate is still required.

Unfortunately, there seems to be no information on defecation rate for water deer in this country; a further potential problem is that buck water deer produce more pellet groups, each with fewer pellets, during the rut (Sun et al. 1994). Also, where water deer occur, there are likely to be muntjac, and droppings of the two species often cannot be readily distinguished. Therefore, recording water deer dung cannot currently be used to estimate abundance, but it can provide information on when the rut is occurring and on the position of territories, as bucks mark the edges of their territories by defecating.

There are published figures for muntjac defecation rates (Mayle et al. 1999; Chapman 2004), but these are substantially less than those of other species and are only available for penned animals: these figures may be lower than would occur in the countryside and their use would therefore overestimate muntjac density. With this in mind, I have counted muntjac dung in plots or along routes in different situations for different reasons, but have typically used the data as an index for comparison between locations with similar habitat or to monitor the same locations over several years (Figure 13.4).

7.2.2   Counting paths

Frequency of deer paths crossing wood boundaries has been used as an index of the density of roe and fallow deer in woodland (Mayle et al. 2000). Deer paths were counted in and out of a sample of woods in south and east England varying in size up to 30 ha. Relationships were found between amounts of dung along transects in the wood and number of deer paths per 100 m of woodland perimeter. Thus, path frequency around the edge was considered to give an indication of deer density. Muntjac was the most widespread species in the sample of woods – its dung was recorded but nothing was said about the frequency of its paths.

Between 2002 and 2005, I counted muntjac paths at Monks Wood, Raveley Wood and at Marston Thrift in Bedfordshire (Cooke 2006). Deer paths (Figure 2.5a) crossing rides at the edges of the woods were generally fewer than those crossing more central rides. Applying the relationships found by Mayle et al. (2000) to numbers of paths at the edges of the woods gave densities of about 5–15 per km²,

Figure 7.1   A path well used by water deer at Woodwalton Fen out on to adjacent farm fields.

well below the probable densities of Monks Wood and Marston Thrift. I concluded that the published relationship gave unreliable estimates for muntjac density. Nevertheless, I continued to record deer paths along one edge of Monks Wood as the muntjac density decreased because of prolonged stalking. Number of paths and the proportion of well-used paths both declined as stalking progressed, indicating that such a technique may be useful in monitoring change in density (Figure 13.2).

Paths made by water deer have been recorded in recent years along 540 m of the western edge of the Woodwalton Fen reserve where deer have commuted to the adjacent farmland (Figure 7.1). The total number of well-used paths decreased from 59 in 2011 to 12 in 2017. Path counts reflected the general decline in the population during this period but were not a good guide to year-to-year changes in numbers counted at dusk in spring grazing on the farm fields. As with muntjac paths, some water deer thoroughfares can remain in use for several years, despite their tendency to become overgrown in summer in wet habitats. In South Korea, Rhim and Lee (2007) and Son et al. (2017) used counts of water deer paths as indices of abundance to judge the relative usage of different types of forest habitat.

7.2.3   Camera traps

Camera traps have been employed in ecological research with increasing frequency over recent years (O'Connell et al. 2011). These devices are particularly valuable in studies with mammals because many species tend to be nocturnal and relatively large. Such cameras could almost have been designed with tigers *Panthera tigris* in mind. The tiger is a large animal living at low density and with an individually striped coat. This means that individuals can be readily recognised and populations estimated using 'capture–recapture' techniques. Deer are less easy to work with because individuals live at higher densities and are harder to recognise. However, population size in white-tailed deer *Odocoileus virginianus* has been estimated in the United States using camera traps (Hamrick et al. 2013). The method depends on being able to differentiate between bucks, does and fawns, and to identify all of the bucks as individuals, primarily on the basis of antler size and structure. Cameras are deployed in a specified fashion. A population factor is calculated for the bucks (the total number of bucks identified divided by the total number of photographs of bucks). This factor is then used to estimate the number of does and fawns from the number of photographs of them. There is also an extrapolation factor that takes into account deer that were missed if cameras were deployed for less than two weeks.

Unfortunately identifying individual muntjac or water deer bucks from camera trap images is much more difficult than for species with complex antlers. At a site like Woodwalton Fen, there have been too many deer living at too high a density in recent years to have any chance of identifying all individual bucks. Some water deer bucks are apparently easy to identify on the basis of damage to their ears caused by fighting (e.g. Figure 19.3c). However, one can never be certain that seemingly unique injuries are not replicated in other individuals; furthermore, ear damage in an individual can worsen over time. Many water deer lacking obvious tusks seem to look alike especially when viewed on night-time images taken with infrared light.

Muntjac deer are rather easier to distinguish from one another by virtue of their facial markings (Demidecki and Demidecki 2016), particularly if there are few in the population. Although it is possible to overestimate the number of individuals present if an individual looks different on different occasions, a simple addition of deer believed to be present is perhaps more likely to under-estimate the total as some individuals may not be recorded. This can be minimised by having a long recording period, but then immigration and emigration are more likely to complicate the picture. I have used this type of technique inside both of the large deer fences at Monks Wood. In November and December 2012, I deployed cameras in nine locations for 11–21 days each inside the 10.6 ha fence in the south-west of the wood. Four individual muntjac deer were identified, but only two were believed to be present for the whole period because of movement through a gap under the wire fence. During the summer of 2015, cameras were used at two places inside the 6.1 ha coppice fence for 54 days each. This time six individuals were identified; images of four of these deer are shown in Figure 7.2. Identification of bucks was made easier because they were regrowing their antlers at that time, so differences in antler growth were probably more obvious than differences in final antler structure would have been. The images shown are not photographs but frames from videos. Although frames have much lower resolution, each ten-second video provided 100 images, so finding one with a deer motionless, close enough and at the correct angle was often successful. Distinctive nicks and notches in ears can be useful in identification, as can other injuries or abnormalities such as limps, scars or broken antlers, and I have even videoed a female without a tail. In the field, if muntjac can be seen sufficiently closely and frequently, it can be possible to recognise individuals with a reasonable degree of certainty by their appearance and behaviour (Blakeley et al. 1997).

A completely different method of determining population size models the process of animals being recorded by camera traps, thereby eliminating the need to be able to recognise individuals. This is the Random Encounter Model of Rowcliffe et al. (2008). Cameras are deployed as randomly as is practical, and their zone of detection should be known. In addition, the group size of the species being studied and how far animals range in a day need estimating. Coincidentally, the initial study using this method was undertaken at Whipsnade Zoo in 2005 on four species including muntjac and water deer, and was found to give results for the two deer species that agreed well with census counts. Humphries and Dutton (2015) tested the method on muntjac in the Forest of Dean and derived an estimate of 164 per km². This appears to be the highest density ever reported for a muntjac population in this country. A few years before, Gill and Morgan (2010) had spent 16 nights with a thermal imager in the forest without encountering a single muntjac; and in 2006, I accompanied George Peterken to Lady Park Wood in the forest, but failed to find any evidence of muntjac. Among other things, these results demonstrate the difficulty in assessing deer densities. However, the trail cameras registered even higher numbers of fallow deer and wild boar in the forest, and Humphries and Dutton considered that 'an issue with the camera calibration may have over exaggerated the density estimate by underestimating the speed of muntjac'. Nevertheless, on the Ards Peninsula in Northern Ireland, colonising muntjac were detected by this method at an estimated density that was 'within the expected range' (Dick 2017).

In the Forest of Dean, muntjac were estimated to have travelled an average of 380 m per day, whereas the figure for muntjac at Whipsnade was 8.3 km per day (Rowcliffe et al. 2008). Difficulty in measuring this critical piece of information

Figure 7.2   Frames selected from camera trap videos of four muntjac:
(a) buck with widely-spaced, symmetrical antlers; (b) buck with asymmetrical
antlers; (c) doe with broad facial pattern; (d) doe with a narrow pattern.

may be the Achilles heel of this method. Distance moved per day by muntjac in the King's Forest in Breckland was found to be in the region of 1 km (Chapman et al. 1993). Thus there may be considerable variation between situations. A further complication was that male muntjac in the King's Forest travelled further than females. Water deer at Whipsnade moved an average of 1.2 km per day (Rowcliffe et al. 2008). Xiao and Sheng (1990) found that water deer at Poyang Lake in China moved an average of about 110 m per hour in the hills in the flooding season and about 550 m per hour across the grassland in the dry season.

Because of such issues I have tended to use cameras in a simple way to provide an index of deer activity in specific locations. Thus, in Monks Wood, activity of muntjac and roe deer has been recorded in several summers to determine how it changes over time for each species. Without information on deer behaviour, it is not possible to translate activity into density, and comparing abundance of the two species requires other information including distance moved. In the King's Forest, daily distances travelled by muntjac and roe were roughly similar in spring but muntjac moved more than 50% further in early summer (Chapman et al. 1993). Placing cameras at random would be recommended for focusing solely on monitoring changes over a period of years. The disadvantage with this method is that deer tend to move along fixed paths and cameras placed away from such paths may not be very effective at detection. I have tried to maximise effectiveness because I have also been interested in obtaining information on diurnal activity patterns, on whether natural vegetation is eaten and if possible on recognising individuals. In an attempt to achieve this level of detection, frequency and usage of paths is checked in specific woodland compartments, a camera being positioned on the path that appears most used in each compartment (see Section 4.1). If the number of registrations changes appreciably in the following year(s), this may suggest a change in overall deer activity, depending on factors such as whether specification of cameras has altered or whether there are changes in the use of paths. Thus, if conditions stay much the same from one year to the next, but encounters per day tend to decrease, then an overall decrease in deer activity may be inferred. It is often found that deer use the same main paths for years, so camera positions may not change from year to year.

7.2.4   *Deer and damage scores*

Another index that I have used frequently for more than 20 years is the deer score. This is used in conjunction with the woodland damage score. Scoring signs of muntjac and their damage is a technique that I developed in the 1990s for my own use, but it has been adopted by other individuals and organisations, sometimes in a modified form. The complete method will be briefly described at this point although only the score based on deer signs is relevant to recording or monitoring the deer population itself; for descriptions and illustrations of signs see Section 2.6. The method was designed to provide an assessment that could be obtained during a single visit, enabling comparisons to be easily made between woods or over time at the same wood. Experience and/or training are required to use any of these methods effectively and consistently. I have undertaken all assessment of scores reported in this book.

Although scoring was initially developed for muntjac, it is applicable to other deer species with minor modifications. Where more than one species of deer is present, attempts can be made to separate signs of their presence, but this is more difficult for signs of damage. Typically a wood is visited for 1–2 hours, although woods larger than about 50 ha may need to be split up and each part scored separately. When a wood is being monitored annually, each recording visit should preferably be at the same time of year because season may have a minor influence on score. Fuller description of the original method is given in Cooke (2006). The Deer Initiative has run training courses for a number of years in the use of an improved version of the technique (Section 12.5.4).

For the deer score, the relative abundance of each of four variables (the deer, their slots, droppings and paths) is scored 0 (if absent), 1, 2 or 3 (if at maximum abundance, as was found in Monks Wood in the mid-1990s). The overall score can therefore vary from 0 to 12. Muntjac damage scores are based on recording the extent and frequency of five variables (browsing on woody vegetation, breakage or fraying of woody stems, browse lines and grazing on ground flora), each again being scored 0, 1, 2 or 3. Damage score can therefore vary between 0 and 15. Numerical scores are accompanied by written descriptions of observations. If used annually in a wood, the method ensures that features are observed and recorded on a regular basis (see Appendix 2 for a breakdown of scores at Monks Wood).

Here the word 'damage' is used to indicate that signs of feeding or other activity on vegetation are recorded. Damage will equate to 'impact' if it is sufficiently severe

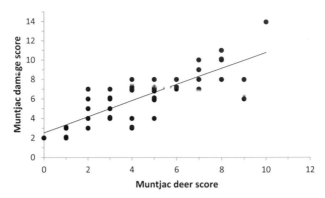

Figure 7.3   The relationship between muntjac deer score and damage score for 50 sites in Cambridgeshire. None of these sites had fallow deer.

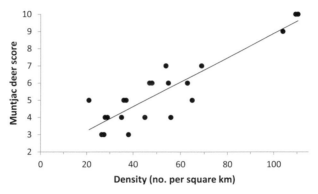

Figure 7.4   The relationship between muntjac deer score and estimated muntjac density at Monks Wood, 1995–2016.

or long term in nature. The relationship between deer score and damage score for 50 sites in Cambridgeshire is shown in Figure 7.3. For sites that were scored in more than one year, the year with the highest scores was selected. As a general rule, damage score increased as deer score increased – evidence that helped to confirm that muntjac were responsible for the observed effects.

Muntjac deer score is a simple index of density. How density has been estimated at Monks Wood is described below (Section 7.6.1). The relationship between deer score and density at Monks Wood (Figure 7.4) means that in future a single scoring visit could provide an indication of density without the need for deer surveillance. The relationship may differ in other woods, but as a means of deriving approximate muntjac densities more generally, I convert muntjac deer scores as follows:

- Scores of 8 or more correspond to densities of roughly 100 per km² or 1 per ha.
- Scores of 5–7 correspond to densities of roughly 50 per km² or 0.5 per ha.
- Scores of 2–4 correspond to densities of roughly 25 per km² or 0.25 per ha.
- Scores of 0–1 correspond to no deer.

Later in this book, I have cautiously applied this conversion to scores in other woods to provide estimates of muntjac numbers where no better information is available.

7.3   The method of counting deer

If simple counts are to be used to monitor changes in population, it is necessary to know how numbers seen vary through the day and through the year, as it is best to avoid times and seasons when few deer can be seen. It would also be an advantage to have an idea of the percentage of the total population that might be counted during a single visit under certain conditions. When deer can be seen is of course of interest to anyone who wants to find these animals, and this has been covered in Chapter 4.

Lynne Farrell and I began investigating water deer at Woodwalton Fen in June 1976. The basis of the long-term surveillance was to cover the reserve with dusk walks as uniformly as possible throughout each winter, with about 30 visits being undertaken (Figure 7.5). The number of deer seen per hour was calculated for each visit and the average calculated for the whole winter period – the winter sightings frequency. We also needed sufficient close observations of individual deer to enable estimates of recruitment and survival to be made (see below). The time commitment was not inconsiderable, but Woodwalton Fen is a perpetually interesting reserve and the monitoring never became onerous or tedious. The main aims of the surveillance were to determine the stability of the population and find out how and why it fluctuated – in the early years, there was concern that the population might be dying out. There was, however, no conservation intervention on our part, our role being to observe.

In 1980, muntjac began to appear at Woodwalton Fen and we simply recorded the muntjac in addition to the water deer. So we were active in the field during the muntjac's initial colonisation.

At this time, I was being kept informed of the muntjac's colonisation of Monks Wood by the warden, Jeremy Woodward. From 1977, he spent most of his time working in the wood, and mapped all of his sightings of muntjac. He copied his maps to me and I tallied up the numbers. In 1985, his muntjac sightings suddenly, and quite literally, went off the scale. At the same time, a plot of hazel coppice failed because of deer browsing, and, sadly, Jeremy left. This meant that there was a large

Figure 7.5   Lynne Farrell on a surveillance walk at Woodwalton Fen with the setting sun causing a few problems.

and damaging population of muntjac that were no longer being monitored, so Lynne and I began muntjac surveillance there in 1986. The warden's sightings had shown highest numbers from December to May with a peak in April. We selected the period January–May for our surveillance visits. Each surveillance visit took in sunset, number seen per hour was calculated and the average sightings frequency was calculated for each year. The principal aim was similar to that with the water deer at Woodwalton Fen – to follow and try to explain any changes. When we began, it was unclear whether the dramatic increase would continue or whether the population might crash. Initially we could only manage to fit in six visits per spring, but from 1995 I was able to increase this number to at least ten, and maintained this level of input up to 2016.

7.4   Monitoring recruitment and survival

7.4.1   Water deer

The surveillance study on water deer at Woodwalton Fen has usually resulted in a relatively large number of sightings, the most logged in one winter being 632 in 2010/11. It was realised early on in the study that many deer could be seen sufficiently clearly to determine whether they had long canine tusks and were therefore adult bucks. Initially, a telescope was used for observations, but binoculars were later found to be good enough, especially those with 10×50 specification. In the autumn it was also possible to identify young of the year with their lighter build and shorter muzzles. However, to help understand why the population size changed from winter to winter, a technique was needed that could provide a measure of the first-year young in the population, and by midwinter it was more difficult to distinguish them from older females solely based on their appearance.

The method adopted relied on summing numbers of sightings of deer with or without visible tusks (Cooke and Farrell 2000). Deer can be divided into three types, based on tusks seen in the field:

- Long tusks (LT): these are adult males in at least their second winter (Section 2.4).
- No visible tusks (NT): these are females and younger males. Female water deer have small canines which are not visible under field conditions.
- Short tusks (ST): these are young males with developing canines. When calculating recruitment, it is assumed that all short-tusked deer seen at Woodwalton Fen in winter are first-year deer rather than slow developing older animals. Less than 1% of deer were classified as having short tusks from 2010/11 to 2015/16, so in practice any error caused by this assumption will be slight.

A potential source of error is that some older bucks have lost tusks (Figure 4.6) so might be classified as having no tusks. However, adult females have a more graceful appearance than adult bucks, and any deer that looks like a buck, but initially appears to lack visible tusks, is scrutinised with extra care.

The recruitment index (R) is the percentage of first-year deer in the winter population. Assuming (1) equal ease of seeing and identifying deer in the different age and sex classes and also (2) that the sex ratio does not differ markedly from 1:1, it can be described by:

$$R = 100(NT + ST - LT) \text{ divided by } (NT + ST + LT) \text{ per cent.}$$

So are these two assumptions reasonable? Sex ratio of adults has been found to vary between populations but not in any consistent manner; the overall average sex ratio reported was females:males 1:0.97 (Section 3.2.1). The second assumption that sex and age classes can be seen with equal ease is unlikely to apply fully as, for example, older bucks become more obvious during the midwinter rut. However, pattern of visits and manner of recording have remained as constant as possible throughout. So while a recruitment index for a single winter should be regarded with some caution, comparison of changes in recruitment through the 42-year study should be more reliable (Cooke 2009a).

To derive survival data for water deer from one winter to the next, sightings frequency is subdivided into first-year deer and older deer using the recruitment index for each winter. Then the sightings frequency for older deer in the second winter is compared with the overall figure in the first winter. To give an example, in the first winter of the study, 1976/7, 6.4 deer were seen per hour. In the following winter, this figure was 4.6 with a recruitment index of 8%, indicating that the contribution from older deer was 4.6(100–8)/100 = 4.2 deer per hour. Annual survival is the percentage of deer from 1976/7 appearing as older deer in 1977/8 = (100 × 4.2)/6.4 = 66%. This measure of survival takes into account all 'losses' including dispersal away from the site. The winter of 1976/7 was hard and this figure of 66% for survival in 1977 was one of the lowest recorded. Survival has never exceeded 100%, which it could do if, for instance, detection of deer became much easier during one particular year.

7.4.2   Muntjac

Surveillance of muntjac began several years after the water deer study, and a recruitment index (R) was adopted based on sightings of deer with and without antlers or pedicles for muntjac at Monks Wood. This again represented the percentage of young deer in the population. In a young male muntjac, pedicles begin to grow at 5–6 months and antlers at 8–11 months (Chapman and Harris 1996). By the time a subadult male is about eight months old it should have easily observable pedicles and be recorded in that class. The recruitment index is calculated for a period of time taking into account cumulative counts of deer with pedicles/antlers (A, which are adult and subadult males) and without such adornments (NA, females and younger males):

$$R = 100(NA - A) \text{ divided by } (NA + A) \text{ per cent.}$$

As with water deer, to provide an unbiased index of recruitment for muntjac, there would need to be an even sex ratio and the age classes and sexes should be seen with equal ease. These conditions are unlikely to apply as sex ratio appears to vary from situation to situation (Section 3.1.1) and behaviour may differ between the sexes (Chapman et al. 1993). Also, while young deer may be less conspicuous on average, newly independent individuals are likely to be at greatest risk of starvation during hard winters and may, by foraging more, become more conspicuous than adults (Figure 20.2). Several of these complications only came to light after I had been calculating recruitment indices for a number of years.

Better knowledge of the composition of populations and the behaviour of sex and age classes is fundamental to understanding, managing and monitoring muntjac, and both topics require further detailed study in a range of situations. Use of the index has been restricted in this account to following trends over lengthy periods of time at Monks Wood, where large numbers of deer could be observed, especially prior to stalking beginning. Annual survival of muntjac at Monks Wood was not calculated as it was for water deer at Woodwalton Fen because muntjac breed in all seasons on a cycle of less than one year and the cut-off point between young males and older males is about eight months of age.

7.5   How well does the counting method detect and measure change in deer populations?

The simple surveillance and monitoring technique of counting deer at dusk has been used to calculate the average number seen per hour during a defined period of time. But, when repeated for a number of years, how well might this method describe changes in population size? Problems arise if the ease with which deer can be recorded changes between years. When detection rates are considered, it must be borne in mind that they might change even without noticeable variations in amounts of available cover. For instance, if recorders change, then results might reflect their relative skill or experience rather than real alterations in the deer population. Alternatively, the ultimate cause could be lack of food or other resources, or disturbance leading to behavioural changes in the deer. If detection rates do change, then at best the method will be misleading in a quantitative sense – in other words, it could under- or over-emphasise any change in population. At worst it will be qualitatively misleading, indicating a change where none has occurred or vice versa. Behavioural changes can be unexpected. The message seems to be that, if there is a sudden, significant and unforeseen alteration in sightings, this may indicate something other than a change in deer numbers. Such dramatic events are, though, unusual.

7.5.1   Muntjac

Behaviour of a population can be modified by management, such as the intro-duction of stalking. When stalking started in Monks Wood in the winter of 1998/9, the average number of muntjac seen per hour at dusk fell from 17.0 in the spring of 1998 to 5.7 in the spring of 1999, a reduction of 66%. As part of a project on ranging behaviour, Ian Wyllie of ITE had fitted several muntjac with coloured plastic collars or radio collars in 1996 and 1997. My observations of deer in the wood with and without collars suggested the population decreased by about 50% between 1998 and 1999 – the reduction in sightings may have been greater because surviving muntjac tended to be more difficult to see after stalking began.

An extra degree of wariness could be explained by individual deer changing their behaviour to become less active generally or less active in more open habitats or at dusk (see Table 12.1) when the surveillance walks were undertaken. For instance, in New Zealand, heavy stalking pressure on introduced red deer led to a marked change in behaviour, with deer avoiding habitats where they were more likely to be shot (Putman 1988). Changes in behaviour associated with hunting have been recorded in a range of situations, such as for white-tailed deer in Florida (Kilgo et al. 1998), elk *Cervus canadensis* in Alberta (Ciuti et al. 2012) and roe deer in Poland (Sönnichsen et al. 2013). However, there is another possible explanation for observations in Monks Wood involving the most conspicuous muntjac being shot first. The Albertan elk had individual 'personalities'; those that were bolder displayed increased use of open areas and greater movement, and were therefore more likely to be shot than less conspicuous individuals. In an extension of the study in female elk, it was demonstrated that learning contributed to a more cautious strategy in older individuals (Thurfjell et al. 2017). Females that reached ten years of age were described as 'almost invulnerable to human hunters'. This fact could be significant for muntjac stalking here as females can live beyond that age, and the young of cautious mothers are likely to learn from her to be wary themselves.

Muntjac display individuality. In Monks Wood, I tracked eight deer with radio collars on about 20 occasions in 1996 and 1997 and each time attempted to see the deer. The most conspicuous individual was seen on 43% of occasions whereas the least conspicuous one was not seen at all until it reappeared once several years later, long after its radio had stopped transmitting. If these eight deer were representative of the Monks Wood population, it is possible to calculate what might have happened to sightings if deer were shot in sequence from the most to the least conspicuous (Cooke 2006). Thus, for example, shooting the most conspicuous 25% of the population in the wood during a single winter should have reduced sightings by 46%, assuming behaviour of the survivors was unaffected; and shooting the most conspicuous 50% would reduce sightings by 78%. In reality, some of the warier deer will have been shot and the reduction in sightings will be less, but this may eventually be countered by selective pressure from stalking over a period of years leading to a more pronounced effect.

In addition, however, the behaviour of individual muntjac was, and is, also being modified by stalking and other disturbance in Monks Wood (the public has free access to the wood). This fact was brought home to me during the foot-and-mouth epidemic in 2001. A stalking ban was in force from 23 February 2001 until late May. Other visitors were also banned from the wood, and the reserve staff kept away, but I was allowed to continue with my visits providing that I took hygiene precautions. Although lack of management meant the rides became overgrown, my muntjac sightings increased significantly from an average of three per hour at dusk during January and February to nine per hour in March–May. Such a dramatic change has not been seen in other years and probably resulted from the deer becoming less wary because of the almost total elimination of disturbance.

7.5.2 *Water deer*

This species has never been shot inside the reserve at Woodwalton Fen and I have not monitored what happens to frequency of sightings when stalking of a water deer population begins. However, sightings of the Woodwalton population increased for more than ten years from 1996 without any deliberate human intervention. This provided an opportunity to test the underlying processes (Cooke 2009a): did it represent a real increase in population or perhaps simply that the deer had become easier to see because of management changes inside and outside the reserve? From the late 1990s, there was clearance of trees and scrub within the reserve to create reed beds, fen fields and grassland.

What happened was resolved indirectly with data collected on the level of young deer in each winter population and survival of older deer from the previous winter. As described above, recruitment index was independent of sightings frequency. Had the increase in sightings been caused by an increase in detection probability, recruitment index should not have changed, but the figure for survival would have been expected to increase. But this did not happen. From 1996/7, recruitment index increased significantly whereas survival decreased slightly (Cooke 2009a; Table 8.1), indicating the increase in sightings frequency was caused by an increase in population size – production of young had increased and/or dispersal of young was reduced. A visit on a calm spring evening in recent years has sometimes resulted in a count of water deer similar to the estimate for the entire population in some of the earlier years of the study.

7.6   Using counting to estimate density

7.6.1   Muntjac

From the information in Section 7.5.1, it has probably become apparent that while number seen per hour usually changes in the right direction if the population changes, the extent to which it changes may differ from reality. Whenever possible, density has been estimated (1) to have a better idea of the real extent of changes, and (2) to relate to impacts. A few years ago, I made a tentative attempt to construct a graph showing how sightings frequency changed with density using information that was to hand (Cooke 2006). Since then, thermal imaging and wildlife camera trap studies have been done in the wood, and the graph has been updated (Figure 7.6). Disproportionately fewer deer are seen at lower densities. The graph cannot be applied directly to sightings in other sites because their detection probabilities may be different from that in Monks Wood.

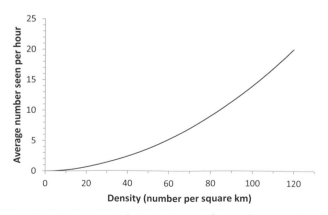

Figure 7.6   Muntjac at Monks Wood: how sightings frequency changed in relation to estimated deer density (updated from Cooke 2006).

7.6.2 *Water deer*

During the first three winters of our study on water deer at Woodwalton Fen, 1976–1978, Lynne Farrell and I took advantage of the generally uniform distribution and territorial nature of the deer during the rut each December to attempt to determine the number based within a study area. This was 39 ha of mixed fen and sallow carr and comprised 19% of the reserve's total area of 208 ha. Working out the territories proved easier than was initially anticipated. The area was separated by ditches into ten rectangular compartments of roughly equal size, and deer could usually be identified with a telescope on the basis of presence/absence of visible tusks, amount of white on the nose, general build, coat colour, scars and torn ears. Similarly, Sun and Dai (1995) were able to identify 23 individuals at Poyang Lake in China. Observations on the deer at Woodwalton Fen were supplemented by a study of signs. Deer and signs were recorded on routine surveillance visits to the reserve and on visits of 1–3 hours specifically to search the study area more thoroughly, including walking through the sallow carr and reed to flush deer. Numbers in the study area were estimated to be 18 in December 1976, 11 in 1977 and 13 in 1978.

Our ability to estimate numbers was tested in two ways. First, estimates for single compartments were well supported by numbers flushed in separate netting operations to tag deer, including when a team of dogs was used. In four compartments netted, we had estimated a total of seven or eight, and seven were found during netting. Secondly, observations on tagged and untagged deer in the whole study area in the early months of 1978 gave an estimate of 13 deer which agreed well with the December estimates of 11 and 13 for 1977 and 1978 respectively. Estimated density and average sightings per hour were highest in 1976/7 (at 0.46 deer per ha and 6.4 deer per hour respectively) and lowest in 1977/8 (at 0.28 deer per ha and 4.6 deer per hour respectively). A linear relationship was found between sightings per hour and density for those winters. Later in the study, approximate density figures were derived from this relationship, extrapolating the trend line if necessary. This was done principally to provide an indication of the densities of water deer associated with damage to conservation features on the reserve (Chapter 19).

7.7 Percentage of deer seen during surveillance

A question that has intrigued me since the start of these studies on deer is what percentage of deer might be seen on a single visit or during a single hour at dusk? During winter surveillance visits in the late 1970s, an average of 7% of the water deer population was estimated to have been recorded per hour of observation. This figure will be site specific and method specific. Longer visits at dusk will detect more deer, but number recorded per hour will eventually decrease because fewer deer are seen prior to 90 minutes before sunset.

Percentage of muntjac recorded at dusk at Monks Wood varied depending on population size and history of stalking. In 1998, just before stalking began, an estimated average of 10% of the population was seen per hour at dusk. In 2013, when the population had been reduced by 15 years of stalking and any survivors were likely to be wary, this figure was only 3%.

7.8   Conclusions

Walking around these sites in daytime at random will allow detection of a small percentage of the deer population. This percentage can be increased by visiting at dusk in those months when the deer are most conspicuous. Effectiveness, i.e. recording as many as possible without the number seen per hour decreasing significantly, can be achieved with dusk walks of 1–2 hours. However, the number of deer seen during a walk does not just depend on number present on the site. Factors such as seasonal changes in amounts of cover or deer behaviour, disturbance, previous and current weather conditions, as well as pure chance, will all influence how many deer are seen. Some of these factors can be controlled by avoiding counting in high winds or heavy rain, for example. Other factors, such as seasonal variation, can be controlled if the intention is to compare counts over a period of years, by repeating the pattern of counting each winter. Thus at Holme Fen, where there is appreciable variation in numbers seen each month from October through to May, roughly half of the counts have been done before the end of February each winter and half during March, April and May.

One should also bear in mind any new events that can change the detectability of deer. These will include clearance of cover from areas that remain attractive to deer and changes in recorder. All that can be done is to try to keep conditions as consistent as possible and record changes in factors that might affect detectability. Then if numbers do appear to have changed, a judgement will be needed about whether this is a real change in population size or just a result of conditions changing. In such circumstances, a second measure that is unaffected by the same change in detectability would be valuable.

How many counts should be made and how frequently depends on the nature of the overall aim. The variability in counts means more will probably be needed per winter if it is important to determine the difference between two consecutive winters than if the intention is to undertake open-ended surveillance over an undefined number of years. Just a single visit each year will provide some information, but it may be some years before it provides the basis for firm conclusions, even if substantial population changes are taking place.

The crucial question is of course does this straightforward method work? Can population changes be detected? Evidence provided here shows that it can work qualitatively – but the method is less reliable quantitatively. Had we set out to begin in 2016, rather than in 1976, we may well have tackled surveillance differently, but it is unlikely that it would have provided as much relaxation and pleasure as has been derived from what now amounts to approaching 2,000 walks around the local NNRs.

7.9   Postscript

Many years after we began counting deer at Woodwalton Fen to produce an index of abundance, other methods were developed along similar lines. A measure known as the Kilometric Index (Vincent et al. 1991) has gained some support (Putman et al. 2011b). This method is used particularly for roe deer which, like muntjac and water deer, is a non-herding species with quite small home ranges. Fixed transects are walked a number of times during January, February and March and the number of deer seen is divided by the distance walked to give the transect index. An index is calculated for all transects in that area at that time, and an annual index worked out using all the area indices calculated that year. Whereas transects may be fine in woodland, however, it is very difficult to walk far in a straight line at a wetland site such as Woodwalton Fen – unless you are on an open grassy ride or you have a boat!

In 1995, in response to the rarity of reliable national monitoring information on mammals, the BTO extended its Breeding Bird Survey to count individuals of relatively large species of mammals seen on field visits. Volunteers annually surveyed approximately 2,600 one-kilometre squares in the British Isles (Wright et al. 2014). Changes in abundance indices, 1995–2012, were published for nine well-recorded mammal species. The technique demonstrated a statistically significant rise in the national muntjac population during that time. Palmer (2014) used the BTO's data for roe to select sites with a range of deer densities to help determine whether this species had affected vegetation structure, abundance and composition in English woodlands.

CHAPTER 8

Colonisation, population stability and change in uncontrolled populations

8.1 Introduction

This chapter looks in detail at specific populations of the two species in order to understand how they functioned when relatively uncontrolled by stalking. Attempts are made to determine how long it took to colonise a site fully and also to begin to understand population dynamics and how each species responded to environmental change. Spatial colonisation of areas of various sizes has already been dealt with, but this is a different type of colonisation relating to what goes on in a site. Few sites have been studied long term, and every population is likely to differ to varying degrees in how it responds to environmental factors. Nevertheless, these are important issues with regard to understanding how we arrived at the current situation in Britain and for predicting how populations and impacts might change in the future. So it is worth examining in some detail what information is currently available. Most of this chapter is based on studies at Monks Wood and Woodwalton Fen, which, as far as I am aware, have been recorded continuously for longer than any other unmanaged population.

8.2 Muntjac

8.2.1 Colonisation
Muntjac began to colonise Woodwalton Fen in 1980, four years after the study on water deer began – so the entire process has been recorded in terms of numbers seen per hour in each winter (Figure 8.1). It took many years for serious impacts to develop and stalking in the reserve started in the spring of 2011; the graph stops at that point. Numbers seen per hour peaked in the winter of 2009/10, but hard weather that winter caused the deer to be more obvious (Figure 20.2). Density was at its highest during 2007/8–2010/11. It needed nearly 30 years of colonisation to

reach this level and, had stalking not intervened, density might subsequently have increased still further, so further lengthening the colonisation period. Conversely, they may have reached a peak in the mid-1990s had there not been habitat changes starting at that time. How long colonisation took depends on how the term is defined and what data are used.

Another example of the difficulty in defining period of colonisation comes from the King's Forest in Suffolk (N.G. Chapman et al. 1985; Blakeley et al. 1997). The first record of muntjac was in 1963. By 1985–1987, when the subpopulation in a study area began to be surveyed, numbers changed little suggesting a colonisation period of less than 22 years. Then, however, the storm of October 1987 changed the habitat to such an extent that the subpopulation had virtually doubled by 1993. There had been some culling of deer in the forest up until 1987, but Norma Chapman has assured me that cessation of culling was not a significant factor in this later increase.

It is possible to be more precise about the colonisation period at Monks Wood and Holme Fen. Muntjac started to colonise Monks Wood in about 1970 and became extremely numerous by 1985 or 1986, while at Holme Fen, they first appeared in 1980 and plateaued from about 1998. Around Grafham Water, Keith Mason and I recorded any muntjac seen or heard during the monthly wildfowl counts each winter from 1971 until 1998. A buck's skull had been found near the shore of Grafham Water in 1970, but none was seen alive until 1982 and the number recorded was still increasing in 1998. So the colonisation process lasted at least 16 years or 28 years, again depending on how colonisation is defined. It is not unusual for a solitary animal to be recorded in a new site some distance from the nearest known population, but then no more are seen for perhaps a decade until permanent colonisation gets under way (Chapman and Harris 1996). White et al. (2004b) computed that, in the absence of any emigration or immigration, a muntjac population would take five years to develop from a level of 25% of a stable population up to 50%.

The evidence suggests that, under favourable conditions, muntjac populations may reach a peak in 15 years or more from the start of permanent colonisation; but these are all comparatively large sites in the region of 50–250 ha, and colonisation would be predicted to be faster at smaller sites. One of the reasons for the relatively

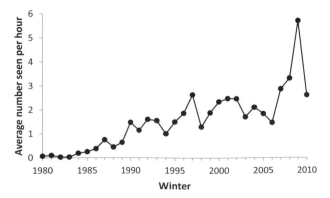

Figure 8.1   Muntjac at Woodwalton Fen: average number seen per hour during dusk visits each winter, 1980/1–2010/11.

rapid build-up at Monks Wood may have been the lack of resident foxes at that time (Cooke 2014b). Foxes appear to be important predators on muntjac fawns and a flourishing fox population is likely to slow the rate of colonisation. Foxes have been more numerous in Monks Wood since 2008, and this has been associated with lower muntjac recruitment.

Peak winter sightings frequencies of muntjac at the three NNRs were Monks Wood 23 per hour, Holme Fen 9 per hour and Woodwalton Fen 6 per hour. These should be compared with caution as detection probabilities will vary between the sites, and attempts have never been made to estimate density at Holme and Woodwalton fens. However, Monks Wood's peak density of more than 100 per km² was almost certainly higher than at the other two sites. Large areas of Holme Fen are dominated by bracken, which offers little to muntjac apart from cover. And Woodwalton Fen is a predominantly wetland reserve with a high density of competing water deer. As is discussed later (Section 10.3), large woods tend to have higher densities of muntjac. As Monks Wood is a relatively large site with suitable habitat, few woods would be expected to match its frequency of sightings prior to stalking starting.

8.2.2   Monks Wood, 1986–1998, the 'muntjac slum'

What then happened at Monks Wood during this time when the population peaked before stalking started inside the wood? How stable was the population? The situation is summarised in Figure 8.2, which shows numbers seen per hour and recruitment index for the period 1986–1998. The immediate answer, after a glance at the basic raw data, is that population size was fairly stable overall, but fluctuations occurred from year to year. Instant answers lacking due consideration can, however, be misleading, and looking at all available information in greater detail reveals a more complex story.

Jeremy Woodward's observations during his wardening duties had indicated that muntjac numbers increased steadily up to 1984, and then markedly in early 1985. Attaining peak population size coincided with two of the worst winters of recent times, 1984/5 and 1985/6 (see Table 8.2). There was no unusual level of mortality

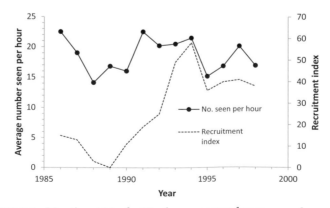

Figure 8.2   Muntjac at Monks Wood: average number seen per hour and recruitment index (% young deer in the population), January–May, 1986–1998, a time of peak population density (adapted from Cooke 2006).

during those winters, presumably because the wood was still in a comparatively unbrowsed state. Although it provided the deer with sufficient food, they had to forage further to find it. In the early months of 1986, when I first started monitoring muntjac in the wood, two of the six visits were made when it was bitterly cold with snow on the ground; on those visits, the deer were mainly congregating around bramble bushes, and 30 and 44 sightings per hour were logged. Over the next five years, average sightings frequency decreased by more than a third with very limited recruitment of young deer (Figure 8.2). This was not surprising as a population that had reached saturation point was likely to produce fewer young.

In the early months of 1991, however, the new warden, David Massen, reported a number of dead deer in the wood. That winter was not unusually cold, but there was a period of snow cover in February and the wood was probably starting to show signs of damage. Sightings frequency was high and recruitment index increased from its previously low level. I was not able to study the die-off that year, but the most plausible explanation for such mortality was that the population bred better and recovered to its former peak, prompting a reduction in winter survival. In 1992 and 1993, the die-off was not repeated, sightings remained quite high and recruitment index increased still further in response to the die-off in 1991. Foraging on the grass fields immediately to the south of the wood (Figure 1.6) had been comparatively low up 1991, but increased in 1992 and 1993. By this time, there was clear evidence that the fabric of the wood had been affected by the muntjac which, in response to a diminishing food resource, were foraging more frequently outside the wood.

During February–April 1994, I was coming to the end of a year's intensive study of the wood's muntjac, and I was ideally placed to observe at first hand the major die-off that occurred (Cooke et al. 1996). February witnessed a number of muntjac deaths and very high totals of live muntjac seen on daytime and dusk surveillance walks, with unusual numbers being seen outside the woodland boundary. In March and April, sightings frequencies decreased and most of the casualties discovered were judged to have been dead for more than a week. A check in March of 1.7 km of roads close to the wood located ten traffic casualties; that this total was more than has ever been recorded before or since reflected the wider foraging taking place. Nationally, thousands of muntjac are killed on our roads each year (Chapman and Harris 1996; Langbein 2007), but the toll around Monks Wood is usually low as less than 1 km of its edge abuts roads.

Sightings frequency for January–May 1994 remained relatively high (Figure 8.2), despite the exceptionally low number of sightings in April and May, because deer became more conspicuous in and around the wood. Twenty-three freshly dead individuals were found in the wood during February and March and taken for post-mortem investigation. Average body weight for the age/sex subgroups was 30–45% below that of muntjac collected from elsewhere in southern Britain, January–April, 1977–1994 (Cooke et al. 1996). It was concluded that all of the deer collected in Monks Wood in 1994 exhibited signs of advanced starvation, with pneumonia being the ultimate cause of death in several cases. Of this sample, 48% were first-year deer aged 20–40 weeks, 22% were second-year deer and 30% were adults. Deer that were newly independent of their mothers seemed especially vulnerable. The recruitment index was 58%, the highest figure ever recorded – this was probably largely due to increased visibility of the hungry young deer (Figure 8.3).

Figure 8.3   (a) A young muntjac crosses a ride in Monks Wood during the die-off in March 1994, when there was very little food. (b) The same view, March 2018, with no muntjac to be seen but with much low bramble.

Although there had been little change in sightings frequency from 1986 to 1994, a picture was emerging of a population that was highly stressed and living on the edge. Up until then, I had assumed fencing was all that was required to solve the problems being posed by the muntjac in the wood. But, in the space of two or three months, I came to appreciate that the population required active management to prevent die-offs from occurring again. At the height of the die-off, I was entering Monks Wood expecting to find dead deer. This should not happen in a lowland wood during a period of unexceptional weather. In all, 46 muntjac were found dead or dying, and the population was estimated to have crashed by roughly 50% (Cooke et al. 1996).

Average frequency of sightings in 1995 was about 30% down on the figure for 1994 and was only about 10% less by 1997 (Figure 8.2). The recruitment index remained high throughout – the wood's muntjac population had demonstrated it was capable of making a recovery, at least in terms of numbers seen, from the traumas of 1994.

Muntjac have been shot just outside Monks Wood since 1984 (Peter Green, personal communication). Such shooting is likely to have increased during the 1990s as word spread about the size and habits of the population, and as deer grazing on adjacent cropped land was perceived by the farming community to be more of a problem. Members of the local Deer Management Group (DMG) shot 27 deer emerging from the wood during the early months of 1995. This activity was in part to provide a sample for comparison with the deer that died the previous spring. Of the sample shot in 1995, eight (30%) were in the 20–40 week age group, confirming good recruitment, but only one was in its second year indicating that few first-year individuals survived the die-off in 1994. The remaining 18 deer (67%) were all adults, some being at least ten years old. These older, experienced animals were far more capable of surviving die-offs.

A further 125 muntjac were shot outside the wood by DMG members from 1996 until spring 1998. The only demographic detail available was that the sample shot in early 1996 comprised 66% adults, virtually the same as in 1995. The other reason for shooting outside the wood was to stop or discourage deer from feeding on arable fields along its eastern and northern edges. From 1990 to 1994, 4% of all muntjac sightings on surveillance walks were on arable fields. This halved to 2% during 1995–1998.

This shooting, however, had no marked effect on overall sightings in the wood. Statistical analysis of the sightings data in Figure 8.2 reveals only a very marginal decrease during the whole period. One thing to bear in mind, though, is that virtually the entire understorey of the wood disappeared down hungry throats during the 13 years while the observations were being made, and individuals should have become progressively easier to see. So, if the population really had remained more or less stable, sightings frequency would have been predicted to increase during the later years. That this did not happen suggests that some decline in numbers and density occurred. Any suggestion that the population struggled for food latterly is supported by the fact that, whereas 5% of all deer seen on surveillance walks, 1986–1991, were outside the wood on the grass fields to the south, this figure increased to an average of 17% during 1992–1998. Oliver Rackham (2003) referred to Monks Wood as a 'muntjac slum' and commented that they had done more damage than the clear-felling during the early part of the twentieth century. The other side of the coin was that the deer had to live in the slum they had created and so suffered similar hardships to slum-dwellers everywhere. A muntjac in Monks Wood during the 1990s will have had a tough existence.

So, what might have happened if stalking had not started inside the wood in the autumn of 1998? It seems that the stalking outside had some effect in that it reduced numbers on arable land and there was no further die-off during 1995–1998. The level of stalking was not inconsiderable, amounting to shooting 0.24 deer per hectare of woodland per year. This represented shooting about 20% of the resident population every year. It may sound enough to have had an appreciable impact, but at that time Monks Wood was still functioning as a source of muntjac, with emigration far outweighing any limited immigration (Cooke 2006). Thus, many of those shot were probably potential emigrants. Continuing beyond 1998 with just stalking outside would, at best, have led to a slow reduction of muntjac density. Many muntjac may live for years without venturing outside this 157 ha wood. Had there been no shooting at all outside the wood, it is likely that there would have been further periodic die-offs.

Previously, major mortality incidents have only been observed in this country during periods of unusually severe weather, principally the prolonged deep snow of early 1947 and the persistent cold, frosty weather of 1962/3 (Chapman et al. 1994a; Chapman and Harris 1996). Die-offs were sufficient to deplete or even wipe out populations in whole forests. Muntjac can withstand brief periods of starvation, but persistent bad weather is sufficient to kill them. In China, winter snow cover is believed to have prevented muntjac from colonising further north (Sheng 1992a).

The situation at Monks Wood was clearly exceptional with its very high deer density causing chronic starvation problems with occasional die-offs even during fairly typical winter conditions. Finding another similar example to begin studying now is difficult. Monks Wood was then a large, quiet deciduous wood with few deer of other species and with hardly any foxes. What has tended to happen over the last 20 years in areas where muntjac occur is that their populations are controlled before or soon after they reach peak density.

I have regularly studied one small wood where there has been no stalking at all: Lady's Wood, about 4 km from Monks Wood (Figure 1.4). In this wood, self-regulation of the muntjac population appeared to occur (Section 13.1.2. and Figure

13.6), but in a much less drastic fashion to what occurred at Monks Wood. An adult deer forced to move away from a small wood is likely to be familiar with part of the surrounding landscape and should have a good chance of survival. However, in a large wood, a stressed animal may move out of its normal range but only encounter further tracts of the wood with the same problems. Leaving the wood totally would be a dangerous step for a young animal with no experience of the world outside; but staying inside the wood is a no better option. This might be described as being on the antlers of a dilemma!

8.3   Water deer

8.3.1   Colonisation

The capacity for water deer to build up quickly in numbers was observed at Whipsnade Zoo following their initial introduction from Woburn Abbey (Middleton 1937). During 1929 and 1930, 32 water deer were liberated on to about 60 ha of 'undeveloped pasture' on the edge of the Whipsnade Zoo estate. They shared this area with a range of other ungulate species. By late 1933, the total number was estimated at 200. Despite a number of deaths being recorded, this corresponded in broad terms to a doubling of the population each year. This rate of increase proved unsustainable, and about 140 deer died during the winter of 1933/4, possibly due to enteritis associated with a heavy burden of nematodes.

Information on the colonisation process at Woodwalton Fen is patchy. The population has been reported to have 'originated from a small number of Woburn stock released in the vicinity sometime between 1947 and 1952' (Chapman (1995) reporting a statement by Guy Thornton). It was, however, November 1962 before tracks of small deer were first noticed inside the reserve by Nature Conservancy staff. The founder population evidently survived the freezing conditions during the winter of 1962/3, as single fawns were noted in the reserve records in 1963 and 1964. The population built up and became well established by the late 1960s. At the time, the deer were believed to be muntjac, and it was January 1971 before a dead individual and live animals were correctly identified by Raymond Chaplin. It is probable that all records of small deer in the reserve up to that time related to water deer. In 1972, Raymond Chaplin (personal communication) estimated that their population might number in the region of 50–75, based on (1) knowledge of densities elsewhere, (2) an assessment of the type and extent of habitat available in the reserve and (3) the frequency of pathways through the undergrowth. In the early part of our study, Lynne Farrell and I estimated the population to number about 100 in the winter of 1976/7 and 50–70 during the next three winters (Cooke and Farrell 1981). Thus, using 1962 as the first record, the population was considerable within ten years and peaked after 14 years.

I am not aware of other examples of colonisation in the wild in this country where it is possible to estimate the length of the period of build-up. However, if opportunities arise in favourable habitat, colonisation may be rapid. The first record in the Norfolk Broads was in 1968 and was considered to be a local escapee (Section 5.3.1). Twelve years later, they were the most commonly seen species of deer in the Broads. And that was just the start of their expansion.

8.3.2   Woodwalton Fen, population data, 1976–2018

The population at Woodwalton Fen seems to be the only one to have been studied in depth over a sufficient period to provide information on population dynamics and long-term stability or change. Surveillance results are summarised in Figure 8.4. Total height of each 'bar' in Figure 8.4a represents the average number of deer seen per hour for that winter, the sightings frequency. The recruitment index has been used to subdivide each bar into first-year and older deer.

The recruitment index (Figure 8.4b) varied between zero and 46% of first-year deer in the winter population, with an average of 22%. In terms of number of young per female, 46% equates to 1.70 and 22% to 0.56. In a study on coastal populations in Jiangsu Province in China, production of young per adult female to August was 1.53 (Xu et al. 1996).

Change in sightings frequency from one winter to the next at Woodwalton Fen was positively related to recruitment index in the second winter (Figure 8.5), that is if recruitment was high in the second winter, the population was more likely to have increased. There is of course nothing surprising about this result, but as sightings frequency and recruitment index were derived independently of one another, their clear relationship provides confidence about how data have been collected and interpreted. No change between successive winters corresponded to a recruitment index of 19%. Figure 8.5 also reveals the frequency with which annual changes of different magnitude occurred. Thus, during the 41 years, change was 0–19% during 26 years, 20–39% in 12 years, 40–59% in a single year and 60–79% in 2 years. The range was from –36% to +77%. All three of the largest changes were increases. Despite having the capacity to change rapidly, this rarely happened with the population at Woodwalton Fen.

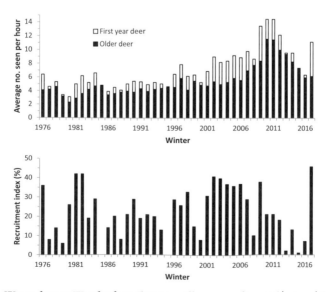

Figure 8.4   Water deer at Woodwalton Fen surveillance results, 1976/7–2017/18: (a) average number seen per hour; (b) recruitment index (% of first-year deer in the winter population).

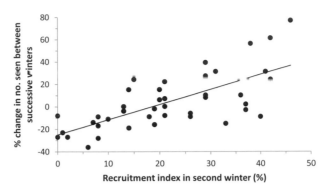

Figure 8.5   Water deer at Woodwalton Fen, 1976/7–2017/18: the relationship between the change in sightings frequency between successive winters and the recruitment index in the second winter (updated from Cooke 2009a).

In Figure 8.4a, comparing height of the bar of the 'older deer' component in a winter with total height of the bar in the previous winter gives a measure of survival during the intervening year, providing detection probability remains unchanged. Over the whole study, survival during a year was, as with recruitment, positively related to the change in sightings frequency from the first winter to the second. Again, it is no surprise that a population increase was associated with good annual survival. Analysis indicated that the tipping point between likely decrease/increase corresponded to survival of 76%.

Average annual survival for the whole period was 79%. Here, survival refers to deer remaining in the population; deer that fail to 'survive' may have dispersed rather than died. It was lowest in 1998 (52%) when severe floods at Easter forced most deer to vacate the reserve, which is used as a water storage area following periods of high precipitation (Cooke 1999c). Frequency of sightings of young and old deer in a winter was always greater than that of older deer in the following winter, so the calculated figure for survival was always less than 100% (but it was 99% in 1978 and 1996).

8.3.3   *Woodwalton Fen, 1976–1996, a time of stability*
Focusing on the period 1976/7–1995/6, when winter sightings frequency fluctuated but did not change overall, the highest value (6.8 deer seen per hour) was only just over double the lowest (3.1). Recruitment index averaged 19% and survival averaged 80% (Table 8.1). As with the whole data set, both recruitment index and survival were negatively related to sightings frequency in the previous winter. In other words, if there were few deer present in a winter, then this situation tended to be followed by good recruitment and good survival, and vice versa. This suggested that density-dependent regulation occurred. Such stability might not be intuitively predicted for such a prolific species of deer. The underlying mechanism may have centred on the finite and fairly stable resources that the reserve had to offer this territorial species. As was seen at Whipsnade Zoo, water deer have the potential to double their numbers in a single year. While this may only happen in exceptional circumstances, they would be expected to produce enough young in many years to compensate for losses of adults to factors such as inclement weather. If

Table 8.1   Average recruitment index and annual adult survival for water deer at Woodwalton Fen for three periods of differing population change.

Period	Change	Average recruitment index in the rut (%)	Average annual survival (%)
1976/7–1995/6	No change	19	80
1996/7–2011/12	Increase	28	78
2012/13–2016/17	Decrease	8	78

there are surplus young, then there will be some dispersal – especially of young in the autumn, but some older deer may also move out. However, if winter numbers are lower than normal, there will be more resources for each deer, the deer may reproduce better and there will be less pressure for them or their young to disperse before next winter. In this way, any annual fluctuation is kept relatively slight. The number dispersing will, however, vary from year to year.

8.3.4   Woodwalton Fen, 1996–2017, a time of change

Having spent 20 years studying a population that did not change appreciably, we might have stopped – but we carried on, and the population first became less stable and then increased perceptibly before decreasing to more or less its original level. Many of these changes were probably related to large-scale alterations to the local landscape. During the mid-1980s and late 1990s, there had been a certain amount of clearance of trees and scrub from the reserve in order to create reed beds, fen fields and grassland. Although this will have led to short-lived increases in detection probability, it soon became as difficult to see water deer in a reed bed as in sallow carr. Such changes were unlikely to have rendered deer easier to see in the longer term but would have been expected to increase the carrying capacity for a species that generally favours more open wetland over woodland and scrub. There were signs of an increase in 1996/7 and 1997/8 with the latter winter recording the highest sightings frequency to date. Any immediate further increase was, however, nipped in the bud by the floods of Easter 1998 forcing deer to disperse far and wide, and the sightings frequency of 1997/8 was not exceeded until 2002/3.

The early years of the twenty-first century saw significant changes, both inside and just outside the reserve:

- There was major tree and scrub clearance in the reserve, 2002–2004.
- In 2004, the last arable crop was grown on Darlow's Farm to the north (Figure 1.4), the land being seeded with grass soon after.
- The last crop on the adjacent Middle Farm fields (Figure 1.8) to the west was in 2007, with grass seed being sown in 2007/8.
- About 25 ha of elephant grass was planted on Castlehill Farm to the south-west in early 2007 and destroyed in summer 2011. These fields were within 800 m of the reserve.

Some changes outside the reserve were driven by the area being within the 3,700 ha of the 'Great Fen', an initiative aiming to restore wildlife habitats across the landscape. All of the above changes (apart from destruction of the elephant grass) seem to have increased prime cover and feeding-grounds for water deer in the local area, and from 2002 to 2011 the population increased. The recruitment index was unusually high, whereas survival changed little (Table 8.1). Not only did clearance in the reserve increase favourable habitat, but the conversion of adjacent arable land to grass provided both better feeding in the early months of the year and a more stable habitat for year-round occupation.

Annual average counts of water deer on nearby and accessible farm fields are shown in Figure 8.6. Usage of fields on Darlow's Farm increased from the winter of 2004/5, and escalated further from 2007/8 when both Darlow's and Middle Farms were under conservation management. Then in 2009/10, there was a massive increase in numbers on Castlehill Farm. The elephant grass that had been planted there in 2007 had been maturing and, by the autumn of 2009, had developed into ideal deer habitat with dense cover and a flora of edible weeds that was luxuriant in places. In addition, weedy stubbles were left throughout the winter in fields between the reserve and the elephant grass. Counts of 10–20 water deer were commonplace on the Castlehill Farm fields that winter. Although the deer were counted from the reserve boundaries, many were considered resident on the farmland rather than being based in the reserve and going out to feed. Decisions on residency were aided by observations such as locations of deer and the directions in which they moved if disturbed. Unprecedented numbers were by then living on the farmland (Figure 8.7). The hard winter of 2009/10 had little or no immediate impact. Survival was high in 2009 and recruitment was high in the winter of 2009/10, leading to an increase in sightings frequency of 56% (Figure 8.4a).

The elephant grass was harvested for the first time in March 2010. When their cover and food suddenly disappeared, many displaced deer seemed to return to the reserve. From there, they foraged out on to the Middle Farm grass in the spring. It could not have worked out better if the sowing of grass on the arable fields of Middle Farm had been planned to provide food for hungry deer from the areas of elephant grass. But it was of course completely fortuitous. The elephant grass soon regrew and its metapopulation of water deer was back in place by the winter of 2010/11, when the number of farm residents was again high (Figure 8.7). However, they were

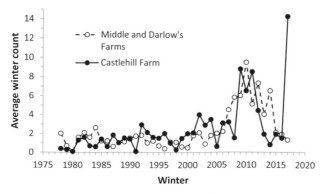

Figure 8.6   Average winter counts of water deer on adjacent farms to Woodwalton Fen, 1978/9–2017/18: Middle and Darlow's farms to the west and north, and Castlehill Farm to the south-west (updated from Cooke 2009b).

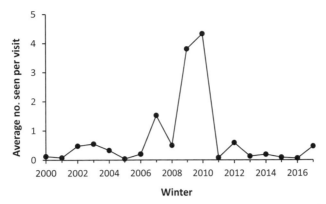

Figure 8.7 Water deer at Woodwalton Fen: average winter counts of deer considered, from their positions and behaviour, to be resident on adjacent farmland, mainly to the north and west, 2000/1–2017/18. Elephant grass was planted on the farmland in early 2007 and destroyed in summer 2011.

displaced once more when the elephant grass was first partially cut in February 2011 and then totally destroyed by spraying with herbicide later that summer.

In 2011, the deer were not so lucky. The population had expanded to take advantage of the elephant grass, which suddenly disappeared. Population density, estimated at about 125 per km^2 based on the area of the reserve, was unsustainable. Sightings frequencies for the winters of 2010/11 and 2011/12 were virtually identical (Figure 8.4a), but deer distributions were very different. Few deer were left living entirely on the farmland by 2011/12 (Figure 8.7), most apparently having returned to the reserve once again. They were unintentionally helped through the winter by oilseed rape being sown beside the reserve in autumn 2011 on Castlehill Farm. So many deer grazed this crop that the average winter count on the farm fields was comparable to those when the elephant grass was mature (Figure 8.6). The difference was that these deer were living in the reserve by day and going out in the evening to feed. Grazing damage to the rape was obvious in 2012 (Figure 11.3). In addition, high numbers grazed on the Middle Farm grass and on mown areas inside the reserve.

After 2011/12, sightings frequency declined by a total of 56% in five years back down to pre-elephant grass levels as a result of poor recruitment (Figure 8.4a). Although this confirms the instability of the situation after destruction of the elephant grass, other factors contributed to the decrease, as described below. Table 8.1 indicates that when the population increased then decreased, it was the recruitment index that changed, not the survival of adults. The dramatic recovery in numbers seen in 2017/18 is covered briefly in Section 8.3.10. The changes seen between 1996 and 2017 clearly owed much to alterations in the nature and extent of the habitat being utilised, but what were the mechanisms and other factors that might have been involved?

8.3.5 Factors and mechanisms of change, 1996–2017
Increasing the amount of good habitat led to a rise in numbers while loss of habitat had the reverse effect. The detailed three-year study on a population of water deer

in the Zoological Park at Branféré in France (Dubost et al. 2008) provided some information on mechanisms. Caution is required in comparing observations made in captivity with those in the wild because, in captivity, food should be adequate throughout the year, veterinary care may be provided, density is high, there is no migration and genetic diversity is likely to be low. However, the population at Woodwalton Fen probably has similarly low genetic diversity and both populations can apparently follow their origins back to Woburn stock.

The population at Branféré was established in the early 1970s and by 1994 numbered about 60 animals living in a 12 ha enclosure. The study took place from late 2002 until late 2005. At the end of 2004, 86 water deer were counted. Nearly 50 were transferred to other parks during 1994–2004, and the average increase in population was 8% per annum over those ten years. During the study, the population increased by 14% in 2003, by 7% in 2004 and by 1% in 2005. This then was a population that had increased in the longer term, but in which the rate of increase was slowing down during the three-year study. It was found that overall litter size at Branféré was 2.0 compared with 2.5 in a Chinese study (Sheng and Lu 1984), the difference being attributed to the much higher deer density in captivity. Litter size at Branféré decreased progressively from 2003 to 2005 and this turn led to fewer young deer in the midwinter population. This seemed to be the mechanism for the density-dependent regulation that was slowing population growth. At Woodwalton Fen, alterations to litter size may be one mechanism that moderates increases and, conversely, aids recovery. It is of relevance to increases or losses of habitat. If good quality habitat increases in area, productivity would be expected to increase and dispersal of young may decrease. This is likely to be what happened between 1996 and 2011 when recruitment of young deer was high and the population increased.

After the elephant grass disappeared in 2011 and the population started to decline, decreased productivity and increased dispersal probably played some part in reducing recruitment. Dispersal of young deer in autumn will be important as a factor that reduces recruitment to the winter population at Woodwalton Fen and in other dense, free-living populations. Mothers and young can often be seen grazing on the farm fields outside the reserve on evenings in early autumn, and this will be an introduction to what the farmland has to offer a juvenile deer. If there are more deer than the reserve can support, young rather than adults are likely to be pushed out, and perhaps more young bucks than young does. Thus, of the 16 water deer culled on farmland up to 2 km away by Martin Guy and his sons during November and December 2014, eight were young bucks. There were evidently more young deer in the early autumn of 2014 than in the following two years: greater numbers of dispersing individuals were reported by the stalkers, and the recruitment index was higher in the winter (Figure 8.4b).

Nothing is known about how fawn predation may have changed from year to year at Woodwalton Fen, but a number of species that occur in the reserve have been observed or suggested as predators of water deer fawns (Harris and Duff 1970; Chaplin 1977), including fox, stoat, weasel *Mustela nivalis* and carrion crow. Foxes have never been controlled on the reserve and are considered to have slowed the colonisation of muntjac (Cooke 2014b). Gnawed legs of water deer fawns can be found in the summer, and occasionally foxes have been seen with fawns in their mouths. Scandinavian studies have shown that foxes were able to kill 10–50% of roe

Figure 8.8   Stock grazing has deterred water deer. (a) Cattle have been used for management purposes inside the reserve throughout the study. (b) Sheep grazing on Middle Farm, outside Woodwalton Fen, February 2010. The damper area of Darlow's Farm and the trees of Holme Fen can be seen beyond the grass fields. This view can be contrasted with Figure 1.8, taken eight years earlier when all fields were used for arable crops, and with Figure 10.1 when lack of livestock allowed large numbers of deer to graze.

deer fawns born each summer and have a significant effect on recruitment rates (e.g. Linnell and Panzacchi 2006). Roe, like water deer, are seasonal breeders. This research indicated that, where roe deer density was high, foxes switched from their main prey of small mammals to hunt fawns actively as soon as the birthing season began. Where roe lived at low density, active hunting by foxes did not begin until later in the season. The lifestyles of roe and water deer are similar, and in sites like Woodwalton Fen, where there are high densities of both water deer and foxes, predation by foxes on water deer fawns is likely to be an important factor. Activity of badgers *Meles meles* has increased in the reserve since about 2010; signs were relatively rare at that time, but by 2013 were common throughout the reserve and just outside too. Badgers will take deer fawns (e.g. Jarmeno 2004) and could also be implicated in the low recruitment rates observed in recent years.

While the recent decrease in the winter population may have been driven initially by loss of habitat, other factors are also likely to be involved, including unsuitable periods of weather in both summer and winter (see below). Since 2016, there has been a marked decrease in water deer numbers on the Middle Farm fields, which may have become less attractive to deer. Number counted on the fields was not related to number seen in the reserve, but was associated with the presence of livestock on the fields. For the period 2009–2018, there was an inverse relationship between the extent of livestock grazing on the nine study fields and number of deer counted. For instance, there was no grazing in February–April 2013 when the average count of water deer was 27, but 4.3 fields were grazed on average during February–April 2017 when the water deer count averaged only three.

Water deer are displaced when livestock are introduced, both inside and outside the reserve. In fields inside the reserve, deer were present on 22% of winter surveillance visits prior to cattle being introduced to manage vegetation; this decreased to 4% while cattle were present, but recovered to 18% after cattle were removed (Cooke 2009a; Figure 8.8a). Deer are probably deterred by both the presence of livestock and their impact on the habitat. The effect of sheep on the farm fields was studied during the winter of 2010/11; they were moved around from one field to another, spending about two weeks on each (Figure 8.8b). Average count of water deer per field was 1.1 during the month before sheep were introduced, zero while the sheep were present, 0.4 during the month after sheep were removed, and 2.6 during the following month. Numbers of grazing deer increased as they responded to the growing grass. Water deer are, however, occasionally seen with sheep – one apparently spent more than a week with a flock on a farm close to Holme Fen during the winter of 2016/17.

The winter grazing afforded by the new grasslands on Middle and Darlow's farms has been crucial in supporting the higher population of deer based in the reserve in the early years of the twenty-first century. Any factor that reduces the suitability of those fields may prejudice the situation. In this respect, I have investigated whether the increase in visitors showing an interest in the farm landscape may be detrimental. Various observers, including myself, have described how many water deer could be seen from the western flood-bank on spring evenings. In addition, this has become a well-known spot for viewing birds such as short-eared owls *Asio flammeus* and common cranes *Grus grus*. The likelihood of encountering visitors on the flood-bank while I was undertaking dusk surveillance walks increased fivefold during 2005/6–2016/17, although springs with high visitor numbers did not have lower numbers of deer on the grass fields. Nevertheless, even a single visitor can affect deer numbers on a particular occasion.

8.3.6   Effects of inclement winter weather on adult deer, 1976–2017
How weather affected the population throughout the whole study is considered in this and the following two sections. This section examines how severe winter weather affected survival and recruitment during the following year.

A list of the harshest winters was prepared by using data collected since 1976 by the meteorological station at the National Institute of Agricultural Botany in Cambridge and from weather observations made during surveillance visits to the fen. Winters were provisionally selected for inclusion in Table 8.2 if they had a total of more than 40 days with air frost. Then the list was restricted to winters in which

Table 8.2   Descriptions of the six most severe winters (October–March) during the study at Woodwalton Fen, 1976–2018, with figures for survival of water deer during the following year and recruitment index the following winter.

Winter	Days of air frost	% visits with snow cover	Rainfall (mm)	Flooded	Survival next year (%)	Recruitment index next winter (%)
1978/9	60	29	291	Yes	60	6
1981/2	45	19	228	No	72	42
1984/5	55	23	269	No	73	0
1985/6	56	21	223	No	70	14
2009/10	48	17	329	No	86	21
2012/13	53	9	347	Yes	77	2

Figure 8.9   Flooding at Woodwalton Fen, April 1980, a view
from the entrance down the main cross ride.

deer at Woodwalton Fen may have suffered the most from snow cover, high rainfall and flooding. To put these winters in perspective, the earlier winter of 1962/3 had 89 days of air frost.

The most severe winter on the list was 1978/9, when it was unusually cold and numbers of dead deer were found (Cooke and Farrell 1981). During that winter, the reserve was used for water storage and, at peak flood, much of it turned into a lake up to a metre in depth. The combination of snow and flood from early January to early April 1979 affected deer distribution across the reserve and resulted in the deer moving to live on or close to the western flood-banks where they had ready access to the farmland beyond.

One particularly memorable visit was on the morning of 3 February 1979 when I put on waders and pulled my rowing boat containing my two sons, then aged eight and seven, across the reserve to the flood-banks. We counted 12 deer on the farm fields and nine on the banks. This was an atypically large number to see, especially on a bright morning. We found a freshly dead buck just inside the reserve, lying in knee-deep water and surrounded by ice. I collected it for post-mortem examination and put it in the boat with the boys – it is a trip we all still remember. An example of flood conditions in the earlier years of the study is shown in Figure 8.9.

Reasons suggested for the observed mortality in the early months of 1979 included starvation, exposure and perhaps reaction to the stress of being compelled to move beyond their normal range – deer were more wary on the farmland than in the reserve. Whitehead (1950) noted that a few bucks died at Woburn each winter after the rut was over 'for some unaccountable reason' that was not related to starvation. The appreciable mortality of adults in early 1979 at Woodwalton Fen will have contributed to the virtual absence of first-year deer in the population during the following winter. Other reasons for the breeding failure may have included an increase in stillbirths or a reduction in the fitness of fawns.

The first four winters listed in Table 8.2 were during the first 20 years of the study when the landscape and the population remained fairly stable overall, whereas the final two were during a period of change. So effects of severe weather were more likely to be apparent in the first four winters. There was no universal recruitment failure – indeed, the figure for the winter after 1981/2 was one of the highest indices recorded. As regards adult survival in the following year, all four of these severe winters were characterised by low values, with an average of only 69%. Average survival for the remainder of the years in this period was 83%, so these severe winters reduced adult numbers.

Deer in the winters of 1978/9 and 2012/13 had both cold and flood conditions with which to contend. During those winters, I found walking through parts of the reserve difficult and tiring at times. If the water was shin-deep and covered by thin ice, it was possible to cross compartments – but exhausting, as the ice gave way at each step. I expect the deer found it similarly unpleasant, quite apart from the fact that their usual grazing areas were under water and ice. It is not surprising that survival and recruitment were low in 1979 and 2013. Flooding in the winter of 1978/9 started in January, so there was an opportunity to mate in December; in the winter of 2012/13, however, flooding began in December and continued until February, thereby disrupting the rut. Other winters suffered varying degrees of flooding, but none had cold conditions as severe as in these two winters.

8.3.7   *Effects of mild winter weather on the rut*
It is possible that if a winter is too mild, this may also disrupt the rut, as Dubost et al. (2008) found that most mating occurred during the coldest weather. The number of days with air frost was calculated for each winter during November–January (i.e. just before and during the rut); a weak positive relationship was found on recruitment the following winter. Two mild winters stood out from the rest: 2015/16 with only six days of frost and with none in December; and 2013/14 with seven days. December 2015 was the warmest December in central England since records began in 1659 (Sparks et al. 2017). Camera traps on one of the main rutting stands showed a general low level of rutting behaviour in both winters (Section 3.2.1). The following winters had relatively low recruitment of 7% and 13% respectively. The four winters just discussed with highly atypical weather (1978/9, 2012/13, 2013/14 and 2015/16) had an average recruitment the following winter of only 7%. The fact that three of these winters occurred during the recent five-year period of decline suggests that weather contributed significantly during this spell.

8.3.8   *Effects of summer weather on recruitment*
Elsewhere, fawn mortality increased because of heat stress in open habitats in hot summers (Chaplin 1977; Dubost et al. 2008). At Woodwalton Fen, however, fawns tend to be left in dense ground vegetation, and hyperthermia is unlikely to be a problem. In their Chinese range, summers are generally hotter and more humid than in this country (Zhang 1996). At Woodwalton Fen, the relationship between average temperature during the first few months of a fawn's life (May–August) and recruitment index the following winter was investigated for each year apart for the four years when very atypical weather conditions were associated with poor recruitment (1980, 2013, 2014 and 2016). A relationship was found with higher summer temperature usually being associated with higher levels of recruitment. No effect of summer rainfall on recruitment was apparent.

8.3.9   *Shooting, poaching and road mortality*

Another mortality factor to be considered for water deer in the local population is shooting outside the reserve. Up until 2008, stalking was seldom recorded on Castlehill Farm. In contrast, there were periods of quite intensive shooting on Middle and Darlow's farms. In the early years of the study, of 37 deer known to have died during 1976–1980, nine (24%) were shot (Cooke and Farrell 1981). Most of the stalking on Middle and Darlow's farms was undertaken from straw bale hides in fields next to the reserve and occurred during the periods 1982–1989 and 1997–2007. Information from some of the stalkers and from Stewart Papworth, the farmer at Darlow's Farm, indicated that 5–10 deer may have been shot each winter. Up to 2008, stalkers were recorded on those farms on 50 dates, 6% of the 804 occasions on which the farms were scanned. However, stalking will have occurred more frequently than this as both farms were relatively rarely scanned entirely and stalking was known to occur on winter mornings too. The fact that numbers of deer seen on these farms did not increase until fields on Darlow's were converted to grass and stalking stopped may have been due in part to stalking pressure and the presence of stalkers; on the then unstalked fields at Castlehill Farm, numbers showed a long-term, if erratic, increase (Figure 8.6). Although the deer shot on Middle and Darlow's farms will have been mainly based in the reserve, culls were probably insufficient to have affected the reserve's population. Indeed, 1997–2007 saw a marked increase in the population (Figure 8.4a).

The stalking situation now has changed. The Deer Act 1991 was amended in 2007 thereby curtailing the shooting season to just five winter months, November through to March. Since 2008, no shooting of water deer has been observed on Middle or Darlow's farms, but 15–30 were shot each winter up until 2015/16 by Martin Guy and his sons on Castlehill Farm and other land to the south and south-west of the reserve. As this stalking took place up to 2 km from the reserve, relatively few of these deer will have been reserve residents, but such a level of culling may reduce both density on the farmland and potential for dispersal.

Poaching has been a serious threat to water deer in China (Zhang and Guo 2000; Harris and Duckworth 2015). In Britain, poaching has been, and still is, a problem, despite being targeted by a number of campaigns, such as the BDS's 'Shine a light on poaching' in 2009. Water deer, with their propensity to feed on open fields at certain times of year, are more likely than muntjac to be sought out by poachers and coursers (e.g. Childerley 2014). I have heard reports of contests with dogs being live-streamed to public houses in London for gambling purposes. And there have been instances of abandoned dogs going on to kill more deer. Incidents of semi-organised poaching of water deer have occurred in western Cambridgeshire, and some of the more accessible fields near Holme Fen have been targeted. Woodwalton Fen is, however, nearly 2 km from the nearest through road and the nearby fields may be less attractive to law-breakers from further afield. Nevertheless, there was poaching recorded in the winter of 2015/16, and it is possible that the combined effects of legal and illegal culling have had some effect on deer density in the reserve by reducing numbers of deer on the farmland available for replacing losses of deer in the reserve.

Lack of roads in the immediate area means that deer based in Woodwalton Fen are safe from being victims of road accidents, although dispersing individuals are at risk (Figure 11.8). Further afield, many water deer are killed on the roads. It was estimated that during 2003–2005, several hundred were killed on roads in England, the number being equivalent to about 10% of the national population (Langbein 2007). Water deer are not the most astute species at crossing roads, and dead deer are becoming more common along roads in western Cambridgeshire.

8.3.10   Recovery in 2017/18

During the last winter before going to press, a remarkable recovery in numbers occurred. The increase during 2017 of 77% in sightings per hour was the largest change seen between winters (Figure 8.4a). The recruitment index in 2017/18 was the highest recorded (46%) and survival was also unusually high (98%). Circumstances proved almost optimal for recovery. With 30 days of air frost, the winter of 2016/17 was neither atypically cold nor mild, so the rut should not have been affected by weather. The summer months were the warmest in the last ten years, thereby aiding survival of fawns. The grass on Middle Farm was well grazed by stock during the winter of 2017/18 and was relatively unattractive to the deer. However, this was more than compensated for by the growing of a large area of oilseed rape adjacent to the reserve on Castlehill Farm, which was grazed by large numbers of deer throughout the winter (Figure 8.6). These crop fields apparently deterred much of the dispersal that might otherwise have occurred.

8.4   Comparisons and conclusions

Individuals of both species may turn up at suitable sites years before colonisation really gets under way. Once a muntjac population begins to develop in a wood of 50 ha or more, it is likely to be at least 15 years before peak population is reached. In smaller sites, this period may be less and peak densities somewhat lower. Muntjac breed on a seven-month cycle whereas water deer breed seasonally but have multiple births. Both strategies enable populations to build up or recover quickly. Water deer have the potential to double their population each year. In reality this is unlikely to occur for very long, even in captivity. Nevertheless, there are indications that water deer populations can sometimes increase even faster than those of muntjac. At Woodwalton Fen, the population apparently reached a typical long-term density only ten years after the first signs were seen, and this was despite early colonisation coinciding with the most severe winter weather of the last 60 years.

That is not to say that water deer are immune from the impact of severe winters. They suffered badly at Woburn during the desperately bad winter of 1946/7 when many were found dead despite food being available (Whitehead 1950), and appreciable mortality occurred at Woodwalton Fen during the less severe weather of 1978/9. Nevertheless, long cold winters may have a more dramatic effect on muntjac, whose feral populations suffered major mortality in 1946/7 and 1962/3 (Chapman et al. 1994a; Chapman and Harris 1996). At Woodwalton Fen, the less severe winter of 2009/10 led to increased foraging in young, newly independent muntjac, but not in water deer. At Monks Wood, just as muntjac were completing their colonisation process, the consecutive bad winters of 1984/5 and 1985/6 were associated with the same behavioural change in the early months of 1985 and 1986.

Weather conditions affected reproduction in the water deer population at Woodwalton Fen. When there was flooding in the reserve as well as prolonged cold spells, recruitment was poor in the following winter, as it was after extremely mild winters; it is likely that both sets of conditions inhibited rutting behaviour. Warm summers were associated with good recruitment during the following winter, but summer rainfall had no effect. As our climate warms, therefore, mild wet winters may on balance have a detrimental effect on rutting, whereas hotter summers are likely to improve fawn survival.

The more highly territorial nature of (buck) water deer may afford them protection against overcrowding by leading to greater dispersal in autumn. Furthermore, they readily take advantage of seasonal grazing outside their core area. Both of these traits may be related to their willingness to venture on to open ground rather than keep in or near cover. Muntjac will range more widely if forced to, as was reflected in the numbers killed on the roads beside Monks Wood in 1994. But newly independent muntjac are at a serious disadvantage in hard winters. A five-month-old muntjac caught in a freezing and snowy spell in February has little experience of fending for itself or what the world has to offer beyond its mother's home range. In contrast, a young water deer in February will be about nine months old and much more worldly-wise, having probably been introduced to the wider countryside by its mother during the previous autumn and having survived the rigours of its first rut.

At Monks Wood, the die-offs of muntjac in 1991 and 1994 were associated with unsustainably high densities and diminishing food resources rather than with poor weather. At Lady's Wood, which is less than the size of the home range of a single muntjac, no dead deer were noticed, but deer activity lessened in response to food and cover being lost. No die-offs of free-living water deer appear to have been reported in similar circumstances to those at Monks Wood, perhaps in part because they live in more robust habitats that are less prone to be damaged by their own feeding activity. There was the die-off at Whipsnade in 1933/4, but this involved captive animals held at an unnaturally high density on grass fields.

The study at Woodwalton Fen has demonstrated that a population of water deer can persist in a reasonably stable state for 50 years, despite their potential for fluctuation. Annual change in sightings was less than 20% for most years. One would expect the Broadland population to be at even less risk of extinction or major decline because of the much larger area of wetland that has been colonised. In contrast, populations on farmland are more dependent on the aims of the owners and managers, so numbers are likely to be more variable in the longer term. Deer at Woodwalton Fen demonstrated that the species can be quick to take advantage of favourable alterations to the landscape.

Recently Waeber et al. (2013) have drawn attention to the degree to which large, dense populations of deer such as muntjac can be significant sources of emigration to the wider countryside. The muntjac population at Monks Wood was such a source of muntjac up until the time that stalking was sanctioned in the wood. Similarly, Woodwalton Fen has been and continues to be the local source for water deer.

Estimates have been made of the percentage of the English national deer population lost to traffic collisions each year, 2003–2005 (Langbein 2007). For muntjac and water deer, these estimates were almost identical at about 10% However, this remarkable similarity probably masked different scenarios for the two species. Muntjac are much more abundant and widespread, including in many suburban areas. The majority of muntjac may encounter roads relatively more frequently than most water deer. If the two species were similarly adept at crossing roads, I would expect a higher percentage of muntjac to be killed. However, I have driven past scores of muntjac crossing roads or standing on verges, only rarely feeling a collision might be imminent; whereas a surprisingly high proportion of the much smaller number of water deer that I have negotiated in my car seemed to have had a death wish. In several areas of East Anglia, warning road signs have been erected specifically for water deer. Road mortality of 10% per annum is unlikely to have any measurable effect on national populations, but there may be local effects around hot spots that will contribute to overall mortality.

CHAPTER 9

Interaction between deer species

9.1   Introduction

This chapter focuses on how muntjac and water deer may interact and affect one another or interact with other species of deer. The most likely interaction is over competing for resources, particularly food. For there to be interaction via food resources, there must be some overlap in the niches occupied by two or more species – in other words, the species must occur in the same habitats in the same parts of Britain and, to some extent, forage for the same food items in the same places. In order to understand better where there is potential for interaction to occur, it is necessary to consider where each of the other four species occurs (Smith-Jones 2017) and what it may feed on. There are, though, other forms of interaction such as the larger deer species destroying vegetation needed by muntjac as cover. As all species of deer expand their ranges across lowland England, instances of inter-action may become apparent in the future.

Roe deer are found in all regions colonised by muntjac and water deer, but have only recently reached parts of central England. They occur in dry, dense habitat similar to that preferred by muntjac, and also in open or wet habitats that might hold water deer. Roe are mainly selective browsers, but also graze, particularly on herbs (Fawcett 1997), so will take the same food items as the two smaller species in appropriate situations. They feed to a slightly greater height than muntjac and water deer, but there is clearly potential for interaction.

Fallow deer generally occur in the same areas of England as muntjac, but are absent from some of the water deer's strongholds, such as the fens of west Cambridgeshire and parts of the Broads. They are predominantly bulk grazers, feeding mainly on grasses, but also take a range of woody species and herbs (Langbein and Chapman 2003). High densities of fallow are capable of impacting woodland composition and structure and it would be surprising if there were not at least local effects on muntjac populations.

The Sika *Cervus nippon* is the species of deer least likely to be currently encountered in this country by muntjac or water deer, but there are a few localities, for instance in southern England, where sika and muntjac populations coexist. Sika are typically found in areas of conifers and heathland. Their diet varies between sites, sometimes being largely composed of grasses and heather, but sometimes being more varied (Putman 2000). Opportunities for niche overlap between sika and muntjac appear limited.

Red deer are now found through East Anglia and here they are most likely to come into contact with muntjac and water deer. Thus, in Norfolk, they are numerous in Thetford Forest, where they live alongside muntjac, and in the Broads which is the national stronghold of water deer. Grasses are particularly important throughout the year, but a range of other graze and browse species may be taken (Staines et al. 2008), so there is potential for impact on the smaller species of deer.

9.2   Interaction between muntjac and water deer

I am not aware of interactions being reported between muntjac and water deer in England other than in Cambridgeshire. At Woodwalton Fen, water deer were well established throughout the reserve by the time muntjac started to colonise in 1980. Muntjac tended to build up most strongly in the drier southern third of the reserve, which is dominated by mixed woodland, sallow carr and grass-heath fields (Figure 1.7). In the early years of their colonisation, they apparently partially displaced water deer from this area (Figure 9.1). Equilibrium was established in the early 1990s, and from then until muntjac stalking began in 2011 roughly similar numbers of the two species were counted. Muntjac continued to show a preference for this area of the reserve; during the winter of 2009/10, when their maximum numbers were recorded, 45% were seen in the south, whereas only 8% of water deer sightings were recorded there. Although muntjac seemed to affect water deer distribution, they

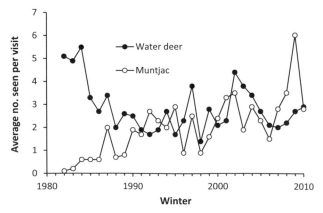

Figure 9.1   Average number of water deer and muntjac seen per visit during dusk surveillance in the drier, wooded southern part of Woodwalton Fen, 1982/3–2010/11. Muntjac began to colonise the reserve in 1980, and muntjac stalking began in 2011 (updated from Cooke 2009a).

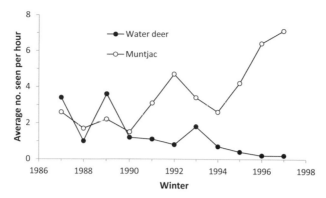

Figure 9.2   Average number of water deer and muntjac seen per hour during dusk surveillance in Holme Fen, 1987/8–1997/8 (updated from Cooke 1998c).

had no noticeable effect on numbers of water deer seen elsewhere in the reserve or on the adjacent farmland.

Events at Holme Fen were similar to what happened in the south of Woodwalton Fen (Cooke 1998c). Paul Burnham was reserve warden during the early and middle part of the 1970s. He occasionally saw deer signs, such as hair or droppings, but never saw the deer themselves. Roger Boston took over as warden in 1977 and began to record sightings of water deer. He first saw a muntjac in 1980, and the same year Peter Carne also recorded one in the reserve (Carne 2000). Up to the end of 1986, Roger Boston logged a total of 349 sightings of water deer and 71 of muntjac. In 1987, I began dusk surveillance and results are summarised in Figure 9.2 up until the winter of 1997/8, which was the last full winter prior to a break in recording. During this period, water deer sightings decreased and those of muntjac increased. When surveillance started again in 2003, sightings frequency for muntjac remained high whereas water deer were absent or rare. What happened after 2005 when muntjac were culled is discussed in Section 17.3.

At Monks Wood, muntjac began to colonise in about 1970 and water deer were first definitely seen in 1977. A small resident population of water deer had developed by the late 1980s. Roughly two-thirds of all sightings were in the north of the wood, being concentrated on two fields enclosed within the wood – these were areas less favoured by muntjac (Cooke 2006). Average numbers of water deer seen per hour on surveillance walks, 1986–1998, decreased from a peak of 1.6 in 1987 to zero in 1998. This decrease coincided with a period of peak sightings of muntjac in the wood with annual averages ranging from 14 to 23 seen per hour (Figure 8.2).

In all three situations, water deer decreased. At Holme Fen, which is dry birch woodland, they became very rare as muntjac increased. At Monks Wood, they hung on for a time despite the very high density of muntjac. However, their population decreased relatively little in the south of Woodwalton Fen. This was probably because the area contained a number of open fields as well as woodland and because of interchange with the adjacent wetter parts of the reserve. I have been told by both Endi Zhang and Min Chen that similar interactions between the two species have also been noticed among populations in their native range in China.

So what is happening to cause such changes? It is likely that several factors are involved to varying degrees. For instance, muntjac appear to tolerate human disturbance much better than water deer, and increasing leisure use of Monks Wood and Holme Fen by people, especially dog walkers, may have had a greater effect on water deer. But this cannot explain the decrease in the south of Woodwalton Fen which is the least disturbed part of that reserve. I have often wondered whether competition for bramble in winter may be important. Both species have relied heavily on bramble to see them through the mid- and late-winter months. In Monks Wood and Holme Fen, bramble became defoliated up to a 1 m browse line in the 1990s as deer populations built up, and substantial die-back occurred. This did not happen in Woodwalton Fen until later, so simple shortage of bramble leaves did not cause the decrease in water deer seen there in the 1980s. However, recent work with wildlife camera traps suggested that water deer may be deterred from feeding where muntjac had previously browsed (Section 19.3.1). So it could be the *presence* of muntjac and not simply loss of bramble that affected the water deer. There was after all sufficient bramble bushes for the muntjac population to thrive at Woodwalton Fen and Holme Fen as the water deer were declining. At Monks Wood, however, declining food resources eventually resulted in muntjac die-offs in the early 1990s.

9.3   Interaction with roe deer

Chapman et al. (1993) studied ranging behaviour, social organisation and activity of muntjac and roe deer in the King's Forest in Suffolk during the 1980s. They found that roe ranged more widely out on to agricultural areas in winter and suggested that, if food and other resources were limited, then roe rather than muntjac would be more likely to move to exploit new habitats. This was supported by an increase in muntjac numbers after the main study finished, associated with roe declining in localities where muntjac increased most and increasing their dependence on areas less favoured by muntjac. These authors pointed out that their radio-tracking study may have lacked the spatial precision that could be afforded by dung counting. This second method was later used in Thetford Forest by Hemani et al. (2004), who found greater habitat selectivity in muntjac. Niche overlap was highest in winter when both species tended to focus on bramble, indicating interspecific competition at times of food scarcity.

A possible case of muntjac having a considerable effect on a population of roe deer may have occurred at Chippenham Fen in Cambridgeshire. Roe deer were frequently seen there in the 1980s, but since muntjac started to colonise in the 1990s, roe have become much scarcer. The reverse effect of colonising roe deer on resident populations of muntjac does not appear to have been studied. By the time roe began to build up noticeably in Monks Wood, muntjac had already been controlled by stalking for a number of years and the habitat was not limiting populations. Natural England considered it necessary to sanction stalking of roe in 2011, 18 years after I noticed the first colonisers.

In my three local NNRs, roe have been slowest to colonise Woodwalton Fen. This may be because it has had high densities of muntjac and particularly water deer. Alternatively, it may be because Woodwalton Fen is much wetter – roe deer are even rarer in the wetter areas of the reserve. So far, they have been largely restricted to appearing in small numbers, mainly in winter, in the woodland in the drier south of the reserve. This is the area previously initially colonised by muntjac and where muntjac have had an appreciable impact on regeneration. Roe, being larger animals, can forage higher on bramble bushes and other browse species, but reduced regeneration from ground level has affected the overall amount of low food and may have delayed roe becoming year-round residents, as they are in Monks Wood and Holme Fen. Since the winter of 2013/14, camera trap studies in the southern woodland at Woodwalton Fen have suggested that roe may be occurring more frequently as muntjac decrease because of stalking (Table 19.2). So far, I have no evidence of roe being deterred from browsing bramble by the prior feeding of muntjac, as may have happened with water deer.

The situation in and around Wicken Fen is very different. Roe deer are abundant especially on adjacent Burwell Fen, while muntjac are locally common in the scrubbier areas, indicating considerable spatial separation. Water deer are occasionally recorded, but have apparently never become established, possibly because of the presence of the other species. However, their lack of success in the extensive reed beds and other wet areas is a surprise and I wonder whether they have been sometimes wrongly identified as roe deer. If there is no control of other deer species at Wicken Fen, water deer may continue to struggle to gain a toehold.

9.4   Interaction with fallow deer

Fallow deer are often found in the same woods as muntjac. Langbein and Chapman (2003) stated that, if fallow deer deplete a site's woody understorey and herb layer, it may become less suitable for muntjac (Figure 9.3a). They suggested that the competition is unidirectional in that fallow may affect muntjac but muntjac are far less likely to affect fallow. That being the case, effects on muntjac densities may be detectable in woods harbouring significant densities of fallow deer.

Looking at my data set of woods in eastern England with fallow deer scores of five or more, the relationship between muntjac deer scores and wood area up to about 100 ha in size was virtually identical to that for woods without fallow (Figure 10.2), that is the muntjac deer score tended to increase as wood area increased. However, for bigger blocks of woodland with a fallow presence, the muntjac score was lower than would be expected on the basis of woodland area. Such situations existed in Wakerley Great Wood in Northamptonshire, in Bourne Woods in Lincolnshire and in Hatfield Forest in Essex. The explanation for these observations may be that, in larger woodlands, suitable patches that are unaffected by fallow deer are scarce within the range of resident muntjac, whereas in smaller woods they have easier access to potential feeding areas outside. An example of muntjac foraging outside a wood wrecked by fallow deer occurred at Short Wood in Northamptonshire. This 25 ha wood was totally fenced in 1999 and some fallow deer were fenced in. There was a deer-leap but they did not use it. They had a huge impact with a 1.5 m browse line through most of the wood by 2002, and the muntjac appeared to survive by getting in and out under the fence. The fallow deer were eventually 'liberated' early in 2003.

Putman et al. (2011a) suggested that a larger grazer, such as the fallow deer, may compete with a smaller selective feeder, such as the muntjac, and so cause the latter

Figure 9.3   Browse lines of larger deer species reduce availability of food and cover for the smaller species: (a) a fallow deer browse line on ivy at Fowlmere, Cambridgeshire, April 2010; (b) a red deer browse line at Minsmere, Suffolk, September 2010.

to focus more on potentially vulnerable plant species. In this way the impact of a given density of muntjac may be greater than had the muntjac occurred alone. However, to date, I have not recorded an example of this process. Indeed, where moderate or high densities of fallow occur, I have found that muntjac damage scores are consistently lower than might be expected from their muntjac deer scores, based on the relationship in Figure 7.3 for woods without fallow deer. While this may reflect less feeding by muntjac on target plants within the woods, it is possible that fallow damage masked some of that done by muntjac. On the other hand, a modelling study by White et al. (2004b) indicated that, at a landscape scale when muntjac and fallow deer were both present, this led to an unexpectedly high level of damage.

9.5   Interaction with red deer

As with fallow deer, if enough red deer use a wood then they would be expected to render it less suitable for muntjac (Figure 9.3b). I am not aware of anyone having studied such an interaction, but in 2013 I deployed camera traps at the Norfolk Wildlife Trust's new reserve between the villages of Methwold and Hilgay – muntjac were widespread and abundant despite the woodland being heavily impacted by red deer, which occasionally numbered 200. Similarly, water deer coexist with red deer in wetland habitat in some parts of the Norfolk Broads. Conflict is reduced because the red deer often commute some way to feed on farmland, and the Broads retain this country's largest numbers of water deer.

CHAPTER 10

Densities and numbers

10.1   Introduction

Knowledge of deer densities and numbers can prove useful in site management. A historical knowledge of densities can be of considerable benefit in predicting current impacts (e.g. should coppice be protected this year?), as well as assessing future risks (e.g. if density is starting to increase following relaxation of stalking). Some knowledge of numbers is essential to stalkers when planning a detailed culling programme (de Nahlik 1992). And a question often asked by an interested visitor is 'How many deer live here?'

Putman et al. (2011a) collated information on densities and impacts in an attempt to derive threshold densities of deer at which negative impacts start to be seen. Attention was drawn to the difficulties in determining density. They concluded that deer density alone was unlikely to be a good predictor of impact, and management should be based on assessment of both *actual* impact and *apparent* density. In this chapter, density is calculated in several ways, including using simple counts.

In addition to determining densities in specific locations to provide information of relevance to management, there have recently been attempts to derive typical or average densities in habitats of different types (e.g. Croft et al. 2017). If such information is available, it can be combined with distribution data to produce estimates of regional or national population size. This is of relevance to deer management and policy at a broader scale. As yet, however, there is limited published information on densities of muntjac and water deer, and existing data largely relate to high-density populations. Most of the information in this chapter is from high-density populations, but attempts are also made to include data from less dense populations and from landscapes.

There is also relatively little information on densities within the species' native ranges, and the main part of this chapter is devoted to situations in Britain. Estimates of water deer densities in China up until the late 1990s varied between 3 and 90 per km² (Wang and Sheng 1990; Sheng 1992b; Zhang 1996). Some of the lower densities were for landscapes rather than sites, while some of the higher estimates were for aggregations during the rut. Since then, numbers and densities have declined (e.g. Chen et al. 2009). Min Chen provided Richard Fautley (2013) with the following estimates for water deer in China in 2011: total, less than 5,000; Jiangsu Province, less than 400 at Yancheng and Dafeng; Jishan Island in Poyang Lake, less than 500; Anhui Province, small scattered populations; Zhoushan Islands, 2,000–3,000.

In South Korea, densities of water deer varied from 1 per km² in urban areas to 2 per km² in upland areas and 7 per km² in the lowlands (Kim et al. 2011). The population in that country was recently reported to be 500,000–700,000 (Chun 2018). The total area of South Korea is roughly 100,000 km², pointing to an overall national density of 5–7 per km²; the last author referred to increased densities since 2000.

The only estimate of density for muntjac in their native range comes from Taiwan (McCullough et al. 2000): a density of 9 per km² in a study area seems rather low and the situation was unusual in that average range size exceeded 100 ha.

Densities and numbers are considered below at four scales: sub-site, site, landscape and national.

10.2   Sub-site scale

Such estimates usually relate to areas of a few hectares or a few tens of hectares depending on the size of the site. Some sub-sites studied in big sites will be larger than entire small sites. Figures are derived mainly as counts of deer inside an area or emerging from it on to more open ground, and illustrate maximum numbers that can be seen. Once in the mid-1990s, I saw 13 muntjac deer grazing in the evening on a field of newly germinated winter wheat outside Monks Wood, but such aggregations of muntjac are very rare. A remarkable concentration was recorded in 1976 in Maulden Wood, Bedfordshire (Figure 5.2), when 13 were netted on a single occasion in a plot of 1 ha (Anderson and Cham 1988). This probably resulted from them lying up in a small patch of dense cover in otherwise fairly open woodland.

Water deer have a greater tendency to form groups than muntjac (Section 2.7). Aggregations of water deer can be seen during late winter and spring evenings both inside and adjacent to Woodwalton Fen. Many people are puzzled to learn that a non-herding species can be found in such gatherings. But these are not true herds – they are simply situations where individuals and perhaps a few pairs have convened to take advantage of good feeding. They are akin to suburban neighbours descending on their local public house on a Friday night. After last orders, people disperse back to their homes. If an aggregation of water deer is disturbed, they are likely to scatter, there being few, if any, bonds holding them together.

Counts of water deer have revealed their potential to exist and feed at high densities. Within the reserve at Woodwalton Fen there is much cover, even in winter, and it is more difficult to see large numbers. The north end of the reserve is 40 ha of mainly reed bed with some sallow carr and mixed fen vegetation. My highest count of water deer in that area was 21 out on the rides one evening during the winter of 2007/8. This equated to a density of 0.53 per ha, but many deer must have been missed. The most concentrated aggregation of water deer that I have seen inside the reserve was 16 grazing on a mown area of fen vegetation of only 1 ha in February 2011; these deer had paused before moving through wet sallow carr out on to the adjacent farm grassland. Again, many deer were presumably still in cover. The most I have ever seen emerging from this 15 ha patch of fen fields and carr was 45 on one evening in March 2013, giving a daytime density of 3.0 deer per ha. This part of the reserve was beside the best feeding area available.

Figure 10.1   Eleven water deer grazing on one of the Middle Farm fields at dusk in March 2015. Part of Castlehill Farm, including a small copse, is in the background.

In the early months of the year, the grass and arable fields of the adjacent farms are open with largely unrestricted visibility so that it should be possible to count virtually all of the deer out feeding or resting (Figure 10.1). Although the grass fields were part flooded in March 2013, 53 deer were counted one evening grazing over 18 ha, a density of 2.9 per ha. This included 28 on a single field of 4.1 ha (6.8 per ha). The most seen on crop fields was 27 on 16 ha of newly sprouting oilseed rape in February 2012 (1.7 per ha). These though are 'honeypot' densities and will have been appreciably higher than density for the whole site.

10.3   Site scale

With situations such as water deer at Woodwalton Fen and muntjac at Monks Wood, there are two ways of calculating densities. First, there is the number in the population compared with the area of the site. I tend to prefer this method for individuals that spend most of their time in the site because its area is known and fixed. The second method is the number compared with the area over which the population ranges. But this area is not fixed, will probably vary seasonally and may not be known with any certainty. There are also two ways of expressing density – as number per ha or as number per km². The latter is the conventional way of expressing deer densities, especially for larger species that may range over several km². But number per ha seems the logical expression for small patches and this is why it is used under the previous heading. It also has relevance for small sites and where impacts on patches of vegetation are being considered. But from now on in this chapter, number per km² will be used.

A major study was undertaken on muntjac in an area of 206 ha in the south-east corner of the King's Forest in Suffolk from the mid-1980s until the early 1990s (Claydon et al. 1986; Chapman et al. 1993; Blakeley et al. 1997; Figures 1.3 and 4.2a). The number of muntjac using this mainly coniferous area was estimated each February, making allowance for deer whose ranges were partially outside the study area; fawns were excluded from the total. Numbers increased from 31 in 1985 and 1986 (15 per km²) to 56 in 1993 (27 per km²).

Studies at Monks Wood revealed that, in some circumstances, muntjac densities could exceed 100 per km² when based on the area of the wood (Cooke 2006). Initially, I had difficulty convincing some people that muntjac could live at such densities. When the density was of this order, the population ranged over perhaps double the area of the wood. After stalking reduced the population, ranging became much rarer out over the adjacent arable land and grass fields.

Large deciduous woods in regions where muntjac are numerous can have high densities of this species. Figure 10.2 shows information for a sample of 32 predominantly deciduous woods in Cambridgeshire during the last 20 years, plotting muntjac deer score against size of wood. For Monks Wood, deer score has been found to be positively related to density (Figure 7.4), and deer score is used here as a surrogate measure of density. Muntjac shooting had started in a number of these woods, so the highest (pre-stalking) deer score has been used for each site. Woods with fallow deer were excluded from the sample as their presence may affect muntjac density (Section 9.4). There were rapid increases in muntjac deer score as wood size started to increase, but less change occurred with woods above 50 ha.

One of the reasons why large deciduous woods may have higher densities of muntjac is that they tend to have larger woodland blocks than small woods, thereby providing a greater proportion of woodland well away from rides and human disturbance. They also have proportionally less draughty exposed edges, and probably provide more opportunities for young deer to gain a toehold in their natal wood rather than having to emigrate.

Examining the size of woods with different deer scores:

- Ten woods had scores of 2–4, suggesting densities of roughly 25 per km² by applying the relationship at Monks Wood; average wood size was 15 ha with only one being larger than 20 ha. The areas of some of the small woods were less than a muntjac's home range.
- Seventeen woods had scores of 5–7, suggesting densities of roughly 50 per km²; average size was 33 ha and all but one was less than 70 ha.
- Five woods had scores of 8–10, suggesting densities of roughly 100 per km², and all were larger than 40 ha with an average size of 96 ha.

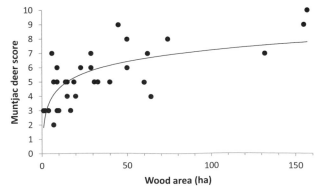

Figure 10.2   The relationship between muntjac deer score and wood area for 32 woods in Cambridgeshire with no fallow deer.

During the first three Decembers of the study at Woodwalton Fen (1976–1978), density of water deer in a central area of 39 ha was estimated to be equivalent to 28–46 per km². If this density could be extrapolated to the whole reserve, the total number would have varied between roughly 60 and 100. In recent winters, the reserve's population increased and peaked in 2011/12 at a density estimated at 125 per km² (Figure 19.1), giving a total of about 260 deer. The population that winter seemed to be boosted by deer migrating into the reserve following the destruction of crops of elephant grass on the adjacent farmland. This proved unsustainable and since then declines have occurred. Again the population ranged outside the reserve boundaries, especially in winter and spring. I am not aware of a wild population living at this density elsewhere in England. When visiting sites with good populations in the Norfolk Broads and marshes, the density of signs has never been greater than in Woodwalton Fen.

Thus there are similarities between the muntjac at Monks Wood and the water deer at Woodwalton Fen, with both populations having had density estimates exceeding 100 per km² when based on the size of the reserve. Likewise, both populations depended heavily on supplementing their food resources by regularly venturing outside the reserve. Muntjac deer appear more reluctant than water deer to move into open spaces to feed and only do so in numbers when they have to. Comparatively few muntjac graze outside Woodwalton Fen despite their population and that of water deer sometimes having been so high. And most of those that have done so have only moved under cover of darkness, as evidenced by camera trap observations. Numbers of water deer have always grazed regularly outside the reserve, even in the early years of the study when the population was smaller.

Very high densities of water deer live within the confines of Whipsnade Zoo in Bedfordshire. For instance in 1994, the estimate for the whole park was about 350 deer at 140 per km², while the subpopulation studied by Zhang (1996) had a density of 240 per km². Those at Branféré Zoo studied by Dubost et al. (2008) were maintained at up to 7 per ha (700 per km²). At Branféré, they received supplementary food.

10.4   Landscape scale

In a forest environment, it is a moot point when site scale becomes landscape scale. In the previous section, I treated the King's Forest study area as a site because it is the same size as the Woodwalton Fen reserve. Since that study finished there have been intensive and extensive studies into densities over forest blocks and entire forests in Breckland. In Thetford Forest, the Mundford block (13 km²) was surveyed in 2001 by thermal imaging, muntjac density being estimated as 20 per km² (Hemani et al. 2007). In 2002, the whole forest, comprising 185 km², was surveyed by counting the standing crop of dung along transects (Hemani et al. 2005). Overall population density of muntjac was estimated as 64 per km², with density in the King's Forest (24 km²) being 60 per km². During 2008–2010, Waeber et al. (2013) undertook a new survey of 12 woodland blocks of Thetford Forest, totalling 132 km², using thermal imaging equipment. Overall, density was 22 per km². Throughout this period, there was heavy culling within the forest. It is noticeable that the survey using dung counting gave an estimate that was roughly three times the other two estimates. The reason for this difference is not known but the drawbacks of dung counting are discussed elsewhere (Section 7.2.1). Dung counting was used to estimate muntjac density in woodland in the southern part of Rockingham Forest, Northamptonshire, at 26–39 per km^2 (Donnelly 1995; Wyllie et al. 1998). In all of these forest areas, muntjac numbers were controlled by shooting. Nevertheless, forest populations can be huge because the areas are so large. In addition, other species of deer are present, including large numbers of roe deer in Thetford Forest and fallow deer in Rockingham.

It is widely accepted that deer management should be co-ordinated on a landscape scale. Stalking of muntjac began in the Holme Fen reserve in 2005, in part to reduce their impact in the reserve, but also to reduce emigration from this large wood to other sensitive areas. Having some knowledge of muntjac abundance in the local landscape would provide an indication of the size of the problem. With this in mind I decided to delineate a landscape-sized area of relatively good habitat and estimate how many muntjac it might contain during 2001–2005 (Cooke 2006). The derived figure can probably best be described as a speculative estimate and the answer has not been tested by other means. The technique might, however, be used with greater confidence to monitor changes in overall numbers in the area, as well as changes in deer density in its woodlands, and this is attempted later in Table 13.2.

The area selected was a polygon of about 130 km^2 encompassing the three large NNRs of Holme Fen, Woodwalton Fen and Monks Wood, as well as other woods, ten villages and much farmland; grid references of the corners are given in Cooke (2006). The polygon incorporated the land shown in Figure 1.4 up to the motorway in the west plus a cluster of woods beyond Monks Wood in the south and extending to the east to include Warboys Wood. To the north and east, open fen meant some emigration would occur but significant immigration was very unlikely. To the west, the motorway formed a partial barrier to migration. Movement to the south was restricted to some degree by two fenced airfields and the town of Huntingdon. There is very little woodland in Cambridgeshire but this polygon is one of the more wooded parts of the county with about 8 km^2 of woods and 1.5 km^2 of wooded fen, totalling 7.3% of the landscape. A view of land between the railway line and the motorway is shown in Figure 4.5b. The landscape between the Woodwalton Fen reserve and the railway line is illustrated in Figure 10.3a; most of the land in the polygon is still arable farmland which is utilised by relatively low densities of muntjac (Figure 10.3b).

Each year a number of woods have been visited for scoring with a focus on those with public access. Based on the information from Monks Wood summarised in Section 7.2.4, muntjac deer scores were converted into muntjac densities of 0, 25, 50 or 100 deer per km^2. When a wood was scored more than once during 2001–2005, the latest value was used. Deer densities could therefore be inferred for each wood scored and an overall mean density estimated, weighted for wood size. The calculated density in the sample was then applied to all 9.5 km^2 of woodland. To estimate the total number of muntjac in the polygon, a figure for density in the wider countryside was also needed, a value of 2 per km^2 being used (Cooke 2006). The calculation has been modified slightly since the original version. The total number of muntjac in woodland is estimated to have been 590 at a density of 62 per km^2, with the overall number in the polygon being 830 at a density of 6.4 per km^2.

Figure 10.3   The area between the Woodwalton Fen reserve and the railway line (Figure 1.4): (a) typical farmland with the escarpment at the edge of the Fens in the background, March 2009; (b) a muntjac buck in a tractor tramline in wheat, Castlehill Farm, July 2013.

Monks Wood and several other local woods have had densities that exceeded the mean densities in the forests, but outside the woods, densities were and still are much lower. Waeber et al. (2013) assessed muntjac density by thermal imaging on the Stanford Training Area in the Brecks. This is a Ministry of Defence training area of 102 km^2 composed primarily of sheep-grazed grass heath, with some arable and woodland. Although this adjoins Thetford Forest, average density was the same as the overall density in the Cambridgeshire polygon at about 6 muntjac deer per km^2, compared with 22 per km^2 in Thetford Forest.

Unlike muntjac, water deer are not forest animals, but their equivalent area to the Breckland Forests is Broadland. In the 300 km^2 of the National Park, there are extensive tracts of suitable habitat. The current population is impossible to estimate with any degree of certainty, but I concluded there were 2,000–3,000 when estimating the national population at 7,000 (Cooke 2010). This translated into an overall density for the Broads of 6–10 per km^2, but densities of up to 40 per km^2 have been reported for wetland reserves (Section 19.1).

Water deer from Woodwalton Fen have dispersed away from the reserve, particularly to the west and the north. To the east and south, large drainage rivers present partial barriers to movement. The area around the reserve is unusual, even for this relatively agricultural region, in having an expanse of 44 km^2 without any through roads. Much of this is now part of the Great Fen, which over a long period of time will see intensively farmed arable land converted into less intensive pasture, reed bed and other wetland habitat (Bowley 2013). The main East Coast railway line forms the western boundary of the Great Fen and bisects this quiet area of farmland. Although some deer may cross the tracks, their slots reveal they often traverse this barrier via various tunnels and culverts. A survey of arable land in 2009 and 2010 suggested an average density of 3–4 water deer per km^2, but two farms probably had local densities exceeding 10 per km^2 (Cooke 2010). The whole landscape of 44 km^2, including the reserve, had an overall average density of about 7 per km^2. This is similar to both the crude estimate given above for Broadland and the overall density for South Korea (Section 10.1).

In the winter of 2009/10, when numbers were relatively high, I estimated totals in 27 tetrads that included the Woodwalton Fen reserve and all connected tetrads where evidence of their presence had been found. Average number per tetrad was 15, ranging from 140 (which included part of the reserve) down to zero (where at any moment in time the species was judged more likely to be absent rather than present). The average density approached 4 per km².

Water deer that escaped from Woburn Park will have found that the immediate Bedfordshire countryside lacked significant areas of wetland and settled into a life on farmland. A pioneering survey in the early months of 1992 indicated a population of 40–100 in roughly 200 km² of farmland around Woburn (Nau 1992), giving a density of about 0.2–0.5 per km². One of the highest concentrations of sightings was on farmland at Potsgrove outside the western boundary of the park (Figure 4.5a). Water deer have now spread through much of Bedfordshire and the quiet, rolling fields of Potsgrove have developed a reputation as home to a conspicuously dense population. The area extends to about 14 km² and a few years ago I was told by the estate's deer manager, Callum Thomson, that the population was maintained at about 120 individuals, a density of 8–9 per km² (Cooke 2009b). Deer tend to be seen on arable land rather than grassland, possibly because of avoidance of fields with stock. Members of the Bedfordshire Natural History Society have counted the deer regularly, the highest number seen being 44 in November 2004 (McCarrick 2006). This count provides some idea of the highest percentage of deer that might be recorded in such a situation – in this case 37%. Another large arable concentration with a density that appears even higher lives within about 3 km of the park boundary to the north of the M1 motorway (Childerley 2014).

10.5   National scale

As the scale increases, so the problems associated with estimating densities and numbers are magnified. In order to have any confidence in a national estimate, where the species occur must be known. In 1972, the BDS organised its first survey of deer in Britain on a 10 km square basis. Since then, information has improved considerably with repeat surveys by the BDS in 2002, 2007, 2011 and 2016, and with county organisations mapping on a tetrad basis. The latter development has resulted in more detailed local distribution maps, either in book form (e.g. Bacon 2005; Bullion 2009), or published in county natural history journals (e.g. Toms and Leech 2005; McCarrick 2007). The NBN Gateway maps incorporate records from many such schemes.

Our knowledge of the distribution of deer has recently become more up to date as well as more detailed. With any species that is expanding its range, however, distribution can never be totally up to date. Estimates of density are also required to calculate national population size, but meaningful average densities are hard to obtain. In my experience, national population estimates rarely meet with ready acceptance – critics usually believe them to be too low or too high for a variety of reasons, but do not often suggest alternative estimates.

The first scientific attempt to estimate national populations of muntjac was by Harris et al. (1995). Half of the records collected by Chapman et al. (1994a) had been from five counties: Berkshire, Buckinghamshire, Hertfordshire, Oxfordshire and Warwickshire. For these counties, adult densities were taken to be 30 per km² in prime habitat such as semi-natural broadleaved woodland and 15 per km² in suboptimal habitat such as plantations. The number estimated was then doubled to give the total for Britain: 40,000. Virtually all of these muntjac deer were in England with less than 50 in Scotland and less than 250 in Wales. Fawns and juveniles were estimated to add about 12,000 to the total, making 52,000. The same authors estimated numbers of water deer by summing estimates from local naturalists for the areas around Woburn (Bedfordshire/Hertfordshire), Woodwalton Fen and Stetchworth (both Cambridgeshire), Minsmere (Suffolk) and the Norfolk Broads. This gave a total for the pre-breeding population of 650.

In 2004, as part of the British Mammal Society's 50th anniversary celebrations, attempts were made to re-estimate the national abundance of all of our mammal species. For muntjac, relevant specialists were consulted about densities in local woodlands and average woodland densities were calculated for each geographical region. Using these densities and the area of woodland in each 10 km square occupied by muntjac in the BDS's survey in 2004, the total muntjac population was estimated at 128,500 (Anon. 2005). Adding up totals for the more discrete popula-tions of water deer gave an estimate of 1,500 animals. Because of the large margin for error, it was considered unwise to compare the 1995 and 2004 estimates.

In 2009, the Deer Initiative estimated how many deer might be in the United Kingdom for the Parliamentary Office of Science and Technology: estimates were more than 150,000 for muntjac and less than 10,000 for water deer. No information was given on methodology, and estimates were described as 'approximate and contentious'.

Also in 2009, I estimated the water deer population in Britain at 4,000 for Defra (Cooke 2009a). This calculation was based on: (1) using tetrad maps for the main concentrations, counting the total number of tetrads and multiplying by the average number of deer likely to be in a tetrad; and (2) estimating how many might occur outside the main concentrations. I was concerned, however, that I had under-estimated both the average number likely to be present in a tetrad and the number present outside the main concentrations. Therefore, much time was spent during 2009 and 2010 surveying farmland in 27 tetrads in the area around the Woodwalton Fen reserve (Cooke 2010). This produced an average figure of 15 deer per occupied tetrad, as stated above, which was then applied to the most up-to-date maps for Norfolk, Suffolk, Cambridgeshire, Bedfordshire and Hertfordshire. The total number in these five counties was estimated at 4,935 deer. In a composite BDS distribution map up to 2007, there were 97 ten-kilometre squares inside these five counties and 39 squares outside. The average number per 10 km square within the five main counties was 51; applying this density to squares outside yielded another 1,664. This method was likely to have overestimated the number outside the main counties where density was probably lower and where a number of colonies had died out (Ward 2005; Smith-Jones 2017). On the other hand, it will have under-estimated the number in the Broads and marshes of East Anglia where there are continuous expanses of good habitat. The total number, 6,919, was rounded off to 7,000 for 2010 (Cooke 2011) because of uncertainties in both the mapping and the calculation.

Croft et al. (2017) undertook a Defra-funded project to estimate the abundance of British mammals by combining publicly available data on distribution from the NBN Gateway with published estimates of density in different habitats. Abundances were expressed as minimum and maximum abundance estimates. For the muntjac these two estimates were roughly 2,000,000 and 5,000,000 respectively. These were believed to be 'unrealistically high' because density estimates were 'unrepresentatively high' due to the published figures for density tending to be for high-density populations. The authors pointed out that studies were required on densities of a number of (common) mammalian species in areas of low abundance. No abundance estimates for water deer were listed because the only density value was for Woodwalton Fen, which was known to hold a high-density population.

The Mammal Society carried out a similar study in 2016 and augmented limited published information on distribution and density by asking specialists, including myself, to provide extra data (Mathews et al. 2018). Considerable problems currently exist for any calculation of this type. These do not just relate to densities in preferred habitats, as both species live at lower densities in other common habitat types. Thus, figures that are used (or not used) for muntjac densities outside woodlands or water deer densities on farmland will make a significant difference to estimated totals. The Mammal Society's estimate for water deer was 3,600, the reliability of which was graded as 'very poor'. The muntjac population was estimated at 128,000 but this was for woodland only and was based on density in a single coniferous site. This was very similar to the estimate for 2004, which was also for those in woodland. In the exercise on my local polygon (Section 10.4 above), the muntjac living outside woodland represented an extra 40%; if this factor was applied to the new national estimate, it would increase to 180,000.

Perhaps not surprisingly, these exercises provide no clear idea of current total numbers of either species. Doubtless, such systematic approaches will form the basis of increasingly accurate estimates in the future. For the present, however, my opinion is that there are likely to be the region of 5,000–10,000 water deer. The total in China was estimated by the deer specialist Min Chen to be less than 5,000 in 2011 (Fautley 2013), so we may have at least as many as are in China. Reintroductions are under way in the Shanghai area (Chen et al. 2016; He et al. 2016) but populations continue to be threatened elsewhere in China: overall population growth is unlikely in the near future. In England, by contrast, the main range of the species is still expanding (Figure 5.5), and sometime fairly soon it should be possible to state categorically that we have more water deer than exist in China. However, the English and Chinese populations are dwarfed by the total of at least half a million reported to be in South Korea (Chun 2018).

CHAPTER 11

Introduction to impacts

11.1 National risk assessments

In 2009, Norma Chapman and I were commissioned by Defra to prepare risk assessments on muntjac and water deer respectively to aid departmental decision-making on alien species in Great Britain. The process involved answering questions on a standard template. The questions only covered negative impacts and did not extend to positive impacts or to management practicalities. Our draft answers went through various review processes before being posted online; the assessments can be accessed via Non-Native Species Secretariat (2011).

Information on establishment, spread and impacts from these assessments is summarised in modified form in Tables 11.1 and 11.2. Some questions have been altered here to reflect the fact that both species have already colonised parts of the country and are causing some issues. It is worthwhile briefly considering the comments.

The first point to make is that there are many similarities between the two species. Both have coped reasonably well in Britain's climate. Muntjac have taken to the eco-climatic zones of southern England although they occur in subtropical habitats in parts of their native range. Deep and prolonged snow is known to affect them in both China and Britain (Sheng 1992a; Chapman et al. 1994a). For water deer, the main differences in climate seem to be the hot and humid summers and abundant late summer rainfall in China (Zhang 1996 and personal communication). The reproductive strategies of the species, although different, are both geared towards producing more than one young deer per annum that are capable of breeding at less than one year of age. Their territorial nature provides dispersal incentives for young deer in particular, although their natural rate of movement is relatively slow, being 1 km or less per annum (Chapman et al. 1994a; Sections 5.4.2 and 6.5). Both species are exposed to competition from other herbivores and to predation, including by motor traffic. These hazards have not been sufficient to prevent establishment and significant spread, but an important factor could be muntjac and other deer species out-competing water deer in some habitats, particularly woodlands and dry sites with good ground cover.

Turning to other differences, the muntjac has shown itself to be adaptable by, for example, colonising suburbia and small pockets of suitable habitat. On the other hand the water deer is less adaptable in this country which may be associated with its low genetic diversity (Fautley 2013). Only 19 water deer were imported to Woburn between 1896 and 1913 (Chapman 1995), and it is possible that no more deer have been imported since then. Fautley et al. (2012) found that whether introduced deer successfully established a population depended mainly on the number of animals released, while further spread depended on breadth of habitat that could be utilised and diet. There has been debate on the degree of genetic diversity of

Table 11.1   Questions relating to establishment and spread
in risk assessments for the two species.

Questions	Muntjac	Water deer
How widely is it spread in Britain?	Very widely	Moderately widely
How easily can it reach suitable habitats?	Very easily	Moderately easily
Can it survive in Britain's climate?	Yes	Yes
Is colonisation affected by competition or predation?	Yes	Yes
Will predation prevent further spread?	No	No
Will competition prevent further spread?	No	To some extent
Will diet prevent further spread?	No	No
Are current levels of control stopping spread?	No	No
Is reproductive strategy aiding colonisation?	Yes	Yes
Is lifestyle aiding dispersal?	Yes	Yes
How adaptable to change is it?	Adaptable	Slightly adaptable
Will low genetic diversity affect colonisation?	No	Possibly
Could an eradication campaign be survived?	Yes	Probably
How rapid is natural spread?	Not rapid	Not rapid
How important is spread by humans?	Very important	Important
How difficult is it to stop further spread?	Very difficult	Difficult

Table 11.2   Questions relating to impacts in risk assessments for the two species.

Questions	Muntjac	Water deer
How important are economic, environmental or social problems in its native range?	Of minimal importance	Of minimal importance
How important is economic loss likely to be in Britain?	Of minor importance	Of minor importance
How important is environmental loss in Britain?	Of major importance	Of minor importance
How important is social harm in Britain?	Of major importance	Of minor importance
How likely is it to be a vector of disease?	Moderately likely	Moderately likely

the muntjac in this country. Williams et al. (1995) found that the original release of eight males and two known females in 1901 comprised a comparatively small proportion of the introduced gene pool – in other words, there must have been other unreported introductions of non-Woburn stock. However, a recent study by Freeman et al. (2016) indicated that the national muntjac population could be traced back to a single founding event consisting of a low number of females. They cautioned that releases of a small number of individuals might result in permanent and widespread colonisation, irrespective of low founding genetic diversity or later genetic input.

The spread of both species has been considerably aided by humans, although for water deer a high proportion of accidental or deliberate releases seem not to have resulted in permanent populations (Ward 2005). Colonising muntjac tend to be secretive, meaning preventing further spread is very difficult. This is compounded by them invading towns where they cannot be shot. Eradication, even over parts of their range, may be impossible. Populations of water deer would be easier to contain and perhaps eliminate, particularly in open arable countryside. However, in situations such as Broadland in East Anglia, they are well entrenched in wetland habitat and a concerted effort would be needed over a very wide area to reduce the population and prevent further spread.

Before discussing impacts, it is worth considering how the two species are viewed in their native ranges, including the benefits that they bring. As in Britain, muntjac deer are much more numerous and widespread in China than are water deer (Sheng and Ohtaishi 1993), but the water deer occurs in Korea whereas the muntjac does not. The muntjac has been highly valued in China for its meat and pelt, with an annual harvest of about 650,000 out of a total population of 2–2½ million (Sheng 1992a). The impact of hunting on its population was lessened by the benefits derived from a substantial decline in predatory leopards and because logging led to an increase in bush and secondary woodland which it preferred to high forest. The fact that it is less at home on open agricultural fields means that reports of conflict with farmers seem lacking.

Robert Swinhoe (1870) considered that, although water deer probably fed on crops such as cabbages and sweet potatoes, any damage must be acceptable because farmers tolerated them in high numbers in various place along the River Yangtze. Since then, the population has suffered massive declines. Water deer have been traditionally hunted for their meat, pelts and for the colostrum in the stomachs of fawns. For example, in the Zhoushan Archipelago from the mid-1970s until the 1980s, about 500 pelts were purchased annually by local companies (Zhang and Guo 2000). As people became wealthier in the 1990s, so demand grew for these commodities. Water deer are still sometimes trapped as an agricultural pest (Harris and Duckworth 2015), but it is unclear whether this is used as an excuse for killing them to eat or sell. Its present rarity in China means that, at worst, agricultural damage is no more than a very local issue. It is said to be 'a low-concern pest of rice fields' in North Korea (Harris and Duckworth 2015). In the past, the species generally declined in Korea because of habitat loss and excessive killing for food and traditional medicine (Won and Smith 1999; Harris and Duckworth 2015). However, serious conflicts between deer and farmers have emerged in South Korea, where it has increased in recent decades (Kim et al. 2011; Jung et al. 2016; Chun

2018). Any agricultural concerns for water deer across its native range seem partially offset by benefits, whereas the muntjac appears to be regarded as wholly beneficial.

In this country, there are also benefits, including financial ones. For instance, foreign and some British trophy hunters are willing to pay for the opportunity to shoot both species. England is one of the few countries in the world where it is possible to stalk wild muntjac and water deer, and there are a number of operations geared towards catering for such clients. Obtaining a 'gold medal standard' head of a buck water deer can be costly. Shooting is a pastime enjoyed by many people. The venison of both species is good to eat but the returns for the stalker are not as good as for larger species. Stalkers such as Charles Smith-Jones have explained to me that many game dealers are reluctant to accept muntjac as they are relatively hard to skin and yield little venison. Currently, joints from a single muntjac from one of my local woods cost £25. Perhaps more establishments should take a lead from an inn at Walberswick in Suffolk that proudly advertises that it can serve locally sourced muntjac and water deer.

Another benefit afforded by these deer is that most people are thrilled to see them in most contexts. I can still remember seeing my first wild muntjac although it was a brief sighting more than 50 years ago. Many people have become aware that our water deer are special and rare on a world scale. Some organisations are now cashing in by offering deer safaris. I have been on such walks led by rangers to see water deer and red deer at sites in East Anglia, and I have led a number of walks myself locally to see deer and talk about their impacts and management.

Finally, it is important to remember that, if the right balance can be maintained, muntjac browsing can improve the floristic interest of a wood by helping to control more dominant vegetation. Viewing all the evidence, however, many people consider that the problems caused by muntjac, particularly in conservation, far outweigh the species' benefits.

As regards impacts, there are many factors that affect severity and geographical range. Two of the most important are numbers and the area over which they occur. The current situation is different for the two species. From information presented earlier (Section 10.5), our national population of muntjac is between one and two orders of magnitude more numerous. Suitable habitats for muntjac providing cover and food are common and widespread in the countryside and even into towns and cities. When dispersing, they can take advantage of corridors such as hedges, or will cross open ground if necessary. They are likely to be most numerous in well-wooded areas and forests with good understoreys. Water deer live at greatest density in wetland areas – habitats which tend to be rare outside the Broads. They have also colonised coastal areas, both in China and East Anglia. And they can flourish on quiet, dry farmland, but at lower densities. The fact that they frequent open fields means that dispersal is not dependent on corridors of cover, although it does place the animals at greater risk of being shot.

Whether the deer impact these habitats depends on other factors such as their diet, how robust habitats might be, threshold densities at which problems occur and seasonal changes in behaviour, as well as a range of site-specific factors such as the availability of alternative food. In Table 11.2, muntjac are regarded as causing major environmental losses. This relates primarily to their effects in conservation woodlands and is discussed at length in the following chapters. Water deer, however, cause relatively minor problems, in part because there are few wetland sites for them to colonise and because these sites tend to be fairly robust. Their impact is considered later by addressing their contribution to browsing and grazing at Woodwalton Fen where significant populations of both species of deer occur.

Economic losses are reviewed below in relation to agriculture and forestry. Road traffic accidents, which are more frequent for the more abundant muntjac, cause both economic loss and social harm. Other social harm comes in the form of muntjac damage to gardens and various non-agricultural situations, and both species have been known to disrupt gamebird shoots by dashing from cover. Their potential involvement in transmitting disease could cause both social harm and economic loss. Most of these issues are briefly considered below. It is clear that muntjac have given rise to more concern than water deer, which was reflected in legislation being introduced for muntjac in 1997 to attempt to stop people moving them to new areas. Legislation is covered at the end of the chapter.

11.2   Impacts on agricultural crops

11.2.1   *Muntjac*
In a major report for Defra, Wilson (2003) concluded that deer damage to cereals and grassland was only believed to be significant in areas with high deer densities. Deer impacts in lowland Britain were reviewed by Putman and Moore (1998). Few incidents involving muntjac or water deer were recorded on MAFF's database, the main culprits being fallow, roe and red deer. Similar conclusions were reached by Langbein and Rutter (2003) across England and Wales, and by White et al. (2004a) who looked at the economic impacts of wild deer in the east of England (Bedfordshire, Hertfordshire, Cambridgeshire, Essex, Norfolk and Suffolk). Muntjac were found to cause some damage to cereals, but had relatively less impact due to their close association with woodland as well as their small size. Their diminutive stature apparently made them less likely to feed on tall cereals during ripening, a time at which most damage from the larger deer species occurred. Muntjac were found to be by far the most numerous species of deer in eastern England, but their biomass was marginally less than that of the much larger fallow deer.

Crops growing outside woods with high-density muntjac populations are situations where impact might occur, and I studied two fields of beans adjacent to Monks Wood in 1998, just before stalking began in the wood (Cooke and Farrell 2001b). The deer ate the stems, leaves and pods of the beans (Figure 11.1), and broke and bent the stems in a similar fashion to how they tackle coppice regrowth (Figure 2.7). Transects into the fields revealed that plant height, pod length and number of pods per plant were affected up to 70 m into the crop, so it seemed reasonable to assume that yield was reduced along the field edges. This was, however, a uniquely severe attack next to a wood with an exceptionally high density of muntjac.

Elsewhere, the presence of other species of deer may make the contribution of muntjac difficult to unravel. For instance, growing numbers of fallow and muntjac deer in Rockingham Forest have impacted agricultural crops and have led to the Deer Initiative co-ordinating a programme of increased culling and monitoring to rectify the issue.

Figure 11.1   Field beans eaten by muntjac outside Monks Wood, 1998.

11.2.2   *Water deer*

Water deer are more at home on open farmland than muntjac, but are their densities high enough to cause even local problems for farmers? White et al. (2004a) considered that water deer had a negligible impact on agriculture in eastern England. However, the species has increased in abundance and range since then. The most likely place where damage might occur is farmland adjacent to wetland sites with high densities of water deer. Because of this possibility, Lynne Farrell and I have always been alert to any issues that might arise just outside Woodwalton Fen. Our early observations made during 1976–1984 raised a few concerns (Cooke and Farrell 1987). Under favourable weather conditions, deer were most likely to be encountered feeding on farmland during the early months of the year when food was scarcer in the reserve and territorial ties were weaker (Figure 11.2). Numbers were generally low, but sightings increased when the population was high or when there was flooding in the reserve or snow cover. Deer were especially attracted to carrot fields. During winter in the late 1970s and early 1980s, most other arable fields were uncultivated, so carrot fields were an obvious target with the green tops being available even when there was deep snow. Moreover, the furrows provided shelter and protection; grazing sessions in the carrot fields were interspersed with periods of sitting in the furrows ruminating. Signs of grazing on green tops and the carrots themselves were noticeable but the damage was slight, with the maximum deer density recorded in a carrot field being 2 per ha. Based on data on food consumption at Whipsnade (Manton and Matthews 1983), food requirement for the larger individuals at Woodwalton Fen was calculated to be 0.5 kg dry matter per deer per day. So the maximum loss from 1 ha of crop should not have amounted to much even if various individuals visited a field each day. And on the farmland, other food was taken, such as potatoes left after harvest or natural vegetation in fields left uncultivated over winter or growing beside ditches.

More recent assessments of the local situation confirmed that deer continued to feed in the highest numbers on farmland in the early months of the year (Cooke 2009a; 2009b). Crop damage included some feeding on sugar beet, both in the field and in storage, and knocking off the ears of barley by running through the ripening crop. But this was viewed by the farmers as of minimal consequence. By this time, grazing on the newly sown conservation grassland had become apparent. Darlow's Farm to the north-west of the reserve had been acquired and seeded for the Great Fen project in 2004/5. This immediately attracted grazing water deer, but it was the seeding in 2008 of fields on Middle Farm to the west of the reserve that drew in unprecedented numbers (Figure 11.2). Close to the reserve, densities on grass fields could exceed 3 per ha, but dung counting showed that deer activity decreased markedly more than 300 m from the reserve boundary. Such grazing was not a problem in this conservation situation, but could be of greater significance if deer grazed pasture prior to stock being turned out on a conventional farm (Putman 2003). Water deer are unlikely to compete for grazing once cattle or sheep are present in a pasture as they seem to avoid livestock (Cooke 2009a). A marked reluctance to associate with cattle or sheep has been noted in fields on Middle Farm; numbers of deer fell as more fields were used for livestock grazing (Section 8.3.5).

By 2009, the effect was readily apparent of the presence of about 25 ha of elephant grass on Castlehill Hill Farm close to the reserve (Cooke 2010). A significant metapopulation of water deer took advantage of the cover and food, in the form of various weeds, provided by this crop for much of the year. It was a refuge outside the reserve for individuals that might otherwise have had to disperse more widely. However, damage to winter wheat growing outside the elephant grass was again minimal, although places could be found where deer had rested in sunny positions. The destruction of the elephant grass with herbicide in 2011 led to a very high density of water deer returning to the reserve by the winter of 2011/12, coinciding with sowing oilseed rape in fields on Castlehill Farm adjacent to the reserve. These rape fields attracted large numbers of grazing deer from November 2011 (and again in 2017/18), with observed densities of up to 4 per ha on individual fields. Grazing effects close to the reserve flood-bank were very obvious

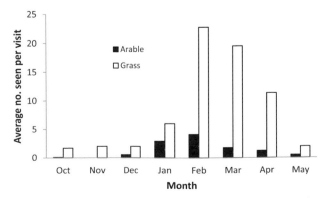

Figure 11.2   Average number of water deer counted on arable fields to the west of Woodwalton Fen, October–May 1976–1979 (Cooke and Farrell 1981) and on newly sown grass fields, 2008/9 (Cooke 2009a).

Figure 11.3   A field of oilseed rape grazed by water deer in 2012.
The reserve at Woodwalton is behind the flood-bank on the left.

(Figure 11.3), but whether there was any effect on yield is unknown. Density in 2011/12, based on the size of the reserve, was about 125 per km².

When Callum Thomson was the Woburn Estate deer manager, he told me density at Potsgrove was kept in the region of 8–9 per km² (Figure 4.5a; Section 10.4): there was some crop damage, but no details were available (Cooke 2009b). In any event the deer were viewed as a valuable financial resource. Close to Aylesbury in Buckinghamshire there is a quiet area of about 5 km² of mixed farming where there is said by one of the farmers to be a density of water deer similar to that at Potsgrove. During a visit in January 2016, I saw a total of 26 in 2–3 hours including a loose aggregation of 12 moving and grazing across a large field of winter wheat (Figure 11.4). Grazing on grain fields so early in the year may not have any marked effect on yield as there is likely to be compensatory growth later (Putman 2003). However, if the deer remain in specific localities, problems could result.

Figure 11.4   Water deer on winter wheat near Aylesbury in
Buckinghamshire, January 2016, with farm buildings behind.

Figure 11.5 Grazing marshes on the Claxton Estate in Norfolk –
water deer favour the ungrazed margins of the fields.

The largest area in the country with significant populations on agricultural land is the Broads. In the past, I have heard from both Rick Southwood of Natural England and James Ellis from Gunton Park that Broadland farmers grumble about water deer, but red deer are a far greater problem. Adult red deer weigh on average as much as 7–10 water deer and their populations in eastern England can number 100 or more. In places, electric fences have been used to try to keep red deer away from vulnerable crop fields, and some populations have been heavily culled in recent years. Some farmers, especially those that shoot, are very happy to have water deer on their land. When I visited the Claxton Estate on the south bank of the River Yare in 2016 (Figure 11.5), the water deer population was held at about 100 – an overall density of about 25 per km². The deer benefited from predator control and the provision of fenced margins on the grazing marshes, where the highest deer densities occurred. There was little or no concern about deer competing with cattle for grass or over crop loss on the grain fields: the latter was said to be similar to that caused by hares and much less than was lost to rabbits in the past.

Agricultural land is a habitat on which water deer can live and thrive to varying degrees. However, the species currently has a restricted distribution in this country and situations where it might have some measurable economic effect are few and far between. Nevertheless, it would be wise to be vigilant about this issue, especially in the light of recent reports of crop damage from South Korea (Jung et al. 2016; Chun 2018). Note that the risk analysis in Table 11.2 was undertaken before these articles were published.

11.3 Impacts in forestry crops

Other species of deer have caused problems in forestry for a great many years, but neither muntjac nor water deer rated a special mention as regards browsing damage on forest crops by Putman and Moore (1998) or White et al. (2004a). This was wholly understandable for water deer as they tend not to be a forest species. However, the muntjac is a woodland species and is also a browser. But muntjac tend to contribute to damage that is being inflicted by other species, particularly fallow deer, rather than be especially troublesome on their own. Until recently, Trevor Banham was the Forestry Commission's Chief Wildlife Ranger in East Anglia with about 30,000 ha of woodland under his control; in an article in 2003, he dismissed concerns about the impact of muntjac on the pine crop, but stressed the problems with ecological impacts at ground level. Charles Smith-Jones (2004) regarded the muntjac as a 'minor offender' when compared with other species of deer, but warned of the potential for large-scale damage as the species continued to increase. There had in fact been concern expressed a comparatively long time ago. Derek Bray (1980) drew attention to the browsing impact inflicted by muntjac apparently during the especially hard winter of 1978/9. He stated that, when snow cover persisted, any density of muntjac was capable of serious browsing damage to all exposed vegetation that was remotely edible. Under normal conditions, the worst threat seemed to be from stem fraying when muntjac densities were high. Up until then, landowners and timber growers had tended to be unwilling to cull muntjac: Derek Bray advocated control of muntjac before such crises occurred again. However, none of the subsequent winters has been as severe as 1978/9 (e.g. Table 8.2). The paradox is that winter weather that might lead to impacts could ultimately result in mass die-offs of muntjac.

In order to understand better the impact of muntjac in the absence of other deer, Keith Wilson (1995) concentrated on the south of the Forest District around Northampton, Milton Keynes and Silverstone. He found that their fraying caused serious damage to few trees and stems in commercial forestry plantations so that overall impact was not great. More problematical was the damage done to tree shelters. Andy Patmore (1995) also reported on muntjac in Northamptonshire, finding them in mixed oak/conifer and pure conifer up to pole stage/second thinning. Bare forest floor, such as under mature Norway spruce *Picea abies* or beech *Fagus sylvatica*, was usually avoided. The only time they were seen feeding on conifers was again during relatively harsh winter weather when they browsed Norway spruce and European larch *Larix decidua* at times of snow cover. Nevertheless, when deer fences are erected around forestry areas, pains are usually taken to ensure that muntjac are kept out as well as the larger species (Downing 2014). Muntjac browsing can impact situations such as pockets of regrowth along the edges of forest blocks intended to improve biodiversity.

Similarly, muntjac contribute to browsing problems faced by farm woodlands, new amenity plantings and the wood-fuel supply chain. The government's Woodfuel Strategy for England has set targets for the amounts of fuel required to help with our energy requirements. If targets are to be reached and production is to be sustainable, then regeneration has to be successful – so browsing by all species of deer needs to be controlled. On the Claxton Estate in Norfolk, browsing on planted trees by water deer was prevented by the use of tubes 1.2 m in height.

11.4   Impacts in gardens and similar situations

Water deer tend to be more nervous than muntjac around people. They have so far failed to colonise suburban areas and are only very rarely encountered in gardens. I have lived in the market town of Ramsey and its attached village of Bury since

1973; despite the fact that a significant slice of the country's water deer population resides only 5 km away in Woodwalton Fen, I am not aware of any evidence of them entering Ramsey or Bury. Muntjac deer, on the other hand, are often seen in gardens here and elsewhere within their range in lowland England. They are very much at home in larger mature gardens with sufficient cover and food. The presence of muntjac within town limits is mainly due to them willingly coming in from the countryside, but to some degree it is the result of towns encroaching on the countryside (e.g. see Figure 11.7). The PTES organises an annual national survey of mammals in the built environment. Over the period 2004–2017, the proportion of sites with muntjac increased, as did abundance of the species (People's Trust for Endangered Species 2017).

The frequency and impact of muntjac in gardens, as well as gardeners' attitudes, were reported by Norma Chapman and Alex and Stephen Harris (Chapman et al. 1994b). Early records of muntjac in gardens are usually of dispersing, transient individuals, but in time they establish home ranges comprising a number of gardens sometimes linked by corridors such as hedgerows or railway embankments, and breeding occurs. These authors gathered information and opinions including during a national survey into distribution and status of muntjac. Of people with muntjac in their gardens, 12% were pleased, 33% were displeased and the remainder offered no view. The degree of displeasure increased in areas where muntjac appeared to be more common. Exasperation was frequently expressed as muntjac ate a wide range of garden plants. Being concentrate feeders on readily digestible and nutritious vegetation, muntjac tended to select flowers and buds. The small size of their mouths limited them to relatively small root vegetables when these were taken. A request for information about damage to market garden crops revealed this to be rare, but when it did occur, it could be severe involving significant economic loss. Cemeteries and gardens of remembrance are other places where muntjac can cause considerable distress by eating flowers and other plants left in memory of loved ones.

Lists of palatable and unpalatable plants have been provided by these authors and by Charles Coles (Coles 1997). However, there are disagreements about exactly which species are eaten. In part this may be due to the preferences of individual deer (Chapman et al. 1994b).

It may be helpful in understanding the fundamental reasons for muntjac causing so much angst among gardeners to describe the situation in my own garden in Ramsey in Cambridgeshire. I first had muntjac in the garden in 1995, 12 years after moving there. That was a young buck, as are many of the early colonisers. Occurrences were sporadic up until 1997, when I noticed the first signs of damage. By the time that my wife and I moved again in 2007, muntjac deer were firmly established locally, albeit at low density. I estimated the average density of muntjac outside woodland in my part of the county to be about 2 per km^2 (Cooke 2006). My garden was within a 50 ha area of low-density housing and grassy fields connected to countryside. So on average, one muntjac might be expected to be present within the area at any one time. The garden itself was half a hectare in size, had many mature trees and shrubs and was bordered by hedges through which muntjac could readily move. It was, therefore, a prime patch within the area.

Each spring from 1997 until 2007, I undertook an ivy trial (Section 12.6.3) with a single group of 20 ivy stems in the same quiet corner of the garden. These results

Table 11.3   Average percentage of ivy stems eaten by muntjac in spring in a garden in Ramsey, Cambridgeshire, 1997–2007 compared with that taken in Raveley Wood, 2000–2006. The distance between the two sites is about 5 km.

Site	Percentage of ivy browsed or defoliated			
	Browsed day 1	Defoliated day 1	Browsed day 7	Defoliated day 7
Ramsey garden	1	0	22	0
Raveley Wood	16	1	80	21

can be compared with those from Raveley Wood where six groups were deployed each year, 2000–2006. By 2000, problems for ground flora were becoming apparent in Raveley Wood, but stalking started and the situation improved. The average figures for percentage of ivy browsed or defoliated (Table 11.3) demonstrated that ivy taken in the garden was much less than the average at Raveley Wood.

In 2006, the garden was scored (Section 7.2.4): one for muntjac deer and three for their damage. Raveley Wood was scored in 2000 as six for deer and eight for damage, decreasing to four and seven respectively in 2006. Had the garden's scores and ivy results been for a wood, then I would have concluded that the muntjac were causing no problems and might indeed be a benefit by browsing the bramble, which flourished in places in the hedges. And this was what would be expected from a density of only 2 per km².

However, direct observation of plants in the garden led to a different conclusion. For example, treasured gifts of roses had their buds stripped off and unprotected tulips were grazed to the ground. In the vegetable patch, lettuce, runner beans and courgettes were eaten (Figure 11.6a). I like deer and I would rather have a few muntjac around and pay the price, rather than have no deer at all. But I can

Figure 11.6   Muntjac like vegetables. (a) Here they have been attracted to flowering courgettes. (b) Young lettuce protected from muntjac in the author's garden.

appreciate that many others would view the balance differently. The question is why is a density of muntjac that might be a benefit in woods a distinct issue in gardens?

The answer seems to be that there are differences between the nature of woods and gardens and between the attitudes and perceptions of woodland managers and gardeners. A woodland manager may notice if recently cut coppice is being badly browsed, but most damage is more subtle and may be overlooked or ignored. Some gardens are packed with edible and reachable fare, and these are likely to be owned by enthusiastic gardeners, who inspect their patches regularly and closely follow the fortunes of flowering plants or planted vegetables. Anything amiss is likely to be spotted quickly. Gardeners are less likely than woodland managers to accept perceived damage, even if the impact on flowering spectacle, survival or crop yield is negligible. Real impact in my garden, as opposed to evidence of feeding activity, was minimal – but I still wanted to stop it if I could.

Deer management in urban and suburban situations poses real and different problems from coping with deer damage in the countryside. Remedial action was discussed by Chapman et al. (1994b) and Coles (1997). Lethal methods in suburban gardens are not an option for safety reasons. Repellents are often suggested but in practice may not work for long, especially in winter when deer tend to be hungrier. They also need approval under pesticide regulations, otherwise their use is illegal. There is currently a product, *Grazers*, advertised for use against deer in a variety of situations from gardens to crop fields. Muntjac may be deterred for some while by human scent: for instance, they seem more reluctant to take my planted ivy if I fail to wear gloves when handling it.

The remaining option is a barrier of some description between the deer and the plant. Here the issues are similar to protecting woodland plants. Just as orchids may be individually protected with netting in woods, so roses can be protected in gardens. I have used wire tunnels to safeguard rows of salad crops (Figure 11.6b). If the problem is on a slightly larger scale, then discrete areas of a few square metres could be caged with netting 1 m high and with mesh size of 50 mm or less; a larger

Figure 11.7   Flitwick Wood in Bedfordshire: gardens adjacent to woodland with a muntjac population may be at risk from raiding deer.

mesh size may tempt deer to try to put their noses through the wire. Some of my small exclosures in Monks Wood, which are 4 × 4 m, have repelled muntjac for more than 20 years. However, if there are severe problems in a garden or if particularly sensitive species are widespread, enclosing the whole garden may need to be considered. In this case, fencing should be at least 1.5 m in height. Unfortunately, even if this is done, muntjac may still find a way in. I have found that I feel better disposed to deer that are habituated to some degree. Those that hide away during the day but raid the garden at night are a problem without giving any pleasure.

Gardens backing on to woodlands are at particular risk (Figure 11.7). If all else fails, an alternative option would be to garden with muntjac in mind and grow plants that they avoid. To give some examples: in spring grow snowdrops, daffodils and hellebores, but not tulips. In summer, species such as foxgloves, peonies, potentillas, hydrangeas and various ornamental grasses are likely to survive intact. The bad news about vegetables is that only members of the onion family are totally safe from being eaten. Charles Coles' book contains extensive and detailed advice, but it would be prudent to experiment with novel species before investing heavily.

11.5   Road traffic accidents

In January 2003, the 'Deer–Vehicle Collisions Project' was initiated in England by the Deer Initiative with lead funding from the Highways Agency. Funding from the Scottish Executive later that year extended the project to Scotland. The project's aims included building a database, estimating the overall toll of collisions, outlining regional differences and identifying current and future black spots. Jochen Langbein oversaw this and subsequent projects and produced two major reports in 2007 and 2011. It is the English data that are of greatest relevance to accidents involving muntjac and water deer.

Langbein's first report estimated that the total number of deer–vehicle collisions (DVCs) in England to have been 34,000 per annum, but due to under-recording the true total may have exceeded 60,000 per annum. These DVCs were considered to have caused injuries to several hundred people each year and tragically resulted in a number of deaths. The minimum cost of damage to vehicles was estimated to be £13.5 million. Muntjac contributed far more to these statistics than did water deer by virtue of their greater population size and wider distribution. In English accidents where deer species were identified, muntjac contributed to about 25% whereas water deer were involved in less than 1%. Muntjac were responsible in 14% of instances where vehicle damage was significant and on 24% of occasions when human injury resulted. In contrast, the respective figures for water deer were 0.3% and zero. The vast number of deer killed or injured was described as 'probably the single greatest welfare issue for wild deer in the UK'. A more important welfare issue than the deer killed outright by the collision was the suffering of injured deer before they could be humanely dispatched or die from their injuries. For both muntjac and water deer, roughly 10% of their national populations were estimated to be lost through DVCs each year.

The main finding in the later report (Langbein 2011) was that total number of DVCs increased from 2001 until 2007, but then began to level off coinciding with a reduction in road traffic from 2008. A number of accident hot spots were identified

Figure 11.8   A young buck water deer beside one of the nearest roads to Woodwalton Fen – killed while dispersing, November 2013.

by the project and mitigation methods were researched and adopted in some situations. Such remedial management included fencing, overpasses, underpasses and signs activated by the deer themselves.

The contributions to total DVCs 2001–2010 were 28% for muntjac and less than 1% for water deer. Muntjac was the species most commonly involved in the East Midlands and eastern England. Several accident hot spots were identified in Thetford Forest in Norfolk. The small number of DVCs reported as involving water deer (total less than 100) were mainly from Bedfordshire, Cambridgeshire and Norfolk. There were also isolated reports from the Cotswolds and the Forest of Dean. It should be noted, however, that some road signs warning of deer in Norfolk have been erected specifically for water deer. In several recent years at least 10% of water deer sightings reported to Norfolk and Norwich Naturalists' Society have been of animals involved in collisions. I have never driven into a water deer, but I have only avoided doing so on several occasions because I have been aware of deer running hard across fields on a collision course with my car. They seemed to have an urge to cross the road before I drove past. Such behaviour is an unfortunate trait that could lead to water deer accident hot spots occurring more frequently as their population continues to increase (Figure 11.8). In South Korea, roughly 60,000 water deer have been reported as being killed annually on roads (Chun 2018), which represents about 10% of the national population – comparable to the statistic for this country (Langbein 2007).

11.6   Deer and disease

11.6.1   *Transmission of disease to humans*
Peter Green is Honorary Veterinary Advisor to the BDS. He has pointed out to me that deer carry a wide spectrum of viral, bacterial and parasitic pathogens and may shed these within the suburban environment as well as in the countryside. Little or nothing is known about specific pathogens being released by different species

of deer, but the potential exists for serious problems to occur, including with *E. coli* 0157, the most important strain of the bacterium causing disease in the UK. This would seem to have more relevance to muntjac than water deer, as they may be visiting gardens regularly without the householders realising.

Lyme disease is the main human condition with which deer have been associated. Species such as muntjac and water deer carry ticks, including the species *Ixodes ricinus*, which is the vector of the bacterial pathogen *Borrelia burgdorferi*. If infected ticks bite people and stay attached for a couple of days, then Lyme disease may result. Should the disease be quickly diagnosed in humans, it can be treated with antibiotics, but if it is not diagnosed it can cause a variety of debilitating effects. The advice is try to avoid being bitten by reducing exposed skin, using a repellent, and checking regularly for ticks and the characteristic signs of an infected tick bite: a circular red rash. If there is any concern, consult a doctor.

The role of deer in this process is complex (O'Connell 2007; Lyme Disease Action 2013; Mysterud et al. 2016). Larval ticks are initially uninfected but may pick up the bacteria from hosts such as small mammals and birds, which tolerate infection without ill effects. Deer are the preferred hosts for adult ticks; they are not reservoirs of the bacteria but help to maintain and spread tick populations. There may be a beneficial effect of deer since larvae that have used deer as a host do not have an infection that they can pass on to humans: this is termed dilution. In a wood where ticks feed solely on deer, there is no Lyme disease. Nevertheless, detailed studies in Norway since 1991 have revealed a weak relationship between deer density and the incidence of Lyme disease. In Britain, the number of cases of Lyme disease has continued to rise over recent decades and some commentators have been quick to blame the increase in our deer populations. Many other factors are, though, believed to be involved, including climate change (warm wet weather favours increases in tick populations), changes in land use, increases in countryside pursuits and improved public awareness of the disease. Interventions based simply on blaming deer could undermine broader attempts to improve public and ecosystem health, and so the BDS is funding research on mechanisms involved in transmission of the disease. In NNRs in western Cambridgeshire, stalkers have noted a significant increase in the number of ticks on deer (Roy Butters, personal communication), and reserve staff are sufficiently concerned to have started recording ticks on skin and clothing. However, the rise began when deer densities were relatively low, and increased numbers of ticks have not been noticed on deer on the local farmland (Martin Guy, personal communication).

11.6.2   *Transmission of livestock diseases*

Deer also have a potential role in transmitting disease to livestock. In 2013, Peter Green reviewed the possible part played by all our species of deer, although the article did not specifically mention either muntjac or water deer. Five diseases were reviewed for which a role for deer was considered: bovine tuberculosis, bovine virus diarrhoea, Johne's disease, Schmallenberg's disease and liver fluke. For bovine TB, fallow and red deer may have had a role in maintaining the infection in some localities. In the United States, white-tailed deer were shown to be capable of infecting cattle with bovine virus diarrhoea. In Britain, some red deer carried antibodies demonstrating they had been exposed to the virus, but the required tests to confirm they were affected were not done. Johne's disease has affected sika in Scotland (Simpson et al. 2012) but little was known about the incidence of the disease in wild deer or whether wild deer shed the bacteria. Schmallenberg virus is of particular relevance to deer in eastern England, having caused deformities in calves and lambs in this region since 2012. Red and fallow deer have been shown to have been exposed to the virus, but whether they act in transmitting it to livestock was not known. Deer in Britain are infected by liver fluke but it was not clear whether they posed a risk to livestock. In 2014, the BDS and the Moredun Institute in Scotland called for faecal samples to test for the presence of fluke eggs (Anon. 2014).

The bluetongue virus is, like the Schmallenberg virus, spread by biting midges. It seems to have been controlled here by the widespread vaccination of livestock; our red deer can become infected but do not become ill (Green 2012). In general, however, as Peter Green (2013) pointed out, it can be easier and cheaper to eliminate deer by shooting than fund and undertake the research needed to inform these issues. With their livelihoods threatened, farmers can hardly be blamed for such a reaction. Another important observation made by Peter Green was that deer in Britain are generally in good health – roughly 20% of the overall population is shot each year and carcases are dealt with by people who are 'trained or experienced in recognising when something is amiss'. In other words, if our deer do begin to suffer widely from some disease, then it should become quickly apparent.

11.6.3   *Chronic wasting disease*

Although our deer population is believed to be generally healthy, there has been much concern recently about chronic wasting disease in deer. This was first characterised by Williams and Young (1980) in captive deer in the United States. Clinical signs were alterations in behaviour, progressive weight loss and death within a few weeks or months; the authors concluded it appeared to be a type of spongiform encephalopathy, similar to scrapie in sheep. The disease was discussed by Alastair Ward of the Deer Initiative at the National Deer Management Conference in 2017 (Anon. 2017). He described how deer could be infected by several routes including via other deer or infected soil or vegetation. As captive deer have been moved, so the disease has spread in North America, including to wild deer, and population declines are expected. Several reindeer and moose in Norway were found to have become infected in 2016. If the disease reaches the UK, the consequences could be severe. The BDS has commissioned research at the Roslin Institute to investigate whether wild or farmed deer in Britain have been exposed to the disease, to assess whether each of our species is susceptible and to ascertain whether there are signs of potential resistance. Samples from water deer outside Woodwalton Fen will be tested.

11.7   Legal interventions

In 1981, the Wildlife and Countryside Act introduced a range of legislation including some of relevance to these two species of deer. Section 14 stated that any person who released individuals of species listed in Schedule 9 would be guilty of an offence. Schedule 9 included those species that were already established in the wild in Britain, and which were considered to be causing problems or might do so in the future. The list included species such as coypu *Myocastor coypus* and Canada goose *Branta canadensis*, but not muntjac or water deer. At that time not many people had even heard of water deer, and Oliver Dansie (1983) described muntjac as 'an almost innocuous asset to the countryside'.

However, opinions about muntjac began to change, and in 1997 the Department of the Environment announced that new controls were being introduced that would make it illegal to release muntjac into the wild. This action followed concern over the increase in numbers and range of muntjac and the damage done to woodland flora and roe deer. This was apparently referring to the studies of Norma Chapman and her colleagues, and also to my own work. The species was added to Schedule 9 later that year with the proviso that licences would be issued to allow a limited release of muntjac provided that deer were taken from one of 12 core counties and were released near point of capture. This was to aid rehabilitation of injured animals or those accidentally trapped. The core counties were Bedfordshire, Berkshire, Buckinghamshire, Cambridgeshire, Essex, Hertfordshire, Leicestershire, Norfolk, Northamptonshire, Oxfordshire, Suffolk and Warwickshire. These were counties where it was believed that limited releases would have no significant impact on future establishment or spread.

Around the same time, mutterings began about whether the Chinese water deer should also be included on the Schedule. There was no getting away from the fact that it was an introduced alien, but there was no evidence at the time that it was increasing rapidly or causing environmental problems. The main difficulty that Lynne Farrell and I had in accepting that water deer should be listed on the Schedule was that it was categorised as 'Vulnerable' on the Red List of the IUCN. We argued that placing it on Schedule 9 would send out the wrong message for an animal that deserved to be viewed in a more favourable light. In 2004, however, Defra and various partners published an action plan for the sustainable management of wild deer populations in England (Defra 2004). One aim was to limit, as far as possible, further spread of non-native species of deer – the muntjac, water deer and sika. One proposal of the action plan was to consider adding Chinese water deer to Schedule 9. In line with the proviso for muntjac, Defra also intended to consider allowing release of water deer at the point of capture if there was no practical means of humane dispatch.

The action plan also recommended that a close season for water deer should be introduced. Until then, it had been legal to shoot them throughout the year. This welcome initiative was set in place in 2007 by an amendment to the Deer Act of 1991. There was a close season from 1 April to 31 October; and it applied to bucks as well as does because the sexes can be difficult to distinguish in the field. So at a time that a pledge had been made to limit the further spread of water deer, the stalking season had been restricted to just five months in the winter and early spring. No close season was afforded to muntjac as they breed throughout the year.

In 2007, Defra released a consultation document on a review of Schedule 9, proposing that the Chinese water deer should be included. They pointed out that 90% of correspondents to a previous consultation supported its addition to Schedule 9. I remember wondering at the time how many of those correspondents knew much about the species other than it came from China. The BDS did not welcome the proposal, stating that until there was evidence of damage that could not be curtailed by normal management there was no need to add it to the Schedule (Anon. 2008). Nevertheless, the Chinese water deer was added to Schedule 9 in 2010.

By 2010, it was clear that adding muntjac to Schedule 9 in 1997 had not stopped the species increasing in numbers and spreading its range. I am not aware of any official monitoring of the effectiveness of this action. However, there was a tacit admission of failure by announcement of a change to licensing policy when Nottinghamshire, Lincolnshire and Greater London were added to the list of core counties, thereby acknowledging the spread that had occurred since 1997. The degree to which the muntjac has spread in Britain since 1997 can be graphically monitored by interrogating data collected by the BDS (Section 5.2) or held by the NBN Gateway. And the water deer is displaying continuing spread since 2010 (Section 5.3).

CHAPTER 12

Processes in impact management

12.1   Introduction

There have been several recent books on deer management (de Nahlik 1992) and the management/stalking of muntjac specifically (Smith-Jones 2004; Downing 2014). In this chapter, issues are considered that have been relevant to my involvement with impact management in conservation woodlands: beginning with types of impact; going on to explore when and how to manage, assess, study and monitor; and finishing with relationships between deer density and impact. The topics covered in this chapter are not exhaustive and related issues are discussed elsewhere in the book, including: the role of foxes in controlling muntjac populations; resurgent deer populations; unexpected cascade effects; and the beneficial effects of low levels of grazing and browsing.

For water deer, emphasis is on learning about its potential for impact. For muntjac, on the other hand, focus is on anticipating and recognising impact, and monitoring the effect of remedial management in conservation woodland. Information on the impacts of deer emanated from many places in the world towards the end of the twentieth century, in particular from North America (e.g. Tilghman 1989; Balgooyen and Waller 1995; McShea et al. 1997; Rooney and Dress 1997; and the review of Rooney 2001). By that time, the muntjac, despite its small size, had been added to the list of deer that could cause problems.

12.2   Types of impact

All our deer species, including water deer, will have some effect in woods, but within eastern England muntjac and fallow deer are the species of greatest concern (Goldberg and Watson 2011). Muntjac will occur in most woods in their main range in England, and evidence of their browsing and grazing will be obvious to anyone attuned to such damage. But in many cases, this will simply reflect their presence and will not be regarded as an impact that may require management. In my immediate area of western Cambridgeshire, muntjac have been by far the most numerous species of deer and have been perceived by woodland managers as being responsible for most of the observed impacts (Cooke 2013b):

- Browsing on coppice regrowth reduces height and can kill stems. Ultimately, stools may die and whole coppice plots become dominated by other species.
- Regeneration of palatable trees can be prevented by seedlings being browsed, broken and frayed. This may result in a shift in species composition with unpalatable species taking over. Similar shifts occur in coppice plots where there are stools of several species of varying palatability.
- These types of effect, combined with browsing on established trees and shrubs, can lead to a marked browse line at about 1 m, below which density of woody stems is much reduced.
- Grazing alters the ground layer, with a loss of palatable herbaceous plants but with increasing cover of non-palatable species and those resistant to grazing, such as some grasses.
- Changes caused by browsing and grazing may profoundly affect the structure and species composition of a wood, so having indirect effects on fauna, most of which will be detrimental to conservation interests.

Another possible impact beyond the immediate reach of deer was detected by terrestrial laser scanning (Eichhorn et al. 2017). This novel technique allows the three-dimensional structure of woods to be described in great detail. Studying woods from the Weald and the Welsh Marches confirmed that those with high densities of fallow, roe and muntjac had less dense understoreys, the average reduction being 68% between 0.5 and 2 m above ground level. In addition, results indicated that deer-impacted woods had taller trees with a different vertical distribution of foliage. These changes at heights above 5 m were unexpected and it was unclear whether they were real effects or artefacts of the method.

Water deer are much rarer than muntjac but inside or just outside their wetland refuges they can attain densities that are comparable with those reached by muntjac in woodlands, so some impact must be anticipated. Fortunately, they often tend to favour grazing, and grass species are both tolerant of grazing and less likely to be of conservation value – although such grazing may be of greater concern in agricultural settings. Adult buck water deer would be unable to chew woody stems to break them because their large tusks would be in the way, and females do not appear to break stems either. Neither is there evidence that water deer fray bark in the manner of muntjac. Within wetlands, high densities of water deer are capable of destroying coppice plots of palatable species by browsing regrowth and are likely to affect regeneration in wooded compartments (Chapter 19). They also focus on certain species of ground flora at specific times of year. As with muntjac, they may, as a consequence of such effects, indirectly affect fauna that are dependent on particular species of vegetation. While impacts from water deer will be very restricted geographically, many sites with the highest densities are nature reserves that are home to specialised and vulnerable plants and animals. Reserve managers need to be aware of the potential for water deer to cause problems, and disseminate any relevant information as widely as possible.

12.3   Management options

The two options most often adopted for reducing vegetation damage by muntjac are culling them and putting a physical barrier between the deer and the vegetation. However, there may be other options, particularly for future use (Putman 2003; Green 2008). For instance, contraceptive techniques hold an attraction for those people who may wish to avoid culling if at all feasible. Simply feeding bait laced with steroid hormone to wild deer is not, however, a practical or safe option. There is no way of knowing the amount consumed by individual female deer, and ingesting the hormone may be harmful to male deer and other wildlife. Defra recommended that novel non-lethal methods of control, such as immuno-contraception, should be kept under review (Wilson 2003). In this method, deer are injected with a vaccine containing protein similar to a natural protein crucial for reproduction; the deer then produce antibodies which temporarily disrupt the reproduction process. Deer may be either caught and injected or injected directly by means of a dart. Such methods can work with captive animals in zoos, and in some park situations – where individual members of the larger species can be recognised – it has proved possible to deliver the vaccine by dart. With an animal as small as a muntjac, however, darting is likely to cause injury; in the wild, individuals would need to be caught regularly and injected by hand. Even without considering immigration of untreated deer, the difficulties involved in catching a significant proportion of the population on a regular basis render this technique inapplicable to muntjac populations, at least within the foreseeable future.

Chemical repellents are of two types: those whose smell acts as a barrier and those that are applied to the vegetation and make it distasteful or smell unpleasant. Such products fall under pesticide regulations. Among various barrier repellents, the BDS recommends diesel-soaked strips of cloth hung from lines of string. Such protection will only be effective for a short time and would need replenishing. It also looks unsightly, but might be used where protection was critical. Spray-on feeding repellents for the protection of trees or crops can be expensive and may need reapplication after rain (British Deer Society 2015). However, *Grazers* is a product that is advertised as being rain fast in one hour and is available for use in amenity situations against damage from deer.

How woodland is managed can also influence the extent of damage caused by muntjac. This is an animal that prefers undisturbed cover. A plot of sensitive vegetation beside a major ride is less likely to be badly impacted than a similar plot deep in quiet woodland. So, for instance, if there are several places where coppicing may be introduced in a wood, then those less likely to be damaged could be chosen. English Nature's guidelines on managing deer in woodlands (Mitchell-Jones and Kirby 1997) stressed the importance of such strategic planning and manipulation, and Defra recognised its benefits by recommending that design of new woodland should incorporate and develop the concept (Wilson 2003). Malins and Oliver (2017) have also proposed that use of stalking and fencing could be reduced if more thought was given to the design and management of modern forests: for instance, impact could be reduced by having larger stand sizes and planting less palatable trees. One of the points stressed by Rory Putman (2003) and other authors is that damage control usually involves more than one type of defence.

12.3.1 *Physical barriers to deer*

Gill (2004) pointed out that an effective fence will protect the trees and other vegetation completely, but listed a number of disadvantages, including they can be ineffective, expensive, unsightly and impractical. They can also force deer to move and cause problems elsewhere. Fencing does, however, have an important role in deer management. Different forms of fencing can be used in different ways. Small cages of wire netting with 5 cm mesh and 1 m width attached to posts have been used in my area to conserve individual rare plants (Figure 12.1). I have deployed similar cages to monitor growth in the absence of browsing, for instance with ash saplings in Monks Wood (Section 14.2) and bramble in Holme Fen (Section 14.3.2). I have also used wire exclosures that measure 4 × 4 m for many years without muntjac gaining access (Figure 18.1) – such a fence could safeguard a small population of a rare or particularly sensitive species. As fences increase in area, however, so muntjac become more determined to gain access, especially if they have been displaced from their home range by the construction process. Larger areas require more secure and robust fences. But no fence can be guaranteed to be proof against muntjac.

Tree guards are another option that can work for small areas. Guards need to be tall enough to protect against the largest species of deer that is likely to encounter them. Recommended height ranges from 1.2 m for muntjac and roe deer up to 1.8 m for red deer (Pepper et al. 1985). Guards of 1.2 m have effectively protected young trees from a high density of water deer on the Claxton Estate in Norfolk. However, Oliver Rackham (2014) observed: 'Deer are not fools. They know that a plastic tree guard means dinner inside.'

Some species of vegetation are unpalatable to muntjac. For example, regrowth of aspen *Populus tremula* or alder should not require any form of protection. The most difficult species to protect are those that are palatable and sensitive but are

Figure 12.1   Cages for violet helleborines *Epipactis purpurata*, Monks Wood, spring 2002. These were in position from the year before and would be repositioned as necessary once the plants began to show above ground.

nevertheless widespread in a wood. Oxlip *Primula elatior* is a classic example. A fence has been erected around about 80% of Hayley Wood in Cambridgeshire to exclude fallow deer (Figure 13.5), while the whole of Shadwell Wood in Essex was fenced to try to keep out fallow and muntjac (Figure 14.4). Although fallow deer were defeated by the fences, muntjac were not and have needed culling inside both woods. Similarly, muntjac could not be eliminated from within the two large deer fences in Monks Wood (see below). The wisdom of constructing fences of more than a few hectares specifically for muntjac must be questioned. If fencing is being employed to protect a plant such as oxlip, several smaller fences are likely to be more effective than a large one covering a hectare or more.

Conservationists often use newly cut mature regrowth as brash to protect coppice stools. This is less unsightly than a wire fence, saves money and removes the need for a bonfire. However, it is less effective than fencing at preventing browsing, and is likely to produce kinked regrowth stems which may be unwelcome if they are to be used for anything other than more brash when the coppice is recut.

Publications on how to erect a fence to protect coppice regrowth from muntjac and other deer species have been issued by the Forestry Commission for more than 20 years. Among those are practice notes and technical guides by Mayle (1999), Pepper (1999) and Trout and Pepper (2006), the last account being especially exhaustive. Fences to protect coppice are usually termed 'temporary' because they are aimed at management problems that might last for a few growing seasons. This distinguishes them from 'permanent' fences that might remain *in situ* for 5–15 years. Two large, permanent fences were erected in Monks Wood in 1999 (Cooke 2006). The smaller one (covering 6.1 ha) enclosed seven coppice plots in the centre of the wood (Figure 13.8a). The larger one (10.6 ha) was erected in the south-west corner where much of the remaining floristic interest occurred.

The Forestry Commission guides provide a wealth of information for the fence builder. For instance, muntjac deer initially look for a way under a fence. To stop them squeezing under, the bottom 15 cm of mesh should be bent out along the ground from the base of the fence. If they fail to get under, they will try to push through. I have seen a determined muntjac repeatedly running head down at the larger fence in Monks Wood – and bouncing off. Their heads are smaller than their bodies and sometimes they try to get through too small a hole and are firmly stuck around their shoulders. The lucky ones might be rescued by people, but many will die. Bucks can pass their heads, but not their bodies, through 100 × 100 mm mesh. When they try to withdraw their heads, their antlers may become snagged, so the recommended maximum mesh size to avoid these problems is 80 × 80 mm.

The fence should be constructed of material that neither they nor other animals that want to gain access can break or tear – various materials are recommended in the Forestry Commission literature. Wire mesh or heavy plastic mesh can be acceptable. If muntjac cannot gain access under or through a fence, they may try to jump it. The recommended minimum fence height is 1.5 m, but muntjac can jump much higher – Raymond Chaplin (1977) referred to a muntjac clearing a 2.4 m wall. With muntjac, and probably with the other species too, there is no such thing as a completely deer-proof fence if an area of a hectare or more is being enclosed. Eventually, they will find their way under, through or over. With smaller areas, they are less likely to try to negotiate the fencing. My exclosures measuring 4 × 4 m are only a metre in height yet have apparently never been entered since they were erected 25 years ago.

Despite available information on how to protect coppice from muntjac, woodland managers still use other, less effective, methods, although this does not happen as much as it did in the past. Reasons for this include a dislike of fencing, lack of time or funding, acceptance of the potential for some browsing damage, some previous experience at keeping out rabbits or keeping in stock, or simply lack of knowledge of how to do it properly. Fencing in deer or rabbits is among other problems that can arise – the chances of this occurring increase as the fenced area grows in size. The observed effectiveness of different forms of fencing to protect coppice regrowth is discussed in Sections 14.1.3 and 14.1.4.

12.3.2   Deer control

I am not a stalker and this is not a book about how to cull muntjac – there are already books covering this subject by Charles Smith-Jones (2004) and Graham Downing (2014). And many stalkers in this country are highly experienced in shooting muntjac. However, I have worked in tandem with stalkers for more than 20 years monitoring what effects they have on reducing local muntjac populations and allowing vegetation to recover. These results are summarised later. I also offer a few observations on stalking techniques and strategies that have come to light during our collaborations.

No comparable account seems to have been written on stalking Chinese water deer. In open countryside, such as on arable fields in winter, they should be relatively easy to cull (e.g. Childerley 2014). There are a number of Internet sites that have videos of stalking or offer stalking facilities including for trophy animals. This and the opportunity of taking an unusual species seem to be the main reasons for shooting them. Few people cull them because they are perceived to be a nuisance by landowners, farmers or managers. I am not aware of any situation where they are being culled for conservation reasons.

The idea of shooting wild animals such as deer is abhorrent to many people. They enjoy seeing deer and argue that deer have the right to remain alive in the countryside. After all, muntjac did not ask to be brought to this country and released. We cannot blame them for being a successful colonist and wanting to eat plants that we cherish. I started studying deer because I found them beautiful and interesting animals that often frequented quiet, wild places. In the 1980s I began to appreciate that they could cause issues in conservation woodlands, but up until the early months of 1994, I believed that problems could be largely avoided by non-lethal means. But then my views changed dramatically. Over the course of a few weeks, roughly half of the population of muntjac in Monks Wood died as a direct result of starvation (Section 8.2.2). I was entering the wood expecting to find dead and dying deer, and doing so on virtually every occasion. It was a distressing situation and finally convinced me that they could be too successful for their own good and, in the absence of enough natural predators, their population needed controlling.

Now that culling is more acceptable in conservation woodland, it may seem strange that it took so long to introduce deer control inside Monks Wood, but it was only after 1993, when I started to look at features of the wood in more detail, that the scale of the impact became apparent and management options were viewed more holistically. It has been claimed that deer numbers were allowed to build up to 'saturation level ... to facilitate studies of their impact at high densities upon a woodland ecosystem' (Carne 2000). However, they reached such a level in the mid-1980s, long before anyone realised what was happening.

Organised stalking started along the edges of Monks Wood in 1995. This was done for several reasons, including to prevent, or at least slow down, population recovery. Around that time, the level of stalking required to protect regrowth adequately was considered to be impractical (Cooke 1995; Cooke and Lakhani 1996), but culling inside the wood was not ruled out as a component of an integrated plan of management. The muntjac population recovered to its former level by 1997, and by that time it had become clear that small-scale fencing was not sufficient to protect other interests, such as ground flora. So English Nature decided that culling should begin inside the wood in 1998 in order to help reduce the population to an acceptable level commensurate with the wood's status as a NNR; large-scale fencing followed a year later.

What happened in Monks Wood in the mid-1990s also illustrated the point that culling some deer may have little if any effect on numbers in a wood because it may simply improve conditions for the remainder and increase recruitment of young, reduce mortality and favour immigration over emigration. It was only in 1998 when stalking was stepped up that there was an obvious population decline and the impacted conservation features of the wood began to recover. It is important that efforts are made to explain to visitors the need for certain management operations, such as culling.

The control of deer numbers for conservation and other reasons should be undertaken on a landscape scale (Mitchell-Jones and Kirby 1997; Goldberg and Watson 2011), otherwise success may be difficult or impossible to achieve. Deer are mobile animals and fencing or shooting disturbance in some woods may lead to other sites being used as refuges, in turn causing them to become more heavily impacted. Although muntjac do not range as far as the larger species, it is possible that a large undisturbed wood might act as a source of deer in an area. There should be close liaison between owners, managers, stalkers and anyone undertaking monitoring so that aims and methods can be agreed at the start and progress can be regularly reviewed and amended if necessary.

12.4   Stalking and deer behaviour

That behaviour can vary between individual muntjac has already been discussed (Section 7.5.1), so if the most conspicuous deer tend to be shot after stalking begins then the reduction in sightings is likely to over-emphasise the real decrease in numbers. But there are general effects on behaviour as was seen in Monks Wood during foot-and-mouth restrictions in 2001, when stalking stopped and sightings promptly increased.

It is well known that disturbance, including stalking, can affect deer behaviour and can for instance make them more active at night (Clark 1981; Smith-Jones 2004; Downing 2010). There are various methods for quantitatively recording diurnal activity. Wildlife camera traps record date and time when they are activated, so can

Table 12.1   Percentage of records of muntjac on camera traps in mixed woodland at Woodwalton Fen during December 2010 to February 2011 before stalking started and during December 2011 to February 2012 after stalking started (adapted from Cooke 2013c). Numbers of records were 177 and 491 respectively.

Period of day (Greenwich Mean Time)	Dec 2010–Feb 2011 No stalking	Dec 2011–Feb 2012 After stalking
Dawn 07.00–08.59	11%	15%
Day 09.00–15.59	37%	21%
Dusk 16.00–17.59	21%	16%
Night 18.00–06.59	31%	49%

be used to piece together a picture of muntjac movement in front of them. The cameras will record specific activities such as feeding or movement from place to place. They provide a different type of information from radio-tracking activity recorders, which will register if an animal is resting or moving.

An example of the effect of stalking on behaviour is shown in Table 12.1 for the main block of mixed woodland at Woodwalton Fen in the winter before stalking began and the following winter. Observations were consistent with the introduction of stalking increasing night-time records. Data indicated that stalking was associated with some reduction in activity at dusk, confirming the need for caution in interpreting changes in observations at dusk following the introduction of stalking.

Muntjac may become less wary when disturbance stops if individuals learn from their own experience that the wood is safer or if there is communication between deer. However, when stalking is introduced, learning solely from experience is a perilous strategy – one mistake and a deer could be dead. It seems likely that they learn much from signs, sounds and smells. A new human activity occurring at an unusual time of day, associated ride management and blood on the ground may affect the behaviour of some individuals, leading to avoidance of previously used areas or paths and increased frequency of barking, foot stamping and tail flagging (Figure 2.10a). In Monks Wood, incidence of tail flagging increased from 73% of deer that saw me during surveillance walks in 1993–1994 to 82% in 2000–2012 after stalking began. These signals would be picked by other deer that then become more wary and also pass on the information. Soon all the survivors would be wary, whereas complacent, conspicuous individuals had been shot. Radio tracking the same muntjac before and after stalking starts could determine the extent to which behaviour changes.

I have no comparable information on water deer to determine whether they are affected by stalking. However, their diurnal behaviour in winter did not change in the mixed woodland at Woodwalton Fen before and after muntjac stalking was introduced. About 50% of water deer activity was at night so, in this situation, their nocturnal activity was similar to that of muntjac after stalking started. Nevertheless, the water deer is a flighty species and I would expect some tendency towards an even greater level of nocturnal activity if a population was being culled.

12.5   Assessment and acceptability of impacts

12.5.1   When is management needed?

One of the most basic questions as regards deer management is when should it begin? In the past, browsing on coppice regrowth was most probably the event that made woodland managers realise they might have a problem with deer. But even that situation did not necessarily warrant management intervention. It may have been sufficiently obvious to be noticed but not severe enough to need fencing or other management. A judgement was made that it was acceptable at that moment in time – in other words, it was slight *damage* rather than worrying *impact*. Another manager, however, may well have decided that any browsing was unacceptable, especially if the regrowth was intended for eventual cropping and putting to some use. Nowadays every woodland manager is likely to be aware that deer browsing can affect the success of a coppice plot, although the outcome may be uncertain if, for instance, coppicing has never been undertaken in that wood or with that species of tree or if the deer density has recently changed. Whether management is carried out will also depend on other factors, particularly the resource implications of the operation.

In parts of the country where muntjac are starting to colonise, however, some managers have a shoot-on-sight policy; no damage may have been seen, but they wish to avoid muntjac becoming established. Here, having muntjac is deemed unacceptable irrespective of whether or not they have caused even minor issues because they have been well documented elsewhere as establishing densities that cause impacts. At the other end of the scale, some managers may go to enormous lengths to avoid culling deer, possibly being influenced by views of visitors, who may be paying members of a conservation society. These are policies with which I would not attempt to argue. It is up to those in charge of woods to decide what management they consider is best in their situation. The approach taken by Natural England locally in recent years has depended on the scientific case for deer control at specific sites. Thus, muntjac control at Woodwalton Fen began after the species was demonstrated to be impacting conservation interests and those impacts were worsening. Below I offer advice to those people who may wish to make judgements based on conservation issues.

Knowledge of impact levels is often considered to be more important than having quantitative information on deer numbers, although some monitoring of the deer population is essential (Putman 2003). It is likely to take muntjac more than ten years from being first recorded to causing real impact in conservation woodland. So there is time to consider what approach might be taken. What is less clear initially is how bad the situation may eventually become if nothing is done, but some general rules can be suggested. Thus, small woods of less than 10 ha are unlikely to be as impacted as much as large ones of greater than 100 ha. Woods that have fallow deer, in particular, should not suffer as much extra impact from colonising muntjac as those without other species of deer. Wet woodland will not be as attractive to muntjac as a dry site, and unpalatable alder coppice should be safe from browsing. In wetland sites with water deer, conservation issues are unlikely to arise unless deer density is unusually high. Such generalities should help to ease concerns in some cases, but vigilance is still necessary to monitor a changing situation. Monitoring is essential for any special species occurring at a site – grazing on herbs can often be easily overlooked. The same applies to impacts on coppice regrowth of palatable species; likewise on tree regeneration, but such an effect may matter less in the short term because of the long life of trees.

12.5.2   What are the aims of management?

The basic question of when to begin management is inextricably linked with two other questions: what is the aim of management and how is success recognised? As regards ground flora, Keith Kirby (2001a) suggested two valid motives for management:

- Evidence that declining species were rarer, more characteristic or more specialised than those species that were increasing.
- Evidence that parts of the wood or the whole wood were becoming less species rich.

He considered that care was needed in deciding what factors had caused the changes and how problems might be remedied. In addition, he pointed out that conservation managers may take a poorer view of the impact of introduced herbivores than they would with the same level of impact from a native species, which is deemed to be part of the ecosystem to be conserved. I would also add that managers look even less favourably at recent introductions from the Far East.

Essex Wildlife Trust set a target figure for oxlips in its woods, stating that less than 20% should be grazed (Tabor 2011). This figure was based on deer activity and grazing levels across a range of woods (Tabor 2002; 2004), and is an example of how survey results can be used to inform management aims in a quantitative manner. The oxlip is a sensitive species so achieving such a target would ensure that other conservation features were safeguarded, but the Trust also had a target of less than 10% for browsing on woody shoots. A target set to protect a species less sensitive to deer grazing, such as the bluebell, would be unlikely to have such a far-reaching beneficial effect.

In general, the aim of deer management is usually summarised along the lines of achieving a recovery of ground flora and woodland processes, especially regeneration of trees, shrubs and coppice. Animal species that have been indirectly affected often seem to be forgotten. Probably the assumption is that if vegetation composition and structure recover, then in time dependent animal species will also return. However, this is not necessarily the case (Chapter 17). Eventually a decision needs to be made about when a wood has recovered sufficiently. This is unlikely to mean that management can stop, but it can be viewed as an end state to be maintained.

Natural England (2013) has inherited a condition assessment process from its predecessors to evaluate Sites of Special Scientific Interest (SSSIs) using defined standards. For woodlands, conservation objectives are tailored to the structural and biological characteristics of individual sites (Joint Nature Conservation Committee 2004). The assessment examines five attributes: extent and distribution of woodland on the site; structure and natural processes such as the balance between canopy and shrub layers and amounts of dead wood; regeneration of trees and coppice; species composition of the trees; and quality indicators which include the composition of the ground flora and presence of rare species. Concern has been expressed that the process may under-estimate impacts on ground flora (Dolman et al. 2010).

For a number of years after deer management started in Monks Wood, the wood was assessed as 'Unfavourable Recovering', which meant that aspects of the site were in poor condition in part because of impacts caused by deer, but measures had been put in place to rectify the situation. Monks Wood was assessed as being in 'Favourable' condition in 2015. Improving the condition of SSSIs as a whole is driven by targets aiming to have a certain percentage in 'Favourable' condition by a specific date. Monks Wood was not unusual as regards deer damage: the Parliamentary Office of Science and Technology (2009) reported that the Public Service Target of having 95% of SSSIs in good condition in 2010 was being compromised by deer impacting woodlands. Other conservation organisations may have similar systems to assess and monitor the condition of woods.

12.5.3   How is information assessed?

As an individual researcher who started studying impacts before any other simple method existed, I have developed my own systems for determining extent of impacts. A method was eventually devised to extend damage scoring and integrate all available information on muntjac impact in a wood so as to derive an overall

Figure 12.2   Three levels of deer impact. (a) Low impact, Lady's Wood, September 2011. Some signs of grazing and browsing could be found, but regeneration was acceptable with no browse line, and ground flora had grown well. (b) High impact, Marston Thrift, April 2004. Regrowth required protection, browse lines were evident, and signs of grazing were easy to find with unpalatable species being common. (c) Severe impact, Holme Fen West Block, June 2006. No regeneration was evident with a general browse line and dead bramble; the ground layer was composed of unpalatable species and grasses.

Table 12.2   Derivation of overall impact stages in Monks
Wood in five selected years. Stalking began in 1998.

Indicators	Impact stages of indicators				
	1995	1999	2003	2008	2014
Unprotected coppice regrowth	Severe	Severe	Severe	High	Low
Tree regeneration	Severe	Severe	Severe	High	Low
Shrub layer	Severe	Very high	Very high	Moderate	Low
Browse lines	Severe	Very high	High	High	Moderate
Stem breakage	Severe	Very high	Low	Low	Moderate
Fraying	Severe	Very high	Moderate	Moderate	Low
Ivy browsed day 1	Severe	Low	High	Low	–
Ivy defoliated day 7	–	High	Severe	Low	–
Ground flora	High	High	Moderate	Slight	Low
Grazing on bluebell leaves	Severe	Severe	High	High	Low
Grazing on bluebell inflorescences	Severe	Severe	Low	High	Low
Grazing on primroses	–	–	High	Low	Slight
Grazing on dog's mercury stems	Severe	High	High	Low	–
Height of dog's mercury (cm)	Severe	Severe	Low	Moderate	–
Overall impact stage	Severe	Very high	High	Moderate	Low

impact stage (Cooke 2009c). This relied on deciding the level of impact on a range
of indicators. The framework I used was based on information regularly obtained
at Monks Wood (Appendix 1) but I have used it routinely elsewhere. It can be
readily adapted to more distant situations or study sites with different deer species
by incorporating new indicators, such as amount of antler thrashing or grazing
levels on orchids. It is intended to record impact by all species of deer, so that at
Monks Wood signs of roe are included with those of muntjac. And I have used it at
Woodwalton Fen where water deer are the most abundant species of deer (Cooke
2009a).

Four levels of impact are shown in Appendix 1: no impact, low, high and severe
impact (Figure 12.2). But I also recognise three intermediate levels: slight, moderate
and very high respectively. This was done because there may be some areas within
a site where signs suggest, for example, low impact on an indicator while in other
areas no impact is seen, and having a level of slight impact allows this to be taken
into account. There are considerable differences between typical examples of the
four categories in Appendix 1, and the extra three levels allow better descriptions of
situations that fall between them.

Impact stages have been derived annually at Monks Wood, and information for five
different years is shown in Table 12.2. Overall impact is typically the median stage for
the individual indicators. If there is any doubt, then extra weight is afforded to the

first four indicators in the list and to grazing on ground flora. Only one indirect effect is included in the framework, namely a change in height of dog's mercury *Mercurialis perennis* caused by grazing reducing plant vigour. So far, effects such as changes in fauna caused by impacts on vegetation have not been included in the framework; there are uncertainties in relationships between cause and effect, and recoveries in animal populations may occur long after reductions in impact on vegetation.

The framework allows an impact stage to be described precisely, and, as shown in Table 12.2, permits recovery to be followed in a wood for each of the indicators. Recovery is not necessarily an entirely smooth process, either for a single indicator or overall for the wood. At Monks Wood, impact was high in 2003, moderate in 2004, high again in 2005, but back to moderate again in 2006. It should also be noted that, in this case at least, stalking had a progressively slower effect on reducing impact to the next stage. Monks Wood had severe impact in 1998, was very high by the next year, reached high by 2001, was convincingly moderate by 2006, but only attained low impact in 2013.

12.5.4   *What is the end product of management?*
So how is success recognised? How much impact is acceptable? I have used a technique of asking what is definitely acceptable and what is unacceptable, then focusing in more detail on where a wood is positioned between the two extremes. Most people would probably agree that what I term 'slight impact' is acceptable and that 'high impact' is unacceptable. Indeed, slight impact where some of the indicators displayed minimal deer grazing and browsing may be ideal for biodiversity and be better than no impact at all. There are often, however, practical constraints that mean slight impact cannot be achieved. For most conservation woodlands, I suggest that low impact is a reasonable goal, providing there are no underlying issues as regards sensitive species or woodland systems.

By 2015, Monks Wood had low impact and was considered by Natural England to be in 'Favourable' condition. Other managers may view their own woods differently. In some woods, moderate impact may be acceptable, especially if resources for management are scarce and important conservation features are lacking or can be well protected with fencing. However, sensitive widespread features could still be unacceptably impacted in a wood that was generally showing low impact, meaning extra management was needed. Returning to the question of when should management begin, letting impact creep past the low level in the face of an increasing deer population is likely to mean that unacceptable effects are occurring and some management should be in place. A situation can be envisaged where, say, unpalatable coppice was not being browsed and everything was assumed to be satisfactory, but if basic monitoring of grazing on ground flora or regeneration of palatable trees and shrubs were introduced, they might indicate that impacts were occurring and management was needed.

Scoring and assigning impact stages were developed for my own use but they have been adopted and adapted by other individuals and organisations. The Deer Initiative (2012) amalgamated and refined my two methods to produce a version that was suitable and sensitive for all species of deer. Experience and training are needed to use any of these methods effectively, and the Deer Initiative has held many training courses. I recommend its version to anyone starting now who requires a method for assessing deer activity and impact. It is based more on quantitative monitoring than my methods and is more applicable to the larger species of deer. I do have some reservations with that method, however. In particular, although their stages correspond closely with mine, some of the stages have different names. For instance, a wood that I would say has unacceptably high impact, the Deer Initiative would describe as having moderate impact. This could signal misleadingly little concern to a manager or landowner, and it may be difficult to convince a reluctant owner of what is described as a 'moderately' impacted wood that remedial action is needed.

12.6   What and how to monitor

The types of monitoring that I have undertaken as regards muntjac damage have been aimed at understanding what species of vegetation and animals are affected and to what degree, and how effects change over time. As regards changes over time, most of that monitoring was in response to management, particularly stalking. Recording and communicating information on the impact of conservation management is especially important (e.g. see Fuller et al. 2016) and has been one of the main driving forces behind making the observations and producing this book.

One of my main roles in woods has been to study the impacts of deer and the results of management, and pass on information and advice to the managers, who then made management decisions. I wanted to work in the field within the constraints of what was seen at the time to be the best management. Occasionally, I asked for management to be modified in order to be able to study situations that were more relevant to acquiring a broader understanding of impacts and remedial management. When I began the study of coppice management in Monks Wood in 1993 (Cooke and Lakhani 1996), I was lucky that there happened to be a relatively large number of newly cut coppice and ride-side plots. Sometimes, however, anticipated management in the wood did not materialise, and on other occasions, management wiped out study plots, the locations of which reserve staff were unaware of for various reasons. Thus plots were flailed, driven over or destroyed by pigs which were once kept to reinvigorate derelict coppice.

Investigating or monitoring deer damage raises a number of questions:

- What aspects or species are of interest?
- How are they sampled?
- What is counted or measured and for how long?

Monks Wood has been one of the best-studied nature reserves in Britain but, when I began my detailed work on muntjac damage in 1993, no one had been recording from that viewpoint. A huge amount of information had accumulated, as summarised in the symposium volumes edited by Steele and Welch (1973) and Massey and Welch (1994). A later bibliography (Gardiner 2005) listed 114 publications up to 1993 in addition to the 47 papers in the two symposium volumes. The papers ranged over history, management, the occurrence of all manner of species, and detailed ecological investigations. Nothing had been published on deer damage, although coppice browsing had been observed and attempts at protection introduced. Nevertheless a number of articles were pertinent to the deer issue, especially those describing abundance or distribution of potential target species, such as Dick Steele's vegetation maps and Terry Wells' comments on flowering plants and ferns in Steele and Welch (1973).

12.6.1   *Monitoring species*

Deciding on which species to monitor in Monks Wood was not a problem, with some being obviously impacted and with more subtle issues coming to light by, for instance, comparisons between vegetation in exclosures and their control plots. Initially, I focused on coppice plots because problems were clearly occurring. Recording methods were simple. In plots up to 0.1 ha in area, I recorded success of every stool of the target species, whereas in larger plots observations were made along predetermined transects (Figure 12.3). Number of growing seasons, incidence of browsing on regrowth on each stool and height attained were apposite details to record.

When I began recording coppice in spring 1993, I noticed that bluebell inflorescences were badly grazed. Plots in which to record bluebells could have been located at random across the wood, possibly being stratified according to habitat. But I was more interested in how muntjac were affecting the remaining major stands in the wood, so set up randomly placed transects within the stands and recorded in ten random quadrats along each transect (Figure 12.4a). This had the added benefit of recording a large number of bluebells with a relatively low amount of effort. All intact and grazed inflorescences were counted, and specific observations, such

Figure 12.3   The author's son recording regrowth along a tape in failed coppice in Monks Wood in 1993. This plot was unusual in having a high density of maiden birch and aspen but few mature standards. Despite the plot being electrically fenced, regrowth of palatable species had been heavily browsed by muntjac and then dominated by pendulous sedge *Carex pendula*.

Figure 12.4   (a) Recording bluebells by the quadrat method, Monks Wood, 1993 (LF). The coppice plot was electrically fenced at that time. (b) Recording bluebells along a tape measure, Marston Thrift, 2003 (RC).

as plant height and whether leaves were grazed, were made on the most central bluebells in each quadrat. Transects and quadrats were then fixed and recorded each spring until 2014. When I started in 1993, I never imagined that I would be repeating the observations for that length of time. This was one of the advantages of being flexible and not being dependent on external funding! Had I been reliant on funding, most monitoring sequences would have been considerably shorter. I needed to be careful not to trample or kneel on bluebells in the quadrats as this could reduce numbers in subsequent years.

Later, in other woods, I adapted the sampling by identifying small sections of ride adjacent to good stands of bluebells and selecting by random means a location to place a tape measure perpendicular to the ride. Then observations were made on grazing and plant vigour on the closest plant to predetermined points along the tape (Figure 12.4b). This method allowed sampling and recording to proceed quickly, but the drawback was that number of inflorescences per unit area was not counted as was done with the quadrat method.

As muntjac were found to be eating an ever increasing list of plants in Monks Wood, so I adapted methods to suit my needs. Grazing on dog's mercury was tackled in the same way as bluebells, and I mapped the distribution of the main stands of both species in the wood by the simple process of walking systematically through the woodland blocks. Abundance of primroses *Primula vulgaris* and amount of grazing was recorded every 2–4 years within 5 m of a fixed route, while orchids were studied in specific areas of the wood where they grew in greatest profusion. Detailed observations of vegetation in Monks Wood provided a list of species to check and possibly monitor in other woods.

Occasionally, monitoring has complemented someone else's study. Thus Martin Baker has monitored the fortunes of early-purple orchids *Orchis mascula* in nearby Raveley Wood since the mid-1990s. He has meticulously counted each plant every spring, recording whether it flowered and whether the inflorescence was grazed. We have combined our two data sets to examine the relationship between grazing on orchids and muntjac presence for a period of more than 20 years (Cooke and Baker 2017; Table 16.2).

Scientific recording in Monks Wood by other people did not of course stop in 1993, and I was able to tap into the broad surveys of vegetation in the wood undertaken by researchers such as Crampton et al. (1998), van Gaasbeek et al. (2000), Sparks et al. (2005) and Brunsendorf (2006). Some of these surveys used plots set up in the distant past or introduced new plots for comparison. In addition, George Peterken and Christa Backmeroff set up four permanent transects in the wood in 1985, which have been resurveyed at intervals (Mountford and Peterken 1998; Tanentzap et al. 2012b). This work has given valuable pointers to recovery of woody species. However, once recovery had started I needed survey data at more frequent intervals, so began recording both the presence of ash seedlings beside rides along a fixed route and also regeneration in specific small patches in the wood.

Browsing on and loss of bramble are classic indicators of deer impact, and I have mainly followed the fortunes of bramble by means of plots along transects. Plots are typically circular and of 2.25 m radius. These have an area of 16 m² and need just a single marker post in the centre. The same plots were used to record deer dung annually at Holme Fen. Some transects were recorded annually until it became impossible to access the plots without damaging both the bramble and myself!

Another question that I asked at an early stage was: are muntjac deer really responsible for the browsing or grazing on certain species? To answer this, I frequently worked with Norma Chapman's captive deer, presenting them with different species to see whether and how they tackled the vegetation, and what signs and clues they left behind. I also noticed the differences in vegetation composition between the inside and outside of (1) paddocks containing muntjac and (2) exclosures without deer in woods. At Woodwalton Fen where water deer were also present, I tried, usually unsuccessfully, to determine which species of deer was responsible for damaging specific vegetation by looking for signs. But it was only when I turned to camera traps that I managed to obtain definitive answers (Chapter 19).

Lynne Farrell and I recorded foxes in the NNRs because we believed they might be important predators on young deer. But species such as brown hares and nightingales *Luscinia megarhynchos* were recorded simply because we liked to see or hear them – and it was little extra effort to note them down. In the early years of surveillance, the indirect effects of deer browsing and grazing on other animals were not of concern. Sightings of foxes, hares and other deer species have been presented as numbers seen per hour of surveillance. Not all observations have turned out to be of use. For instance, for many years, I have recorded numbers of roding woodcock in the spring without finding anything of relevance to deer browsing. However, I did once stumble across a woodcock nest while counting deer dung – even that activity has its highlights and compensations!

12.6.2   *Exclosures*

The name 'exclosure' is often given to a fenced area set up to indicate *what happens* if herbivores are excluded, as well as to a fence erected to *protect* something. An exclosure is the opposite of an enclosure – it is designed to keep specific animals out, rather than in. This definition is somewhat confused by New Forest Inclosures being exclosures rather than enclosures.

In Holme Fen in 2008, three years after stalking began, small wire exclosures were erected to determine what effect muntjac were still having on bramble. This is just one example of the many occasions on which I have found exclosures useful. When recording plant species inside exclosures and their unprotected control plots, I have often used the Domin scale to register the contribution of species in terms of their cover/abundance. I needed to get into some of the exclosures to record vegetation (Figure 12.5a), and recording was usually terminated when bramble and other species grew too densely. Exclosures have often been used by other researchers studying deer impacts in this country (e.g. Putman et al. 1989; Tabor 1999; Morecroft et al. 2001; Peterken and Mountford 2005; Holt et al. 2011). A method similar to that of Fuller and Henderson (1992) was used to monitor long-term changes in vegetation structure in exclosures and control plots in Monks Wood: the degree to which a white disc was obscured when viewed across a plot was assessed in predetermined places at different heights (Figure 12.5b).

Exclosures do have disadvantages: in particular, they do not mimic reality. With few exceptions, studies have investigated the total exclusion of deer and other herbivores that were unable to penetrate the fence. An absence or an extremely low level of grazing and browsing is rarely encountered naturally in woodlands over much of England. Furthermore the vegetation that develops inside may be dominated by bramble and have relatively low biodiversity, so it is not necessarily

Figure 12.5   Recording vegetation inside an exclosure: (a) an exclosure erected in Monks Wood in 1978 being recorded in 1994 (LF); (b) vegetation density being measured across an exclosure erected in Monks Wood in 2004 by estimating how much of a 30 cm diameter disc was obscured. The disc was held in predetermined positions along the edge at heights of 0.5 or 1.5 m.

what managers are striving for in their woodlands. In a situation where a muntjac population is changing, it can be difficult unravelling the exact cause and timing of differences between the vegetation inside and outside an exclosure. Therefore, it helps if exclosures can be erected at intervals of a few years. Nevertheless, exclosures can provide insights into which species are eaten, how persistent grazing can affect the vigour of plants and what shifts in plant dominance might occur. Many exclosures are not erected to be studied, but simply to give a woodland manager insights into what impacts the deer may be having and perhaps then to use them for demonstration purposes.

12.6.3 *Browsing trials with ivy*

With ground flora it can be possible to record annual grazing levels in a location over a long period of time, as was done for bluebells in Monks Wood, because plants regrow or are replaced every year. This is more difficult to achieve for browsing, although I have recorded feeding on successive cohorts of seedling trees, such as ash. To help with the management of roe deer in France, Morellet et al. (2001) and Chevrier et al. (2012) developed indices of feeding activity. The oak browsing index of the latter authors was based on recording feeding on terminal shoots of regenerating or planted oaks up to 1.2 m in height.

As an alternative approach, I developed a technique with ivy (Cooke 2001; 2006). In 1993 in Monks Wood, I was fascinated to realise that I could cut and leave twiggy items for the deer one evening and they might be browsed by the following morning. And I wondered whether I could turn this into a method for quantifying browsing. Ivy proved suitable as it was readily taken by muntjac from February until April. And if it was attacked by rabbits or hares, their signs were easily distinguishable: clean oblique bites to stems and whole leaves left on the ground, sometimes still attached to bitten-off sections of stem. This form of damage contrasted with ivy browsed by muntjac where there were ragged bites on stems and often parts of leaves were left still attached to the stem.

Figure 12.6   Examples of (a) an intact stem of ivy in woodland and (b) a group being set up in a fen location. The stick aids relocation should it all be defoliated.

Ivy for a browsing 'trial' was cut from the tips of rigid stems of densely leaved ivy growing on walls or trees. Cut stems were about 20–30 cm in length and were handled with gloves to avoid passing on my scent. The cut ends were pushed into the ground to a depth of about 2 cm. Stems were arranged in groups of 20 on a 5 × 4 grid with 1 m spacing (Figure 12.6). There were usually at least five groups in a trial, spaced out in a wood or other site. After one and seven days, each stem was checked and recorded as untouched or some leaves taken or defoliated. An advantage of ivy trials over some other monitoring methods is that they only require limited training to produce data on browsing, so can be used in long-term monitoring schemes where a regular turnover of field observers cannot be avoided. Methods such as ivy trials, scoring and impact assessment have been used in many woods in this area by myself and by other organisations such as the Wildlife Trusts.

12.6.4   Fixed-point views
Fixed-point photography can be extremely useful at demonstrating change, although it cannot fully replace recording because many impacts on vegetation will not be obvious in photographs. It is worth taking views whenever remedial management is starting, and the same applies to newly erected exclosures. I often regret not using a camera more in the 1990s when Monks Wood was in such bad shape. Once time has moved on, there is no going back to do something that should have been done before. It is important that detailed records are kept of precisely where photographs were taken. I know to my cost that it can be difficult or even impossible finding the precise location of a view in a woodland block taken 20 years before (e.g. see Figure 18.2). One reason might be that the view has changed so much, making it better to take a series of views over that length of time. And, in this age of mobile phones, potentially important photographs must be stored and catalogued properly.

12.6.5   Duration of monitoring
Intensity and duration of monitoring will depend on many factors, including how urgently information is required and the availability of resources. I have always found starting something new easier than stopping, but continuing to record beyond a possible end point has only rarely produced useful and unexpected insights.

The problem is that the duration of monitoring needed for impact management cannot, in some circumstances, be predicted with any confidence. This does not arise if problems are noticed in a wood as regards browsing on unprotected coppice regrowth, because fencing can be erected and monitoring will soon indicate whether the new protection has been successful. If, however, susceptible, widespread flora are being grazed and culling is introduced, then monitoring may be needed for more than ten years before grazing levels become acceptable. The time to stop might only become obvious once it has been reached.

Detailed monitoring can be discontinued when the problem has receded, but it might be unwise to stop all monitoring at that point in case the impact worsens again. A solution is to undertake less onerous monitoring, perhaps involving annual fixed-point photography, together with some brief notes. Alternatively, recording less frequently or general annual deer and impact monitoring may suffice.

12.7   Relationships between deer density and damage

There may be complex relationships between density and damage in a wood so that reducing deer numbers might not result in any anticipated improvement (Gill 1992a; 1992b; Putman 2003). Factors affecting how well density can predict impact include site conditions, the availability and spatial arrangement of habitats more widely, and the availability and quality of other food sources. Thus, a vulnerable species of plant may be unaffected in a site with ample alternative forage. Often there seems to be a threshold density below which significant damage does not occur, but above which damage can happen, although not necessarily in a simple manner. Putman et al. (2011a) endeavoured to establish these threshold values for different deer species in different situations.

Other authors have reported on the impacts of deer enclosed at different densities (Tilghman 1989; Horsley et al. 2003; Nuttle et al. 2014) or have compared results from various studies at different sites (Gill 1992b; Gill and Morgan 2010; Putman et al. 2011a), but none of these studies has contained specific information on muntjac. Work on muntjac at Monks Wood (and also at Raveley Wood) has focused on changes from year to year as the wood recovered following the start of stalking. This is more likely to show significant relationships between density and damage because intrinsic variability between sites or between plots within a site is no longer an issue. It may, however, be the case that such a relationship for decreasing density differs from that for increasing density. It is also likely that relationships will differ to some degree from site to site. But results from Monks Wood demonstrate the types of relationship that occur – and are reported in several of the following chapters.

CHAPTER 13

Controlling muntjac populations

13.1 Muntjac populations at individual sites

Stalking has been introduced into my three local NNRs to control muntjac, and a variety of monitoring techniques have been used to follow how populations changed. These have usually involved counting the deer, their paths or their droppings; observations in the first section below are restricted to effects on deer numbers and these indicators. Scoring is the principal technique used at other sites, mainly because only one visit is required per year. This method also yields information on damage, and the impact of stalking on damage scores is included for those woods. The chapter ends with discussions of stalking impacts on a broader scale and the possibility of muntjac populations recovering if control is relaxed.

13.1.1 National Nature Reserves

For Monks Wood, sightings frequency data can be converted into muntjac densities using the relationship in Figure 7.6. Deer density data in Figure 13.1 begin in 1995 when muntjac deer were recovering from the die-off in 1994. Stalking started in the wood in the winter of 1998/9 when 106 were culled, equivalent to 0.7 per ha of woodland. Deer density decreased by 43% in the space of a year. The cull rate in Monks Wood averaged about 0.5 muntjac per ha per annum up until 2006, and 0.2 per ha per year between 2006 and 2016. From 2006 to 2016, the number of muntjac shot per stalker visit dropped from 0.5 to 0.1. Estimated deer density was 110 per km² in 1998, but had roughly halved by 2006 and had halved again by 2016, that is the reduction from 1998 to 2016 was 75%. The decrease in density was not the only reason why they were harder to cull. Individuals tended to be more wary (Cooke 2013c) and the vegetation recovered. Muntjac were comparatively harder to see than prior to stalking starting, and therefore more difficult to shoot.

The number of deer paths crossing the woodland boundary was counted each spring, 2002–2016, along 840 m of the eastern edge of Monks Wood where it abutted two arable fields. The percentage of well-used paths cut by muntjac hooves was

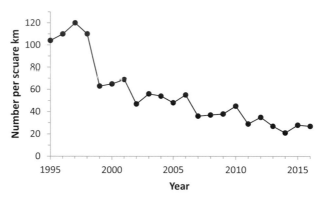

Figure 13.1 Number of muntjac per km² estimated from sightings frequency data, 1995–2016, using the relationship in Figure 7.6. Stalking began in 1998.

estimated to be nearly 100% in 2002 and about 75% in 2003. In later years each path was recorded as muddy and well used or less well used (Figure 13.2). Total number of paths changed little until 2007, since when they have decreased. Well-used paths decreased throughout the whole period. If paths are to be monitored during a potential population decline, it is important to record some measure of usage as well as the total number. Total number of paths was correlated with estimated muntjac density, but not in a linear fashion – abandonment of paths accelerated as the population decreased. By 2011, few well-used paths remained. If the population continues to decline, then total number of paths should decrease further with well-used paths falling to zero. Eventually at very low densities, there would be little or no need for deer to leave the wood and there would be few or no obvious exit paths. An increasing population may display the reverse pattern with an increasing number of faint paths before any become well used; then a rapid increase in total number of paths before finally (virtually) all paths become well used. Evidence of roe deer using the paths was noted 2012–2014, but not in 2015 or 2016 when they were often seen on fields outside the wood on its other sides. The two fields beside

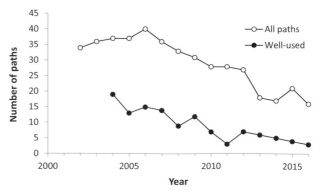

Figure 13.2 Deer paths across 840 m of the eastern edge of Monks Wood, March 2002–2016, showing total number of paths and those well used. Stalking began in 1998.

the eastern edge where the paths have been counted are closer to motor traffic and more disturbed, so may be less attractive to roe.

From 2005, the number of pellet groups seen per hour was recorded during scoring visits. This decreased from 2005 to 2007 and then remained low. I regret not having counted dung pellet groups during the early years of scoring. Amount counted will depend on the route taken and on weather conditions preceding the visit: for instance, heavy rain could result in rapid loss and a low count. However, if the intention is to walk the same or a similar route annually over a number of years, then recording the amount of dung seen per unit of time or distance can be a useful indicator of changes in the deer population.

The study at Holme Fen concentrated mainly on dusk surveillance visits to provide information on the impact of stalking on the deer population. During the summer of 2005, 60 muntjac were shot (about 0.2 per ha) but this had no effect on the number of deer seen the following winter (Figure 13.3). This contrasted with what happened initially at Monks Wood when shooting 0.7 per ha had an immediate, marked impact. About 0.2 per ha were culled again in the summer of 2006 and this did lead to a significant reduction in winter sightings which persisted through to 2014. Numbers shot decreased over time as density decreased and the deer became harder to see, in part because of recovery in the vegetation. Number shot per stalker visit decreased from 0.6 in 2007 to 0.4 in 2011. In the first ten years of stalking, 326 muntjac were shot (0.1 per ha per year). This low cull rate probably had an impact because large areas of the reserve are dominated by bracken where there is a relatively low density of muntjac. So the average annual cull rate was probably in the region of 0.2 per 'useful' ha from the muntjac point of view.

Dung pellet groups were counted from 2005 in plots along five transects spread through the reserve in areas where the ground was dominated by bramble and wood ferns (Figure 13.4). Although these five transects were not representative of the wood as a whole, changes in dung within the plots should be reasonably representative of the best areas of the wood for muntjac. Numbers of pellet groups decreased but the initial drop was less than that which occurred for deer sightings (Figure 13.3). The former is considered to be a truer reflection of the actual decrease for the reasons given previously; the decrease in sightings following stalking is likely to

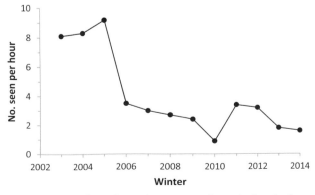

Figure 13.3   Average number of muntjac seen per hour during dusk surveillance walks at Holme Fen, winter 2003/4–2013/14. Stalking began in 2005.

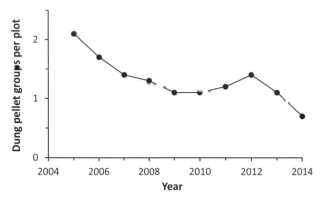

Figure 13.4   Average number of muntjac dung pellet groups per 16 m² plot
along five transects in Holme Fen, 2005–2014. Stalking began in 2005.

over-emphasise the real reduction in density. Comparing the final results in this
period with pre-stalking data, surveillance sightings decreased by 78% whereas
dung counts decreased by 67%.

The situation at Woodwalton Fen is rather different from that in a uniform wood
of a similar size. Muntjac deer are especially attracted to the mixed woodland in
the south of the reserve. They are present at lower density in the rest of the reserve,
and at much lower density on the farmland that surrounds the reserve. So the
population functions much as if it was in a small wood of 7 ha set in 2 km² of
reasonable, but suboptimal, habitat. A combination of winter dusk surveillance,
scoring in April and occasional camera studies have been used to monitor events.
A total of 57 deer were shot in 2011 representing a cull of 8 deer per ha in the mixed
woodland or about 0.3 per ha for the whole reserve. However, stalking was reduced
over the next five years so the average cull rate during the first six years of stalking
was about 2 per ha in the mixed woodland and less than 0.1 per ha in the reserve.
Initially stalking had comparatively little effect on numbers in the mixed woodland
but a larger impact on deer in the reserve as a whole, as deer moved into the
woodland from the sallow carr and reed beds to replace losses. Sightings frequency
for the reserve was 2.6 per hour in 2010/11 decreasing to 0.4–1.2 in the following
seven winters. Further details on impacts and management on the fen are given in
Chapter 19.

13.1.2   *Other ancient woods*
Monks Wood is a remnant of the ancient oak-ash woodland with relevant records
dating back to the Domesday Book (Hooper 1973); the name 'Monekeswode' first
appeared in 1279. In contrast, woodland in the two fen reserves is of relatively recent
origin. Ancient woodland is defined as that which has been under continuous tree
cover since 1600 (e.g. see Peterken 1993; Rackham 2003). Ancient woods are viewed
as being prime candidates for conservation, and inventories have been compiled
(Spencer and Kirby 1992). How the concept of ancient woodland was developed
and protection was afforded to such woods has been recently reviewed in depth by
Derek Niemann (2016). Oliver Rackham (2003) forcefully argued that deer browsing
was the main threat to ancient woodlands in England. He also pointed out that

Table 13.1   Scoring results for muntjac and stalking in nine ancient woods. All are in Cambridgeshire except Marston Thrift which is in Bedfordshire. Decrease = a trend that was statistically significant. (Decrease) = a trend approaching significance.

Wood	Size (ha)	Years recorded	Trends in scores		Stalking	
			Deer	Damage	Start	Cull rate
Aversley	62	2000–2011	Decrease	Decrease	2000/1	Culled 0.4/ha/y in first 8 years
Brampton	132	2000–2011	Decrease	Decrease	2000	Culled 0.2/ha/y overall
Gamlingay	60	2000–2011	(Decrease)	No change	1999	Variable, <0.1/ha/y overall
Hayley	40	2004–2010	No change	Decrease	2004/5	Variable, up to 0.4/ha/y, 0.1 later
Lady's	7	2000–2015	Decrease	Decrease	–	No stalking
Marston Thrift	56	1998–2007	No change	Decrease	1998	Culled 0.2/ha/y to 2004/5
Raveley	6	2000–2016	Decrease	Decrease	2000	Initially 0.2/ha/y, but zero 2005–2014
Waresley & Gransden	50	2003–2010	(Decrease)	Decrease	?	Stalking occurred, but no data
Wistow	9	2006–2012	No change	No change	–	No stalking

most species of deer that damage woodlands are at least partially sustained by foraging outside woodlands, so are especially successful (and damaging) in agricultural landscapes studded with woods. Such landscapes dominate parts of East Anglia.

As stalking started at other conservation woods in this area, I began regular monitoring to study effects on the deer and their damage. Data are assembled in Table 13.1 on nine woods where scoring was undertaken annually over at least seven consecutive years. All of these are considered to be (fragments of) ancient woodland, and the muntjac was the dominant deer species in each one. Seven of these sites had stalking and the other two had stalking at a wood nearby. Regression analysis has been used to determine if significant trends in scoring occurred. The information on stalking summarised in the table was sometimes incomplete or not available. In some cases, records were not kept or were not passed on to the site managers or relayed to me. Seven of the sites are in Cambridgeshire and were managed by the Wildlife Trust, but Aversley Wood and Marston Thrift were managed by the Woodland Trust and Bedfordshire County Council respectively.

Further information on stalking and on colonisation and impact stages (Cooke 2007; 2008; 2013b) is summarised below:

- Aversley Wood was colonised by muntjac during the 1990s. Impact stage decreased from high in 2000, to moderate in 2008–2010 and to low by 2011. Signs of improvement began to show after two or three years.
- Brampton Wood (Figure 6.2) was the first wood in the county to be colonised and by the 1990s impact was high. Impact decreased to moderate during 2005–2007 and to low by 2008. The aim had been to shoot 40 muntjac per year, equivalent to 0.3 per ha per year. However, this aim was thwarted by the

difficulties of stalking in a wood that had large woodland blocks and mainly narrow rides. Signs of improvement started rather suddenly after five years of stalking. Up to that point it seemed that stalking was having no effect on the deer or their damage.

- Muntjac colonised Gamlingay Wood during the 1990s. Impact stage was assessed as moderate throughout the period 2000–2011, although there were some minor improvements. Stalking information was fragmentary.

- Information for Hayley Wood in Table 13.1 was for inside the large deer fence erected in 2002 – this excluded fallow deer but not muntjac (Figure 13.5). The stalker, Bob Smith, reported that muntjac became progressively scarcer and more difficult to shoot, with an estimated density of 0.20–0.25 per ha by 2008. This change did not result in a significant reduction in muntjac deer score, but the trend was downwards. Had observations been made over a period of longer than seven years a significant decrease may have become apparent. Impact stage was moderate or high during 2004–2007 and low or moderate during 2008–2010. Outside the fence, impact from primarily fallow deer was severe in 2009.

- Colonisation climaxed during 1998–2000 in Lady's Wood, when impact was moderate. By 2009, it had decreased to low and stayed at that level through to 2015 (Figure 12.2a), despite no stalking taking place in the wood. There was limited stalking in Raveley Wood, some 600 m away.

- Muntjac had been long established in Marston Thrift when monitoring started in the late 1990s. Impact was high (Figure 12.2b), but decreased to moderate in 2006. During most years, the Chilterns DMG organised a deer count, which involved a line of beaters driving deer across rides, where they were counted. The number of muntjac counted during 1998–2006 did not change significantly.

- Raveley Wood was colonised during the 1990s with the score peaking in 2000. At that time, impact was moderate but decreased to low by 2011.

- Waresley and Gransden woods are adjoined and the Wildlife Trust reserve covers most of the woodland area. Muntjac colonised during the 1990s. Impact was moderate in 2003 decreasing to low in 2009. Stalking occurred

Figure 13.5   Hayley Wood. (a) Ray Tabor at the entrance; the deer fence encloses about 80% of the wood. This photograph was taken in 2002, just after the fence was erected. (b) Two coppice plots inside the fence in 2010.

in Gransden Wood during the study period and game shooting interests might have accounted for some muntjac in Waresley Wood, but no data were available.

- Wistow Wood is separated from the larger Warboys Wood by a narrow lane, which is regularly crossed by deer, as evidenced by their paths. Muntjac colonised Wistow Wood through the 1990s up until about 2005. Impact was moderate during 2006–2012. Stalking took place in Warboys Wood but not in Wistow Wood until 2015.

Andrew de Nahlik (1992) explained how to plan a cull for the larger species of deer but considered that all that was needed for muntjac (and water deer) was control related to 'containment of damage'. The Deer Initiative website has a guide on cull planning (2009), which explains the difference between a reduction cull where the purpose is to lower deer density and a maintenance cull which aims to hold a population at a particular level. An annual maintenance cull for muntjac would be expected to be in the region of 30% of the population (Dolman et al. (2010) gave a rule-of-thumb figure of 25%). So culling more than 30% should reduce the density, which is consistent with a level of at least 35% computed by White et al. (2004b). This means, though, that knowledge of population size is necessary in order to plan a cull. For impacted woodlands, I have tended to assess cull totals in relation to wood size, as described below.

Peak impact from muntjac was assessed as high in four of the woods in Table 13.1: Aversley, Brampton, Hayley and Marston Thrift. Stalking occurred in all four, as well as in three of the other woods where impact peaked at a moderate level. For more than ten years, I have been recommending that in conservation woods a reduction cull should be at least 0.3 per ha of woodland per annum (i.e. 30 muntjac per km^2) to be effective within one or two years. Effectiveness will depend on the proportion of females taken, but the recommendation was initially and simply based on early results from Monks Wood and Aversley Wood where higher levels of overall culling appeared to work, and from Brampton Wood and Marston Thrift where little or no change occurred with lower levels of culling. It eventually became clear at Brampton Wood that 0.2 culled per ha per year could lead to improvement, but at that site it took five years before the first signs of recovery were noted. At Marston Thrift where there was a similar intensity of stalking, damage score stabilised at a lower level, but deer score did not. The Essex Wildlife Trust has confirmed that my recommendation works in practice (Tabor 2011).

If an initial muntjac density was 50 per km^2 (a density likely to trigger concerns), culling 0.3 per ha equates to 60% of the population. If the initial density was higher, then this level of culling would be equivalent to a lower percentage. In the second winter of stalking at Monks Wood in 1999/2000, 92 muntjac were shot, although the total population in the wood was estimated to be 'only' 100 in spring 1999. The population was sustained by breeding and by immigration.

Two of the woods in Table 13.1 had no stalking during the period covered. Wistow Wood may have benefited from stalking in Warboys Wood, which could have helped to keep the overall population in check. At Lady's Wood, improvements occurred despite no stalking (Figure 13.6). It seems unlikely that the limited stalking at Raveley Wood played much or any part in what happened at Lady's Wood. For instance, it is known that no deer were shot in Raveley Wood during 2005–2008, but

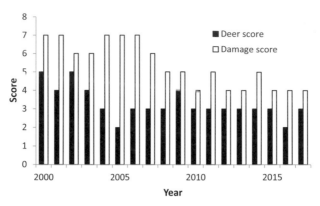

Figure 13.6   Muntjac deer and damage scores at Lady's Wood, Cambridgeshire, 2000–2017 (updating Cooke 2007). There was no deer stalking in this wood.

Lady's Wood was steadily recovering at that time. As stated earlier (Chapter 8), the most plausible explanation for the improvement at Lady's Wood was that muntjac created a browse line, thereby reducing the carrying capacity of the wood, so the population decreased and the habitat recovered. A general browse line disappeared between 2005 and 2011 (Figure 12.2a). Extensive growth of bramble and blackthorn meant that the understorey recovered appreciably and the wood could support more deer again. However, there were no signs of the population increasing.

Since the late 1990s, colonising pressure in this area has presumably decreased as stalking in most local woods lowered the muntjac density. It is also possible that a general reduction of muntjac density in the surrounding countryside contributed to the reduction in density in Lady's Wood after 2002 (Section 13.2). Without this level of stalking in other local woods, recolonisation of Lady's Wood might have been expected until a new equilibrium was established. An additional factor that may have deterred deer from colonising the wood is its popularity with dog walkers. During scoring visits in 2015–2017, I counted averages of three people per hour and four dogs per hour using the wood – and these counts were made on weekdays in September. The muntjac lie up in a small, relatively undisturbed and dense area of dog rose *Rosa canina* in the centre of the wood. I have seen only one deer outside this thicket during the last 18 years of scoring, so they seem to stay hidden during daylight hours. One might assume that the limited stalking at Raveley Wood led to the improvements in that wood. However, trends seen there were very similar to those in Lady's Wood and the same processes may have been involved, although the recovery in the understorey was slower in Raveley Wood and it is not so disturbed by people and dogs.

Natural regulation of a population, where deer leave an impacted wood, is more likely in small sites where deer will be more familiar with countryside outside the wood and may be less hefted to the wood itself. In contrast, larger woods, such as Brampton Wood and Marston Thrift, had significant deer populations apparently living in highly impacted habitat for decades prior to stalking. At Monks Wood, the extremely high density was only reduced by periodic die-offs despite the severity of the impact on the quality of the habitat. Even at the time of peak impact, Lady's and Raveley woods were only moderately damaged. Recolonisation potential may also differ between large and small woods. Thus by 2014 at Holme Fen, the muntjac population was much reduced, but because deer remained dispersed throughout the wood, the population was able to start recovering as food and cover increased and stalking lessened. That population was not dependent on external recolonisation.

Of the nine woods listed in Table 13.1, damage score decreased significantly in seven, whereas deer score decreased significantly in only four with it approaching significance in two others. So it seems, at least for this sample, as if damage score was more sensitive than deer score. In other words, a significant change in damage score became apparent first. But no woodland manager will want to wait for seven years before adjusting recommendations or making decisions. The surveillance summarised in this book has been undertaken in part to determine what happened in the longer term. But throughout, information from these studies has been relayed to managers. It is important that each situation is reviewed regularly, preferably involving managers, stalkers and those people doing the monitoring, so that best use can be made of monitoring information even if it does not demonstrate anything conclusive.

In this respect, stalking records are extremely important. It was unfortunate that information was not available or fragmentary at some of the sites discussed above. However, stalking data for the NNRs has always been good and has routinely included details of sex, weight and age. In recent years, each visit by a stalker has been recorded, irrespective of whether deer have been shot. This enables success per visit to be calculated for a season, so that trends over a period of years can be followed. Such a statistic will vary according to factors such as deer density, behaviour and amount of food and cover. Nevertheless, if stalking methods remain similar over a number of years, success per stalker visit could conceivably be adopted as a surrogate for surveillance, although the approach might need to be more formalised in advance. For instance, a certain number of visits per annum would need to be agreed, as would how a site is covered. Number of deer seen by the stalker per visit (or per unit time) could also be used for monitoring purposes. It would not be necessary for every stalker in a management group to record what was seen, but the role of each stalker should be agreed in advance.

13.2   Muntjac populations on a broader scale

An attempt at estimating number of muntjac in a landscape of 130 km² in western Cambridgeshire for the period 2001–2005 was introduced in Section 10.4. The calculation had two elements:

1. Estimating deer density in a sample of the woods and wooded fen, and applying this to all woodland (9.5 km²).
2. Making an informed guess of density in the remainder of the area and calculating the number in the whole area.

Three other time periods have been examined in order to follow the changes as stalking was introduced to a number of woods. First, 1995–2000 was a period of great change. Muntjac populations were becoming well established in many woods. Stalking was already occurring in some woods with gamekeepers. In the woodlands under conservation management, Monks Wood in 1998 was the first to have stalking – and the effect was immediate. Data for Monks Wood in 2000 was used in the calculation, that is after the population had declined significantly. The two most recent periods were 2006–2010 and 2011–2015. Stalking began in Wildlife Trust woods in 2000. At Woodwalton Fen, stalking did not start until 2011, and at Gamsey Wood it was 2012. Only 5–10% of the woodland area had no stalking and some of those woods will have been influenced by stalking nearby.

Muntjac density in woodland was calculated from deer scores as before. However, whether deer density changed in the remainder of the polygon required consideration. Number of road casualties might provide clues about deer moving around in the general countryside. There are, though, difficulties with using such data – such as traffic flow rate having changed over time – and the probability of a dead deer being seen and reported may also have changed. Locally, the mammal recorders of the Huntingdonshire Fauna and Flora Society, Don Jefferies and Henry Arnold, have listed muntjac road casualties in their annual reports since 1976. Number per annum reported dead on the roads did not change significantly from 2001 to 2015, but was roughly double during the late 1990s. In the absence of other data on muntjac density outside woods, I have used a figure of 4 per km² for 1995–2000 and 2 per km² for the later periods. My incidental records of muntjac locally are consistent with such a change.

Results of the exercise indicated that overall numbers in the polygon were highest in 1995–2000 (Table 13.2). Woodland numbers were highest in 2001–2005, but then declined, associated with increasing amounts of stalking. In 2006–2010, the overall number had decreased to 77% of its level in 2001–2005; and by 2011–2015, the figure had gone down further to 66%. In woodlands, the decreases were even more dramatic: to 68% in 2006–2010 and 52% in 2011–2015. Nevertheless, average estimated deer density in woodland was 32 per km² in 2011–2015 – still sufficient to have some impact on sensitive conservation features (Section 18.5). While stalking will have driven these reductions in deer numbers and density, one should not forget that deer impacts may have played a part by rendering some woods less able to support high numbers.

Two organisations, Natural England and the Abbots Ripton Estate, have been responsible for managing most of the woodland in the polygon. In recent years, roughly 100 muntjac have been culled per annum in the woodland managed by

Table 13.2   Number and density of muntjac estimated present in a polygon of 130 km² in western Cambridgeshire in four time periods. Area of woodland and wooded fen totalled 9.5 km².

Period	Woods and wooded fen		Overall polygon	
	Number	Density per km²	Number	Density per km²
1995–2000	400	42	880	6.8
2001–2005	590	62	830	6.4
2006–2010	400	42	640	4.9
2011–2015	300	32	540	4.2

these two bodies. In addition, small numbers of deer are shot in the remaining woods and elsewhere in the polygon. This represents annual culling rates of at least 33% of muntjac estimated to be in the woodland and at least 19% of deer in the whole polygon. In terms of area, it is more than 0.1 deer culled per ha of woodland (or 10 per km²). It is believed that this landscape approach is mutually beneficial by reducing immigration into sensitive woodland.

During the period 2008–2010, a joint study by the University of East Anglia and the Forestry Commission (Waeber et al. 2013) focused on muntjac and roe in 132 km² of Thetford Forest and 102 km² of Defence Estates' land, the latter comprising grass heath, arable, deciduous woodland and conifer plantations. These authors took into account deer population size, productivity and culling in one year to calculate how many deer could potentially be present in the following year. What happened to the muntjac in the forest area is of relevance here, in part because their study area was a very similar size to the Cambridgeshire polygon. Deer numbers decreased slightly in the forest area, but calculations indicated that this was because large numbers were dispersing. The forest was acting as a source of muntjac, despite culls averaging 23% per annum. To offset productivity, a cull of 53% would be needed. The estimated number in the forest area was about 2,900, giving an overall density of 22 per km²; the mean annual cull would need to be about 1,500.

These are sobering statistics and they show a need to understand population dynamics and not simply concentrate on the site or area where culling takes place. Because of advice from the researchers, Forestry Commission stalkers increased the number of deer culled. Trevor Banham told me in 2014 that the muntjac population in Thetford Forest had declined since the early 2000s because of culling. And Kristin Waeber confirmed in 2016 that the forest was no longer a source of muntjac.

With regard to the Cambridgeshire polygon discussed above, there was no evidence in later years to suggest that woodlands had acted as sources supplying the wider countryside, either within the polygon or beyond it. Currently, the polygon no longer contains woods that stand out as being dominant and significant sources of muntjac. Monks Wood will have been a source in the 1990s. Holme Fen probably became one before stalking started in 2005, as did Woodwalton Fen prior to 2010. The presence of large numbers of muntjac on fields outside Monks Wood and Holme Fen ceased soon after stalking began.

Moving on to consider control on an even larger scale, the former county of Huntingdon and Peterborough covered 1,258 km² of what is now north-western Cambridgeshire; the polygon referred to above comprises about 10% of that area. Of the deer that might contribute significantly to woodland impact, fallow are more or less restricted to the west of Peterborough, where they are controlled, while roe deer are still colonising and have been culled in Monks Wood since 2011 and in Holme Fen since 2018. So over most of the area, muntjac have been responsible for the woodland damage, which would be expected to improve when they are controlled.

Between 1994 and 2016, I assessed muntjac deer and damage scores in 40 woods or sites with trees and scrub across the former county of Huntingdon and Peterborough. Some sites were scored once, but three sites were scored in 20 of the 22 years. From 2000, however, between 4 and 13 sites were scored each year, and stalking became a widespread practice. Average scores for both muntjac activity and damage decreased from 2000 to 2016 (Figure 13.7). I should emphasise that I did not set out in 1994 or 2000 with an intention of monitoring trends in deer and damage scores in this broad area. Consequently, sites were not selected each year on either a random or representative basis: they were visited individually for a variety of reasons, and it was only in 2013 that I realised that it would be interesting to construct such a graph. In terms of these scores, stalking appears to have returned the overall situation to how it was in the mid-1990s. But in the 1990s, some larger woods had high densities of muntjac while smaller woods were being colonised, whereas by 2016 muntjac were being controlled to some extent in most woods.

Although a number of examples have been described of muntjac being well controlled, it would be unwise to become complacent. Currently, stalking has to continue at a high intensity, and some problems have arisen recently that stem from the success of stalking leading to vegetation recovery and thereby allowing resurgence in muntjac populations; two examples are discussed below. Furthermore, stalking over the last 20 years has failed to prevent muntjac deer from expanding their distribution in Cambridgeshire (Hows et al. 2016). Moreover in eastern England generally, fallow deer are a greater problem than muntjac. For instance, a study in Essex woods by the Wildlife Trust found that deer and damage scores were higher in 2011 than in 2002 despite much deer management, primarily because of increases in fallow deer; since then there has been an improvement, although response has varied between woods and over time (David Hooton, Graham Foxall, personal communications). There have been other wide-ranging and long-term investigations in East Anglian woodlands using scoring techniques, but results are generally not in the public domain. This is one reason why I have gone into detail about my own studies. A range of people from supporters of conservation organisations to stalkers need to know whether management is working.

Looking more widely at the national situation, there are relatively few statistics, but it is worthwhile comparing estimates of population size and national cull totals for a time in the past when the muntjac population was expanding rapidly. In the mid-1990s, the population was estimated at about 40,000 adults and 12,000 fawns and juveniles (Harris et al. 1995), while the total cull was estimated at 11,000

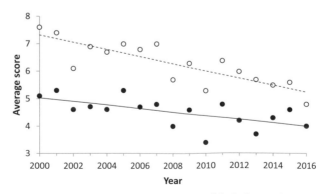

Figure 13.7   Average scores for muntjac deer (black dots, solid line) and damage (open dots, broken line) for sites in the former county of Huntingdon and Peterborough, 2000–2016 (updating and adapting Cooke 2014a).

for the year 1995/6 (Macdonald et al. 2000). The cull represented 28% of the adult population and 21% of the total population. If mortality due to other factors such as predators and road traffic could be added in, the overall mortality figure must have been very high. In the early years of the new millennium, the cull of muntjac may have amounted to 19% (Munro 2002; Wilson 2003). Yet throughout, the national muntjac population has continued to increase and spread. The Deer Initiative (2018) estimated that there were about 2 million deer of all species in the UK and the annual cull of about 350,000 (17.5%) was insufficient to prevent the continuing expansion.

13.3   Resurgent muntjac populations

Until recently, control of muntjac appeared to be resulting in woodland habitats recovering generally within Cambridgeshire. Provided that stalking and monitoring were in place, worsening problems did not seem an issue. However, examples of recovering populations have occurred in both Holme Fen and Monks Wood.

13.3.1   *Holme Fen*
In Holme Fen, earlier success was allowing a problem to develop partly hidden by the recovering vegetation. The increase in food and cover enabled the muntjac population to recover and start to impact bramble again from 2014. A similar situation may arise in woods affected by both fallow deer and muntjac. If fallow are sufficiently reduced so that bramble recovers, this may trigger an increase in the muntjac population, which is then more difficult to control because of the increase in cover. A reduction in tree canopy cover, for instance because of die-back of ash may have the same effect.

The situation at Holme Fen has been considerably complicated by colonisation by roe deer. Camera traps have indicated that while muntjac activity in the most impacted block of woodland increased by about 40% between 2013 and early 2018, roe activity rose nearly fourfold. More than half of 57 bramble patches examined in the area had browsing above 1 m in height when examined in November 2017, indicating roe were making a considerable contribution to browsing. Bramble is viewed by Natural England as a key species at Holme Fen by providing food and cover for a range of wildlife and by protecting tree and shrub regeneration. Muntjac control has been stepped up since 2016 and stalking of roe deer began in 2018 to try to allow the vegetation to begin recovering for a second time.

13.3.2   *Monks Wood*
The situation in Monks Wood is described in more detail as it was caused by muntjac resurgence alone. A dense population developed by 2015 inside the 6 ha coppice fence. This fence was erected in the autumn of 1999 around a block of seven former coppice plots (Figure 13.8a). It was about 1 km in length, the main part being constructed from light, high-tensile deer net 1.9 m high with 150 mm vertical spacing. Attached to its lower part was hexagonal netting, 1.8 m high with 75 mm mesh, the bottom 300 mm of which was turned outwards flush with the ground. The fence has been in place for longer than its expected life of 15 years. Unsuccessful attempts were made to drive out the muntjac before completion of

Figure 13.8   The coppice fence in Monks Wood. (a) The view in March 2000 with the newly erected fence enclosing about 6 ha of former coppice plots, and with a high seat commanding the ride outside the fence. (b) The same view in June 2010. The fence was still in place but dense vegetation growth inside made stalking impossible, and so the deer population built up within the fence. Woodland on the right was also a former coppice plot that was severely browsed in 1986.

the fence and later to remove deer from inside the fence. At least one deer remained within the fence when attempts at stalking were terminated because of the density of vegetation. Although two small areas of coppice were cut in later years, these were not easily accessed for stalking.

Deer and damage scores remained low until 2007, but then increased until 2015 (Figure 13.9). Holes in the hexagonal netting were first seen in 2005, thereby allowing the deer to move through the fence. These were periodically repaired by the reserve staff but deer continued to gain access.

My vegetation records and fixed-point photographs (Figure 13.10) showed that between 2010 and 2015 bramble decreased in abundance and suffered die-back, while young ash in the height range 20–130 cm decreased significantly in number and any survivors were heavily browsed. Dog's mercury disappeared. The fence was supposed to exclude deer but had become the most impacted area in the wood, and

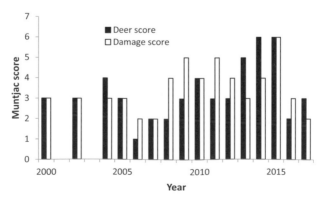

Figure 13.9   Muntjac deer and damage scores recorded inside the coppice fence in Monks Wood, 2000–2017. Scoring was not undertaken in 2001 or 2003. The fence was erected in autumn 1999. Culling in 2016 led to a reduction in deer and damage scores.

Figure 13.10   Fixed-point photographs of a small glade inside the coppice fence in Monks Wood: (a) July 2010; (b) September 2015. An increase in muntjac numbers inside the fence led to impact on the habitat from about 2008.

was therefore reflecting what must have happened generally in the wood in the late 1980s.

In 2015, a camera trap study demonstrated that at least six muntjac were inside the fence, a density of 1 per ha. This meant that comparatively few deer required culling, providing stalking could be started. In spring 2016, two sets of finger rides were cut inside the fence, and the stalkers culled five deer. My camera traps were again deployed and revealed one deer remaining within the fence. Scoring in late summer showed that improvements rapidly occurred (Figure 13.9); scores remained low into 2017, by which time dog's mercury was seen again. The creation of finger rides had allowed the deer to be satisfactorily controlled. However, continuing use of camera traps showed that recolonisation was occurring because holes in the fence had not been repaired. A decision was made by Natural England that the fence would, in time, be dismantled, and stalking would be undertaken as necessary until that happened.

The lesson learnt was that it was a mistake to fence an area of that size lacking internal rides or accessible open areas. No one foresaw in the late 1990s that, if the operation succeeded, stalking would become impossible because of regeneration. Natural England now regards the area as valuable woodland rather than traditional coppice. One reason for a reduction in coppicing potential is that ash is abundant within the fence, and ash regrowth is especially prone to the die-back disease.

13.3.3   Conclusions on muntjac resurgence

Whether resurgence occurs and management becomes cyclical depends on a range of factors:

- The nature of the situation including size of the wood, other deer species present, ability of the muntjac population to recover/recolonise, how quickly and to what extent the understorey recovers, and interactions between muntjac and their habitat.
- How stalking effort, methods and success change in response.
- What is learnt and how this modifies management in the future.

Events at Holme Fen may be a relatively extreme example. It is a big wood, where the deer were well controlled, allowing the understorey to recover rapidly and extensively. The residual population of muntjac recovered quickly in response to the increase in food and cover and the reduction in stalking (and roe deer increased significantly). Now we have learnt from this experience, future fluctuations in the size of the deer population and the condition of the understorey are likely to be dampened down by modifications to management.

Although there was an issue with the coppice fence at Monks Wood, the remainder of the wood has not suffered from muntjac resurgence despite stalking starting seven years earlier than at Holme Fen. This may be because the bramble and the rest of the understorey have been slower to recover and there is not, as yet, so much cover. Slow resurgence might, however, still occur.

CHAPTER 14

Effects of muntjac browsing and improvements in response to management

This chapter describes impacts that browsing by muntjac can have on coppice, tree regeneration and bramble, before ending with a discussion on what information can be gleaned by studying browsing on ivy.

14.1 Coppice

During the latter part of the twentieth century, deer species generally increased in both numbers and range in lowland England (Dolman et al. 2010), and at the same time there was renewed interest in coppicing in broadleaved woods, a practice that had fallen out of favour from the late nineteenth century (Crowther and Evans 1984). Efforts to revive coppicing have been for either economic or conservation reasons. Here I am addressing the latter, but the principles also apply to the former, for which it is especially important to avoid or minimise problems and losses. The essence of conservation coppice is the provision in a wood of a rotation of patches at different stages of the cycle. Different species of plants and animals may be associated with specific stages of coppice regrowth, so a successful coppicing programme will enhance the general biodiversity of a wood.

Browsing by fallow deer was one of the causes of failed regrowth on conservation coppice in Hayley Wood as long ago as the 1960s (Rackham 1975). By the early 1990s, impact on coppice was sufficiently serious and widespread for a number of studies and reviews to be published (e.g. Ratcliffe 1992; Kay 1993; Tabor 1993; Putman 1994). By that time, it was not just fallow deer that were under the microscope, but red, roe and muntjac too. However, muntjac (on their own) had become such a severe problem for coppice in Monks Wood that formal operations were stopped in 1994.

Regrowth on newly cut coppice stools in spring will be optimal fare for muntjac, providing the species is palatable. It is fresh, nutritious and growing at a convenient height in an open plot where the ground vegetation has probably been trampled flat by people. Browsing can lead to several disbenefits for biodiversity. If the new coppice canopy is slow to form or fails to develop, light-loving species, such as grasses, sedges or bramble, may dominate to the detriment of other plants; death of stools may result. In Monks Wood, for example, nearly 80% of newly coppiced unprotected ash stools died between 1993 and 1997 (Cooke 1998a). Alternatively, deer browsing can drive shifts in species composition. Thus, in one plot in the wood, hazel *Corylus avellana* was sufficiently badly browsed for birch to take over as the dominant species in the canopy (Cooke 1994), and birch was still dominant there more than 20 years later. Species of fauna can disappear that depend on certain stages of the coppice cycle or particular species of tree or shrub. In this way, even a relatively short bout of heavy browsing can have long-term consequences, and that is what happened in Monks Wood.

14.1.1   *Browsing on coppice regrowth in Monks Wood*

By the early years of the twentieth century, Monks Wood was still being coppiced on a large scale with about 8 ha being cut annually on a 20-year cycle (Hooper 1973). Much of the wood was clear-felled just after the First World War, and coppice operations only restarted in the 1950s when the wood was acquired by the Nature Conservancy. Coppicing on a much smaller scale was reinstated across 17 sub-compartments, the total area being only 13 ha (Massey 1994). A ten-year rotation was finally adopted. Attempts were made, particularly in the 1960s, to increase the depleted stock of hazel by transplanting and by layering of shoots, but these suffered severe browsing by rabbits and brown hares. There was no long-term monitoring of coppice management to judge its success (Massey 1994).

John Lowday of ITE first alerted me to the problem of muntjac browsing in 1985. The plot cut in 1983/4 had developed satisfactorily and eventually produced a dense closed canopy, but the plot coppiced in 1984/5 had its first-year regrowth badly browsed by muntjac and never recovered, reverting to blackthorn thicket (Cooke 2006). The warden, Jeremy Woodward, became aware of a huge increase in muntjac activity in 1985. The next plot, cut in 1985/6, similarly failed and few of its hazel stools survived by the early 1990s. The transition from success to failure took place over a single year. Such a change was of course not anticipated at the time. But with the benefit of hindsight, coppice operations in any wood with a growing muntjac population are at increasing risk without appropriate fencing.

I assessed the success of each coppice plot until termination of operations in 1994, by asking reserve staff to judge the state of regrowth and by counting regrowth stems that exceeded 1 m in height per unit area. Reserve staff piled brash on stools or enclosed plots with electric fences each year 1987–1994, but success was variable (Cooke and Lakhani 1996; Cooke 2006; Figure 12.3). Plots in quieter areas of the wood were more likely to be badly browsed than those beside major rides. But even regrowth in a ride-side fence was likely to suffer inside its rear edge, with muntjac taking advantage of the cover provided by untouched good growth near the ride to breach the fence where they could not be seen.

Ken Lakhani and I calculated that if English Nature wished to reduce the muntjac density in Monks Wood to a level at which coppice need not be protected, then a cull of about 90% of the deer would be needed. In other words, deer density would need to be kept at about 10 per km². At the time this calculation was made, food for muntjac had been severely depleted. The corresponding figure for deer density when coppice impact might first be noticed in a relatively undamaged wood is likely to be around 25 per km² (Cooke 2006; Putman et al. 2011a). Browsing on coppice regrowth begins at lower deer densities than, for instance, impacts on many ground flora species, and this – together with the clearly visible nature of such browsing – means it is often the first indication of a muntjac problem in a wood.

When recording the success of hazel coppice, I have found it useful to think in terms of how reality measured up to the aim of management. From talking to woodland managers and counting regrowth stems, I developed a concept of an ideal coppice with 400 hazel stools per ha each producing on average 25 stems, resulting in 10,000 mature stems per ha. This ideal enabled me to determine the relative contributions of stool density and stems per stool to overall plot productivity after two or more growing seasons. In particular it permitted assessment of the importance of browsing damage (Figure 14.1).

Coppicing of ride-side strips to diversify structure continued after formal coppicing stopped. These strips may be regarded as surrogate coppice plots and their success can be judged in the same way to determine whether stalking, which started in the wood in 1998, reduced browsing (Table 14.1). In 1993, the average number of stems taller than 1 m in four coppice plots and four ride-side plots, all protected by electric fences, was 14 per live hazel stool (Cooke and Lakhani 1996). This figure reflected the extent of browsing inside the fences. The practice then was to fence virtually everything – there was likely to be minimal regrowth on any hazel left unfenced. By 2005, some hazel was fenced and some was not; an assessment of unprotected stools in six ride-side plots, cut 1999–2004, revealed an average of 13

Figure 14.1   Hazel coppice regrowth heavily browsed by muntjac, Monks Wood, 1993. The stem in the centre of the picture shows a typical ragged deer bite where the incisors in the lower jaw have bitten against the hard pad in the upper jaw.

Table 14.1   Hazel coppice at Monks Wood: number of regrowth stems per stool greater than 1 m, 1993, 2005 and 2009–2011. Stalking began in 1998.

Date	Electrically fenced?	No. of plots	No. of growing seasons	Average number of stems per stool	
				All stools	Live stools
1993	Yes	8	1–2	–	14
2005	No	6	2–6	11	13
2009–2011	No	6	2	19	–

stems per live stool and an average of 11 for all stools (Cooke 2006). The conclusion from these results was that the browsing situation had improved, but browsing was still causing significant losses of stems and was killing some stools. Between 2009 and 2011, hazel was assessed in six ride-side plots after two growing seasons. One advantage of recording second-season growth is that any browsing is the product of deer activity during a relatively brief window of 18 months, so providing a reasonably up-to-date and easily interpreted assessment. The average number of tall stems per stool was 19 for all stools, showing a continuing improvement, but some localised browsing problems persisted, to which colonising roe also contributed.

In the two major surveys of coppice damage by deer in the early 1990s, Kay (1993) studied roe and fallow deer damage in 28 woodlands in East Anglia, while Rory Putman (1994) co-ordinated responses on deer browsing from 106 sites through England. Putman concluded that 'the degree of damage was profoundly influenced by both tree species and deer species present'. Muntjac occurred at seven sites in this enquiry, and results demonstrated the potential for them to cause problems for hazel and ash regrowth. Little information was provided on their browsing on other species apart from lack of interest being shown in alder. Bows (1997) noted relatively low levels of browsing on aspen in Essex woods with muntjac and fallow. Aspen is a species that has benefited from muntjac browsing in Monks Wood by colonising spaces resulting from loss of ash and hazel.

To provide information on the susceptibility of different species in Monks Wood, I studied shrubs and trees in three plots in 1993 (Cooke 1994). All had been cut the previous winter and electrically fenced, but each one suffered significant browsing damage. Privet, dogwood *Cornus sanguinea*, field maple and narrow-leaved elm *Ulmus minor* were the species most severely browsed, while birch and sallow species were least affected. Hazel and ash were intermediate. At first glance, this appears to provide a league table of vulnerability, but this is information from a single site in a single year and few deer will have been involved.

14.1.2   Implications for woodland structure

For many years, Rob Fuller and his colleagues at the BTO have researched the requirements of birds utilising coppice and how those needs are affected by deer browsing. Early work showed the importance of shrub layer density (e.g. Fuller and Henderson 1992), while later work demonstrated that deer browsing reduced both understorey density and breeding numbers of several passerine species (Gill and

Fuller 2007; Holt et al. 2010; 2011). These authors studied woods with mixtures of five deer species, including muntjac, but said nothing about the effects of muntjac alone. Gill and Fuller (2007) found that deer had the greatest impact up to a height of 1.5 m, which is roughly the maximum height at which the larger deer forage, but above the height that muntjac can manage. However, browsing on growing stems lower down by muntjac can ultimately result in reductions at 1.5 m and even higher, and breakage of thin, young stems can have immediate effects on density at such heights. I have been able to investigate the impacts of muntjac specifically in small exclosures erected at Monks Wood.

Using a method similar to that of Fuller and Henderson (1992), I recorded an index of vegetation density at both shrub layer height (1.5 m) and field layer height (0.5 m) in small exclosures erected in 1993 and 2004 (Figure 12.5b). Ten years after erection in coppice plots, vegetation density in the 1993 exclosures was greater at both heights than in the unfenced control plots (Cooke 2006). The understorey community, principally bramble, honeysuckle *Lonicera periclymenum*, hawthorn, privet and rose species, was more abundant and grew taller in the exclosures. When the observations were made in 2003 after ten growing seasons, the coppice structure would have been beyond the stage at which it was especially suitable for nesting passerines. Nevertheless, the understorey was markedly affected throughout the ten-year study, and this may have been one reason for the reduction in some avian species in the wood since the 1980s (Section 17.7). In the exclosures erected in 2004 in ride-side clearance plots, density indices were recorded annually up to 2009. There was no difference in the density at field layer height, but shrub layer density, including coppice regrowth, grew to be much greater in the exclosures than in the control plots (Figure 18.1). It peaked in 2008, when it was more than four times greater than in the controls. So even though stalking had then been undertaken for ten years and coppice regrowth had become less browsed during that time, shrub layer density was still affected.

14.1.3   *What protection works and what does not*
My observations on protection of local coppice operations started with brashing and electric fencing in Monks Wood in the late 1980s. Neither method worked very well in part because of the exceptionally high density of muntjac trying to browse the regrowth. Five-strand electric fences were used but these were regularly breached, particularly in the quieter areas of the wood (Figure 12.3). When deer were seen negotiating the fences, they usually moved through between the first and second strands, which were roughly 15 and 40 cm from the ground (Cooke and Lakhani 1996). This corresponded to nose height for most muntjac. Mayle (1999) stated that muntjac were 'generally undeterred' by electric fences, but Trout and Pepper (2006) claimed that six-strand fences were effective against *low* populations of muntjac for two years.

After the electric fences, metal panel fencing was briefly used in Monks Wood (Figure 14.2). This protected regrowth reasonably well, but difficulty was experienced in filling the gaps under the panels where the ground was uneven. Such fencing is better at excluding fallow deer than muntjac. In the late 1990s, I arranged for a group of young maiden ash to be pollarded at about 1 m, but the regrowth was disappointing, with the muntjac removing all the stems they could reach. With such

Figure 14.2   Metal panel fencing around a ride-side plot in Monks Wood, 1995.

a high density of muntjac, trees would need pollarding above 1 m to be effective. Natural England staff pollarded a small group of ash near the entrance to the wood in 2014/15. These were exposed to a far lesser muntjac presence, but by this time roe deer were contributing significantly to browsing. Those cut at 1 m or higher had acceptable regrowth in 2015, but those cut below 85 cm were unacceptably browsed. The point about pollards and deer is, however, that any regrowth close to the ground is less likely to survive, so pollarding is not an option if dense, low regrowth is the aim.

The Wildlife Trust has undertaken coppice operations in many Cambridgeshire woods. In the early years, only limited success resulted from a variety of protection measures including brash hedging, brash tents around individual stools, chestnut paling or 1.2 m plastic fencing with 50 mm mesh. Muntjac gained access through damage to the plastic fence made either by themselves or by rabbits. Plastic fencing can work well if it is high enough and strong enough, but this material could be readily breached. In Gamlingay Wood, rabbits were fenced inside a plot cut in 2003/4 resulting in high levels of browsing by the following autumn. Since 2004, 1.5 m high wire mesh fencing has been used successfully. For instance, hazel stools in the first wire-fenced plot in Brampton Wood averaged 32 stems reaching 1 m in height, compared with an average of 0.4 per stool inside a nearby plastic fence.

At Hayley Wood, fallow deer were fenced out of about 80% of the wood where all of the coppicing took place (Figure 13.5a). Newly cut coppice received no extra protection (Figure 13.5b). Muntjac remained inside this 40 ha fence and have been culled since 2004. In 2015, it was still possible to find recent browsing damage on hazel although generally coppice regrowth was acceptable.

Much of Rob Fuller's work on the impact of deer browsing on birds has been done in Bradfield Woods in Suffolk, where muntjac occurred along with fallow and roe deer. In the early years of his study, dead hedges were used and gave sufficient protection for adequate regrowth to occur (Fuller 2001). Dead hedges are brushwood fences made by weaving long woody stems between stakes. However, making these was very labour intensive and required considerable volunteer input. In Essex, Tabor (2009) found them ineffective. Later, the coppice plots in Bradfield Woods received complete protection with 1.8 m high steel fences (Figure 20.1).

Chestnut paling found favour with several conservation organisations in the past. The Essex Wildlife Trust used it around coppice in Shadwell Wood to protect regrowth and oxlips after a dead hedge had failed to keep out muntjac (Tabor 1993). Subsequently, it was realised that some of the chestnut pales bowed and muntjac could squeeze through the gaps; the first muntjac got inside after only nine months (Tabor 1999). In Aversley and Archer's woods in Cambridgeshire, the Woodland Trust fenced small areas with paling to which was attached wire mesh. This worked well at Aversley Wood, but not at Archer's Wood where muntjac managed to squeeze under the wire and through the paling.

Various other forms of management were trialled in Essex, such as hurdle fencing using hazel rods woven through upright stakes, but only wire fencing was found to be totally effective at keeping out muntjac and fallow deer (Bows 1997). Some Essex woods also had human disturbance in addition to physical protection, involving charcoal burning or wood turning, sometimes with the extra deterrent of urination on browsed stools; this lowered damage but did not eliminate it.

One innovation trialled in Monks Wood by Chris Gardiner of Natural England was a three-sided fence to offer some protection for ride-side plots. This new style of fence had a back and sides but was open to the ride. The reasoning was that it would deter deer from entering the plot directly from the rest of the woodland block. As it was not critical that such plots should remain totally undamaged by deer, the advantages of a three-sided fence over a complete fence were that it was cheaper and easier to erect, and was not so obtrusive for visitors. Two such fences have been trialled, but both were only erected after one or two seasons of severe browsing on regrowth had occurred. Fencing allowed partial recovery to occur (Figure 14.3). In situations where no damage is essential, appropriate fencing should be in place before the first growing season.

Figure 14.3   A three-sided fence in Monks Wood soon after its erection in 2013: regeneration was reasonably successful after being badly browsed in 2012, but suffered setbacks later as a result of ash die-back and oak mildew disease.

Figure 14.4   Part of the fence around Shadwell Wood, Essex, in 2009.

In 2000, Essex Wildlife Trust erected a high wire mesh fence around the whole of Shadwell Wood to exclude both fallow and muntjac deer (Figure 14.4) – any deer found to have breached the fence were culled. This led to improvements in deer and damage scores and in percentages of coppice shoots browsed and oxlips grazed (site number 3 in Tabor 2004). This is an example of where fencing a whole wood has brought conservation benefits, but generally such action is not recommended (Mayle 1999), in part because it displaces deer. One reason why perimeter fencing worked at Shadwell Wood was because it is a compact site. At 7 ha, it is comparable in size to the permanent deer exclosures in Monks Wood. It has advantages over those structures in that it is separated from other woodland and has rides and open areas where deer can be culled. In such a situation, it is also important to monitor regularly for signs of deer inside the fence and check the fence for holes.

The severity of browsing on coppice regrowth is related both to overall perimeter length of the coppice and to length of the perimeter next to cover (Kay 1993; Putman 1994). Also, plots in Monks Wood were found to be less damaged if they were beside major rides, which were likely to experience greater disturbance from walkers and vehicles (Cooke and Lakhani 1996). Facts such as these might influence management plans in terms of the size, shape and location of new coppice plots.

14.1.4   Conclusions

Muntjac can cause severe problems for conservation coppice, and Monks Wood has provided a vivid example. Browsing regrowth led to a patchy coppice canopy or no canopy at all. Grazing caused a reduction in vernal herbs. Increasing light levels resulted in a proliferation of bramble and pendulous sedge with a further smothering of ground flora and death of coppice stools. Ultimately, browsing of bramble caused loss of thickets and colonisation by grasses and sedges, and the place of mature coppice regrowth stems was taken by birch and aspen. This shift will have had knock-on effects on wildlife dependent on the structure and species composition of the different stages of the coppice cycle.

The solution to a coppice problem centres on placing a barrier between the muntjac and the coppice. Although it would be possible to cull muntjac down to a density at which they no longer have unacceptable impacts on coppice, this might take many years, especially if the problem has been allowed to develop and become severe. Because coppice operations occur over relatively small and discrete parcels of land, they provide an ideal situation for fencing which, if done properly and soon enough, is an immediate and effective solution. To have the best chance of holding muntjac at bay with fencing, adhere to Forestry Commission guidance (Section 12.3.1). Erect a wire fence at least 1.5 m in height with maximum mesh size of 80 × 80 mm at heights that muntjac might try to squeeze through, and have the bottom 15 cm bent out along the ground (and, if possible, buried). Check around the entire fence line regularly for access points, potential weaknesses and trees or branches that have fallen across it. Monitor for signs of muntjac inside the fence, looking particularly for browsing on fresh woody regrowth in summer and for slots, droppings and browsing on bramble and ivy in winter. Have contingency plans for removing muntjac as necessary.

Fences should remain in place for at least two growing seasons. Dismantling fences when growth just exceeds 1 m, the normal browse line for muntjac, is likely to invite browsing on peripheral stems and breakage of leaders. Stems are only safe from breakage when thickness is at least 1 cm at a height of 1 m. In sites where other larger species occur and also have the potential to browse coppice regrowth, use Forestry Commission guidelines to ensure that the fence is suitable for them too – in particular that it is high enough.

14.2   Regeneration of ash and other woody species

This section concentrates primarily on the impact of muntjac browsing on ash in Monks Wood. Ash is a very abundant tree in the wood, making up about 50% of the canopy (Broughton et al. 2011), and ash regeneration has been a good subject for studying both the impact of browsing and the recovery following control of muntjac. Before muntjac browsing became a problem, regeneration was said to be plentiful wherever the canopy was opened (Steele 1973). However, by the time I began investigating impact in the 1990s, the situation had changed radically.

George Peterken (1994) pointed out that change in unmanaged stands in Monks Wood is influenced by (1) growth and interactions between species of trees and shrubs and (2) episodes or events, such as drought in 1976 and 1990–1991 or the increase in deer browsing from 1985. Under his guidance, Christa Backmeroff had set up transects in 1985 to monitor change in four woodland compartments. All trees, shrubs and saplings that had reached a height of 1.3 m were mapped and measured, and saplings and suckers 30–130 cm in height were counted. Part of one transect was reassessed in 1992 (Peterken 1994): a cluster of ash saplings was found to have disappeared since 1985 and mortality of privet and blackthorn had occurred. All four transects were resurveyed in 1996 by Ed Mountford and George Peterken (1998). The clusters of ash seedlings and saplings found in 1985 had suffered high mortality by 1996, and had not been replaced by new individuals. The same was true for other species, most notably blackthorn. Browsing by muntjac was blamed, although it was difficult to be sure because of the influence of natural processes.

An area coppiced in 1992 provided me with much information on the difficulty facing ash seedlings, and not just from muntjac. Four small exclosures and control plots were set up in 1993, and their vegetation was recorded with Lynne Farrell each year until 1997. Beyond that date, bramble was too abundant in the exclosures to continue. By 1997, there were estimated to be 300–400 ash seedlings in a total area of 64 m² inside the exclosures, with a similar number in the unfenced controls (Cooke 2006). Annual survival was 64% inside the exclosures but only 26% in the control plots.

That ash seedlings were not more numerous in the exclosures was probably due to reduced germination because of shading from the faster growing vegetation, particularly bramble. Average height of ash seedlings was, however, consistently greater in the exclosures. Seven saplings in the exclosures exceeded 20 cm by 1997, but none did so in the controls. The fastest growing sapling exceeded this height one year after first being detected, but even this one grew much slower than the bramble. Seedlings and young saplings grow much slower than stems on coppice stools, thereby considerably extending the time that they are vulnerable to browsing. Privet, hawthorn, honeysuckle and other understorey species also grew better inside the exclosures than outside (Cooke and Farrell 2001b; Cooke 2006). Bramble can confer some protection from browsing on seedlings beneath it, but the latter may die if badly shaded. By 2003, a single ash had reached 5 m in the exclosures and a sapling was 1.8 m; no ash was taller than 20 cm in the controls. So, while survival of protected ash was statistically very low, survival of unprotected seedlings was zero in this situation.

Studying the impact of browsing on the early stages of tree regeneration is problematical. In order to have reasonable numbers of seedlings, a good seed source is needed plus a relatively unshaded situation and little competition from other vegetation. Such a combination of factors will occur only rarely, for instance in a freshly cut block of coppice or a cleared ride-side plot. Ideally, studies need to cover

Figure 14.5   Part of the patchy canopy of ash and oak above caged ash and bramble plots in Monks Wood; such a canopy allowed slow regeneration of trees and bramble in the absence of significant browsing.

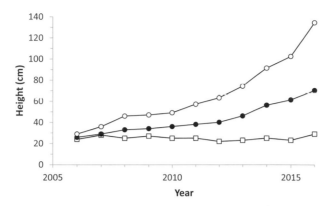

Figure 14.6   Height attained by individually caged ash saplings in Monks Wood, 2006–2016: fastest growing specimen, open dots; average of seven survivors, black dots; slowest growing specimen, open squares.

larger areas than I was able to manage and they need to be long term. Nevertheless, useful monitoring can be undertaken by repeating simple, basic observations. Thus, I monitored browsing on ash seedlings less than 20 cm in height in two areas within one woodland compartment. In 2003, five years after stalking began, 38% were browsed, but from 2005 until 2009 the level of browsing fluctuated between 16 and 28%. By 2015, small ash seedlings were becoming more difficult to find under the bramble and larger seedlings; 15% were still showing signs of browsing, suggesting small mammals were also involved. In one of the areas, ten small saplings in the height range 20–40 cm were individually caged in 2006 and their survival and height was monitored until 2016. These were in a location where there was a patchy ash and oak canopy (Figure 14.5) and bramble height was also recorded. Seven of the young ash survived until 2016 and their growth is depicted in Figure 14.6.

The fastest growing specimen exceeded 100 cm by 2015, but the poorest performer stayed more or less the same height throughout. On average, the sample doubled its height by 2014, eight years after being caged. During the same period, bramble also roughly doubled its height from 47 to 103 cm. Bramble grew much slower than in the more open environment of the coppice plot referred to above, but nevertheless it overtopped many of the ash, including some of those that had been caged. Losses of caged ash were about 4% per annum, that is roughly an order of magnitude less than the loss estimated for tiny seedlings in the exclosures in the 1990s, referred to above. This demonstrated that protected ash had much better survival once they attained a height of 20 cm. Close to the cages, unprotected ash in the height range 20–40 cm had losses of about 6% per annum during 2006–2009.

Peterken and Mountford (2017) reported the long-term fortunes of large numbers of individual ash in Lady Park Wood in the Forest of Dean that became established before and shortly after tree felling in the Second World War. Recruitment stopped around 1950 – the ash developed into crowded stands of saplings, all of similar age. By the time young trees had reached about 32 years old in 1977, 95% of those recorded in 1948 had died. Size in 1977 was an important factor in determining survival to 2013; almost all trees survived if they were of more than 13 cm diameter at breast height, whereas most of those more slender than 7 cm were dead by 2013.

14.2.1   *Recovery in ash regeneration*

In July 2005, I accompanied Ed Mountford to walk along the woodland transect in Monks Wood where George Peterken had noticed the disappearance of young ash in 1992. Superficial examination failed to reveal any ash seedlings that were becoming established. In October 2005, I searched parts of the wood for ash in the height range 20–130 cm. These were young trees that were through the vulnerable stage after germination and were just becoming established. The search included sections of three of the Backmeroff transects. The only young ash saplings seen were inside the large coppice exclosure. Patches of small ash seedlings less than 20 cm in height could be readily found, as could clusters of young trees ranging in height from 2 m up to more than 10 m. On most of the larger ash, there were signs of fraying inflicted by muntjac many years before.

Later, this search was shortened and formalised as a walk along 2 km of rides within Monks Wood counting all unprotected ash of 20–130 cm that could be seen from the rides. This was repeated in late summer in 2006, 2010–2011 and 2013–2017. The first ones of that height were recorded in 2006 and numbers exceeded a thousand by 2015 (Table 14.2). The recovery was not entirely steady year on year in part because some ride-side ash were mown or flailed during management operations, and dead saplings were seen in 2017, having suffered from ash die-back. These data reveal that many years elapsed after stalking began in 1998 before there was any obvious recovery in tree regeneration: eight years before the first seedling was recorded taller than 20 cm and 13 years before any sapling reached 130 cm. It took five years for the first saplings reaching 20 cm to become higher than 130 cm, and then another three years to attain 300 cm.

This walk took place along major and minor rides in the wood. Sides of major rides were always likely to be the first places for ash to recover as disturbance has

Table 14.2   Number of young ash counted in different height classes on a 2 km route through Monks Wood, 2005–2017. No counts were done in 2007–2009 or in 2012. Stalking, which ultimately led to this recovery, began in 1998. Numbers affected by ash die-back are given in brackets.

Year	Number of ash (including number with die-back or dead)		
	20–130 cm	130–300 cm	Taller than 300 cm
2005	0	0	0
2006	20	0	0
2010	179	0	0
2011	663	2	0
2013	453	66	0
2014	791	71	3
2015	1,245	88 (3)	9
2016	1,040 (27)	115 (31)	11 (5)
2017	992 (82 + 38 dead)	233 (99 + 53 dead)	12 (5 + 5 dead)

remained relatively high, light levels are good and browsing pressure lower. In such locations, bramble and pendulous sedge may provide extra protection. That, however, is not necessarily where woodland managers want to see trees growing. Away from the rides in woodland blocks, seedlings are much less likely to survive and eventually mature because of the extra shade and browsing. Nevertheless, there has been partial recovery of seedlings within the Backmeroff transects which penetrate deep into the woodland. Andrew Tanentzap (personal communication) recorded these transects again in 2008 and found some evidence of a recovery in number of ash in the height range of 30–130 cm when compared with earlier data in Mountford and Peterken (1998). This recovery was consistent in both timing and extent with my ride-side observations. Within the transects, Andrew Tanentzap recorded a general recovery in species of woody seedlings and suckers in this height range. Blackthorn showed the greatest change, while hawthorn, dogwood and privet all showed the same pattern. These recoveries occurred although the canopy remained largely closed with canopy gaps changing little between 1996 and 2008 (Tanentzap et al. 2012b). Andrea Brunsendorf (2006) compared species distribution in Dick Steele's maps (Steele and Welch 1973) with her own quadrat survey in 2006 and reported more restricted distributions for dogwood, privet and honeysuckle.

In 2010, I examined 32 diverse locations across the wood and discovered that abundance of ash in the height range 20–130 cm was inversely correlated with a browsing index but not with canopy cover. Most ash saplings were at the low end of this range (i.e. 20–40 cm). Where browsing was sufficient to cause browse lines on the understorey, then ash saplings were only rarely present, but they could be relatively abundant where signs of browsing were absent or minimal. By 2010, young ash trees of 20–130 cm were becoming more obvious in the wood, especially in better-lit locations. Similar observations were made for the community of seedlings and saplings of other species of shrubs and trees, including hawthorn, blackthorn, privet, honeysuckle and dog rose. Several locations were revisited in the south of the wood in 2015 and average height of the tallest young ash in each had increased from 44 cm in 2010 to 110 cm.

14.2.2 *Planted ash*

In the winter of 2005/6, just before the first definite signs of recovery of the species, English Nature staff made the decision to plant ash saplings about 50 cm in height in various localities in the south and centre of the wood. This exercise was undertaken as a trial and also as an attempt to boost regeneration. Most of the saplings were protected with plastic rabbit guards. While the height of these guards gave little or no protection against deer, they did render the saplings easier for me to find, so their survival and growth could be monitored. The planted trees generally fared badly. Outside the deer exclosures, only 20% were still alive by 2009 – they had not been helped by the woodland shade, by particularly dry conditions during 2006 or by the attentions of deer. Average height of survivors decreased slightly from 56 cm in 2006 to 50 cm in 2009. Field maple saplings planted at the same time survived better than the ash but grew slowly.

These ash plantings clearly failed, but of greater interest was the group numbering about 30 planted inside the 6 ha coppice exclosure. Deer activity was low inside the exclosure in 2006 and 2007 and the canopy was patchy. The average height of the planted ash increased to 109 cm by 2009. By then, however, deer activity was increasing (Section 13.3.2), and each year from 2009 until 2015 the percentage of saplings with signs of recent browsing rose progressively from 10% in 2009 to 71% in 2015. Most failed to gain height because of browsing and stem breakage. By 2015, the exclosure paradoxically held the highest deer density in the wood. Roughly half of the young trees planted in 2005/6 survived until 2015. But by this time, only five saplings were well established and in the height range 150–250 cm. All of the others were browsed and/or looked unhealthy and were considered unlikely to flourish in the future. Although the contribution of the planted ash to woodland within the coppice exclosure was not great, they did in this situation provide a useful long-term browsing assay akin to annually repeated ivy trials (Section 14.4). In order to benefit regeneration, small ash would need planting inside a deer-proof exclosure where there was little or no tree canopy or other competing vegetation.

14.2.3   *Interactions with other threats to ash*

Planting of ash might become more important if resistant stock could be developed to help combat the effects of ash die-back. This is a disease caused by a fungus. It leads to loss of leaves, die-back of the crown and can result in death, with young trees being especially vulnerable. The disease was first reported in Britain in 2012 and quickly spread. It was first seen in Monks Wood in 2015, although from the extent of signs on some ash, it probably first occurred in 2014. I saw it on three saplings in the height range 130–300 cm during my count in August 2015 (Table 14.2). Since then it has spread through the wood: by 2016, the best-established planted ash inside the coppice fence all showed signs of die-back; by 2017, a number of trees were dead, including some larger than I record.

How the wood will look in the future is of course uncertain. Few mature ash trees will presumably be left and the canopy will be thinned, but thickets of saplings may develop where equilibrium is reached between germination and die-back rather as happened after Dutch elm disease struck these shores in the 1970s. In addition, as Richard Broughton has pointed out to me, some oaks are showing the symptoms of acute oak decline, which may create even more gaps in the not-too-distant future. The species that will benefit, if there is no intervention, can be tentatively predicted on the basis of what happened when regenerating coppice was destroyed by browsing. If trees grow in place of ash, they may tend to be less palatable species such as aspen, birch or alder, although with the recovery from browsing of other species this is probably less certain than it would have been in the 1990s. Blackthorn thickets could develop, and where trees do not grow, then bramble may flourish in drier conditions with pendulous sedge being dominant where it is wetter.

It seems ironic that so soon after the start of ash's recovery from predation by muntjac, it faces other threats. Furthermore, waiting in the wings is the emerald ash borer *Agrilus planipennis*. This beetle was accidentally transported from the Far East to North America where it has devastated their ash trees and is likely to turn up in Britain eventually (Rackham 2014). Peterken and Mountford (2017) said of the situation in their long-term study site in Lady Park Wood in the Forest of Dean, 'We await developments with interest and apprehension!'

Ash trees may live for 200 years (Rackham 2003), so prior to ash die-back it probably mattered little that a few decades of deer browsing had prevented any significant regeneration. Putman (1998a) commented that so long as there was some regeneration every year or a cohort managed to survive every 40 years or so, the future of a wood was assured. Monks Wood has already survived being clear-felled about 100 years ago – it will survive ash die-back, but its composition may alter dramatically. Deer damage to ash regeneration has been worse in Monks Wood than in any other local wood that I have visited. In some woods, regeneration has remained very satisfactory, at least beside rides. Ash die-back is likely to change all that and affect woods on a broad scale.

14.3   Bramble

Deer relish most species of bramble and can affect its abundance (e.g. Putman et al. 1989; Morecroft et al. 2001). Some rarer forms may be avoided (Chapman 1997b). A study in the King's Forest in Suffolk in the 1980s found that *Rubus* species formed a high proportion of the diet of muntjac throughout the year (Harris and Forde 1986). A year-long camera trap study of browsing on a single bramble bush at Woodwalton Fen revealed that muntjac browsed there fairly consistently at times of year when foliage was available (Section 19.3.3). Other camera trap studies at Woodwalton Fen showed that in winter muntjac returned repeatedly to the same bushes until they had consumed all of the leaves within reach.

Muntjac at a high density produce distinctive browse lines on mature bushes, particularly in late winter. Larger species of deer are usually able to pick off the higher leaves and also break up the thickets. So a browse line at a height of about 1 m around a bramble thicket can usually be taken to indicate the presence of a significant population of muntjac. Figure 14.7 shows a bar chart summarising observations made on 120 visits to 73 woods in eastern England, when browse lines at a height of about 1 m were described as absent, marked or 'intermediate'. As there were increases in muntjac deer score, a measure of their density, so the chances increased of finding browse lines on the bramble. Browse lines are most likely to be apparent in spring at the end of winter foraging.

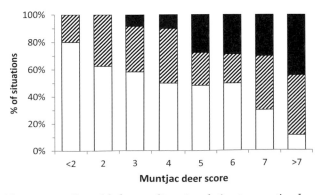

Figure 14.7   The nature of bramble browse lines in relation to muntjac deer score during 120 visits to 73 woods in eastern England (updating Cooke 2006): no browse line (white bars); some evidence of a browse line (hatched bars); marked browse line (black bars).

14.3.1   *Bramble at Monks Wood*

Muntjac will make tunnels through bramble thickets, and, despite their small size, heavy and prolonged browsing can lead to bramble die-back and eventually loss of thickets. In Monks Wood, bramble dominated the ground layer of many compartments in the early 1970s (Steele and Welch 1973), but by the early 1990s, few mature bushes remained and these were suffering die-back (Cooke and Farrell 2001b). Wells (1994) reported an increase in bramble in coppice plots, but Kirby (2005) suggested that this may have been a temporary phase during which the reduction in shading caused by the impact of deer on coppice regrowth outweighed direct browsing on the bramble. Crampton et al. (1998) used a quadrat method to compare vegetation in Monks Wood in the mid-1960s and in 1996 and found that, while bramble abundance had changed little, only small plants and seedlings occurred in 1996. However, exclosures erected in 1978 were still dominated by bramble in 1994 (Cooke et al. 1995).

Bramble also grew well inside exclosures constructed in coppice in 1993, but not in unfenced control plots where it was browsed (Cooke and Farrell 2001b; Cooke 2006). Because small seedlings remained abundant, there was potential for recovery providing the deer population could be reduced. After stalking started in 1998, bramble in the control plots established in 1993 grew faster until 2004, by which time shading from the coppice canopy resulted in die-back.

Bramble height was measured from 2000 onwards in a number of transects in the wood, with recording finishing in 2015. In three transects where growth was continuous over at least ten years, average increases in height were in the range of 7–19 cm per annum, depending on the extent of canopy cover. By 2005, some thickets were beginning to re-form where conditions were suitable.

In 2010, I recorded bramble abundance and maximum height in 32 localities in the wood. Of these, 26 had been photographed in the past and they were being checked for signs of regeneration. Six new areas were included where regeneration was apparent, such as where trees had fallen. At this stage in its recovery, the abundance of bramble was found to be influenced by browsing but not by canopy cover; but height of bramble was affected by canopy cover but not by amount of browsing. Bramble was absent from only two sites – at both, the canopy was dense and there was a general browse line on other woody species. Four of these 32 localities were studied again in 2015. Bramble continued to grow and increase in abundance. Canopy cover changed little during this time.

In the exclosures set up in the wood in 2004, there was no significant difference by 2010 in either bramble height or Domin score between them and their control plots. These plots were located beside wide, major rides where light levels were good and disturbance relatively high. So, even in the control plots in such situations, conditions will have been conducive to bramble growth. After six years, any browsing impact was no worse than the effect of a denser coppice canopy in the exclosures. It was in such locations along the edges of major rides that thicket formation in the wood was most obvious by 2015.

14.3.2   *Bramble at Holme Fen*

Muntjac had a considerable impact on bramble in Holme Fen. In the 1980s, I accompanied Tony Bell of ITE searching for nests of the sparrowhawk *Accipiter nisus* at Holme Fen as part of a national scheme on the effect of pollutants on this species. Traversing the woodland compartments looking for nests was made more difficult and uncomfortable by the abundant bramble. In the winter of 1987/8, I began surveillance of the deer population and muntjac built up steadily through the 1990s (Figure 9.2). The first time I recorded heavy browsing on bramble was in March 1996, when some die-back was also noticed. There was a gap in surveillance between the winters of 1999/2000 and 2002/3, but on an isolated visit in 2002, I noted bramble as being still abundant with no fresh die-back. However, when regular deer surveillance began again, heavy browsing with significant die-back was seen in places, and I was easily able to walk through the woodland. By spring 2004, most bramble was defoliated or dead and it was possible to see 100 m into the woodland from some rides (Figures 12.2c and 16.9a). English Nature staff were concerned at the impact muntjac appeared to be having, and culling began in May 2005.

Average bramble height in six transects at Holme Fen was monitored from 2005 until 2014 (Figures 14.8 and 16.9). It was three years before regeneration really began, but then it progressed fairly steadily apart from in 2012, when no growth was recorded – that spring was cold and winter gales had brought down trees in several plots and affected bramble height. Subsequently, growth rates recovered and showed little sign of stopping in any of the six transects. By 2014, bramble was higher than 1 m in 47% of the plots, thereby being above the usual browse line of muntjac.

In order to test whether muntjac were still having some impact on bramble growth during the middle part of this period, three small plants were individually caged in September 2008 and their growth rate was compared with unfenced controls (for an illustration of similar cages, see Figure 12.1). This was undertaken in the centre of a woodland block where there was a thin birch canopy and light levels were judged to be more than adequate for thicket formation. There was considerable bramble die-back but seedlings and small plants were abundant, although many were browsed. A significant difference in growth duly resulted with the caged brambles being more than twice the height of the controls by 2011.

By 2014, muntjac were creating 'micro-glades' in the sea of new bramble. This was achieved by making small openings and maintaining them via heavy browsing

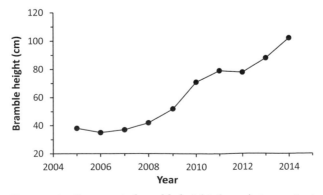

Figure 14.8   Recovery in bramble height along six transects at Holme Fen following the introduction of stalking in 2005.

Figure 14.9   A micro-glade in a bramble area at Holme Fen, September 2014. Muntjac had browsed all bramble in a patch, killing some of it and creating an opening.

of any bramble growing in the centre or trying to reinvade from the edge (Figure 14.9). In this way, deer were helping to diversify what would otherwise have been even denser bramble thickets. This activity may have benefited other wildlife, such as some species of birds. The situation proved to be unstable, however, and a resurgent muntjac population had a more significant impact on the bramble (Section 13.3.1).

Recording varying degrees of dominance by bramble throws up the general question – how much bramble is optimal on any particular site? Bramble can have a smothering effect on other, perhaps more interesting, flora (Tabor 2006). Kirby (2001b) described an experiment at two East Anglian woods where plots cleared of bramble were floristically richer two years later. He considered, though, that it was unlikely that any species would be totally lost from a wood because of bramble. Bramble is a species utilised by many species of wildlife, but too much may be detrimental for biodiversity. It is a case of striking the right balance. A friend who is a specialist on fungi recently complained to me that it is now more difficult to find fungi in Brampton Wood because the increase in bramble renders access to the woodland blocks more difficult. In addition, more bramble and other woody growth may mean reduced soil moisture, which could affect fungal fruiting. It was clear to my friend that removal of conifers in some parts of the wood was leading to a proliferation of bramble but it was not appreciated that deer control for conservation reasons was having a similar effect.

14.4   Ivy

Monitoring grazing on ground flora is often a straightforward process of selecting a plot and recording whether inflorescences or leaves are grazed. Monitoring browsing on the leaves of woody vegetation is not as simple since it can be more difficult to quantify browsing damage, and evidence from previous years may still be visible and liable to complicate the picture. Devising a method involving ivy (Cooke

2001; Section 12.6.3) has provided an index enabling browsing to be quantified and monitored in a regular fashion, and to some extent yield information on wider browsing impact. An ability to quantify level of browsing on a palatable food source means that it is possible to compare ivy taken (1) in the same wood over a period of years and (2) in different parts of the same wood (Cooke 2006). I have also found that, across a sample of 12 woods in eastern England without fallow deer, the amount of ivy browsed after one day was positively related to muntjac deer score assessed in the same year.

All species of deer seem to like ivy. It was an important part of the diet of muntjac from December through to April in the King's Forest in Suffolk (Harris and Forde 1986). Ivy growing up trees in woods often has a browse line, the height of which may reveal the identity of the species responsible; the browse line for muntjac is at about 1 m (Figure 2.6).

Ivy trials were undertaken in the southern part of Monks Wood during March each year, 1995–1997 and 1999–2008, with the same five locations being used. After one and seven days, each stem in a group of 20 was assessed as untouched, bitten or defoliated. Figure 14.10 shows the percentage of stems browsed (bitten + defoliated) after one day and defoliated after seven days. There can be considerable variability in results from one year to the next, so results are better viewed in the longer term. Browsing after one day and defoliation after seven days both decreased after stalking began. The high values in 2001 were associated with lack of stalking and other disturbance during the period of foot-and-mouth restrictions. Groups of ivy were situated close to rides and in 2001 there was more deer activity in such locations. As only five groups of ivy were used each time, the amount of browsing or defoliation was sensitive to changes in the number of groups that were discovered by deer.

The relationships between ivy taken and estimated muntjac density are shown for Monks Wood in Figure 14.11. There was a significant relationship between amount of defoliation and estimated muntjac density in the wood for the ten years

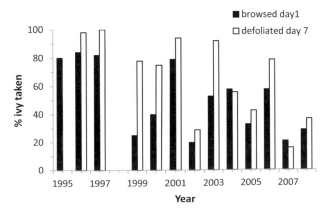

Figure 14.10   Ivy trials at Monks Wood, 1995–2008, using five groups of 20 stems (updating Cooke 2006); defoliation after seven days was not recorded in 1995 and no trial was done in 1998. Percentage of ivy stems browsed after one day is depicted by black bars and percentage defoliated after seven days by open bars. Stalking began in 1998.

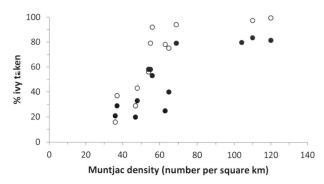

Figure 14.11   Ivy trials at Monks Wood, 1996–1997 and 1999–2008: percentage browsed after one day (black dots) and defoliated after seven days (open circles) in relation to estimated deer density.

of stalking, 1999–2008, and the comparable relationship for ivy browsed after one day approached significance. Data indicated that, below a density of 70 muntjac per km², amounts of ivy taken decreased quite rapidly, especially as regards defoliation after seven days. Above that density, however, amounts taken changed little. This relationship may differ from wood to wood depending on factors such as the availability of natural food and whether the deer population is increasing or declining. The nature of the relationship implied that there may be limitations to the information that ivy trials can impart as ivy taken changed over only a small range of densities. Nevertheless, those densities are frequently encountered in deciduous woods within the main range of muntjac in this country.

CHAPTER 15

Introduction to the impacts of muntjac grazing

15.1 Background

The concept and value of ancient woodland have already been introduced (Section 13.1.2). Ancient woodland in different regions of the country is characterised by different plant species, termed 'ancient woodland indicators' (e.g. see Peterken 1974; Rackham 2003). Among the ground flora, indicator species include bluebell, wood anemone *Anemone nemorosa*, primrose, oxlip and early-purple orchid, all of which are much appreciated by visitors to woods in springtime. Unfortunately, they can all be grazed and detrimentally affected by muntjac. Other perhaps less well appreciated ancient woodland indicators, such as dog's mercury and enchanter's nightshade *Circaea lutetiana*, may be similarly affected. A few indicator species, including ramsons *Allium ursinum*, pendulous sedge and spurge-laurel *Daphne laureola*, are not eaten by muntjac and can benefit from reduced competition. The overall grazing impact of a significant density of muntjac in ancient woodland will, however, almost certainly be negative.

Harris and Forde (1986) reported how the diet of muntjac in the King's Forest in Suffolk varied through the 12 months of the year. Although this was a coniferous forest, results were of some relevance to deciduous woodland because older broad-leaved shelter belts and coverts occurred (Figure 4.2a) and coniferous foliage comprised a relatively minor part of the diet. The work involved examining, identifying and measuring the size of plant fragments in faecal pellets. Ferns were found to be mainly taken in the winter; grass was grazed mostly in April and May when it was growing rapidly; and the peak time for eating herbs was from April until September. However, these observations offer only general guidelines. Jackson et al. (1977) examined muntjac from an estate in Knebworth, Hertfordshire. This was an area with mixed woodland, including coppiced hornbeam *Carpinus betulus*, and dog's mercury, bluebell, bramble and bracken were very common. Most of the deer were shot in February 1967, and all faecal samples contained bramble and herbaceous species, principally bluebell and dog's mercury.

Sometimes muntjac have been found to eat a species in one situation but not in another, and Chapman et al. (1994b) discussed possible reasons for such contradictions. First, muntjac can be individualistic with one deer being attracted to a plant while another animal will avoid it. When I see a widespread plant species in a wood only being eaten in one locality, I wonder whether it is the work of one individual deer. Secondly, poisonous plants may vary seasonally or locally in their level of toxicity, so one study may find a species is grazed while another study at a different season or in a different place may conclude it is left alone. Muntjac, however, often eat plants that other mammals (and maybe other deer) find distasteful. Finally, and perhaps most importantly, if muntjac are starving they may eat plants that they would otherwise ignore – in other words, virtually everything (Figure 15.1).

The studies on faeces mentioned above will have been undertaken in situations where moderate densities of muntjac were living in reasonable balance with the environment and not significantly reducing food availability. In woodland where muntjac are at sufficient density to have depleted their food resource going into winter, it is likely that by February all reachable bramble and ivy leaves will have been stripped off. Bluebell leaves are showing above ground by then and are evidently attractive to a starving muntjac. Honeysuckle leaves also appear early in the year, but the plants may have permanent browse lines because of browsing in previous years. Herbs that start flowering in March, such as primrose, wood anemone and lesser celandine *Ficaria verna*, are at particular risk. Those beginning to flower in April are also at risk, for example bluebell and early-purple orchid. Flowering of oxlips peaks in April. There is something about this species that makes it especially attractive to muntjac, possibly because it can occur in high numbers and densities and each inflorescence provides a highly palatable mouthful. New leaves on hawthorn, blackthorn, bramble and privet are not available in any abundance until April. And new leaves on established coppice regrowth also start to show in

Figure 15.1   An immature muntjac buck grazing in
winter on stinging nettle *Urtica dioica* (MB).

April. But if a wood has been heavily impacted, there may be few leaves on woody vegetation below the browse line. Newly coppiced stools should produce shoots as food for muntjac, but this process does not begin until May. The grazing risk to the general herb community in a broadleaved wood with a dense population of muntjac is probably at its peak during March and April.

Dog's mercury forms new shoots in the autumn that expand in February–March and reach peak height around the end of May (Grime et al. 1988). In spring, it is largely ignored by muntjac despite the findings of Jackson et al. (1977) referred to above. Dog's mercury is poisonous to stock animals (Cooper and Johnson 1984; Grime et al. 1988), but not apparently to muntjac. From summer onwards, muntjac may graze it severely, especially if it is a poor autumn for fruit and nut crops. It may be the only herbaceous species where heavily grazed old stems and pristine new ones can be seen side by side in late winter or early spring. Wood ferns (*Dryopteris* species) are similar in that they grow through the summer and may be grazed from then until the fronds die in winter.

15.2   Observations of grazing and its implications for woodland plant populations

In Monks Wood, grazing pressure probably increased as the muntjac population built up through the early 1980s. But even when the population stabilised in the late 1980s, the level of grazing might have continued to rise as food resources in the wood became depleted. No one appreciated the potential significance of muntjac grazing at this time, so it was not being deliberately studied and only limited information can be gleaned from other research. Thus, no notable changes were seen in ground flora outside five exclosures in the west of the wood in the years immediately after their erection in 1978 (Cooke et al. 1995), so grazing pressure cannot have been serious around 1980.

In 1982, Jack Dempster began a study in the wood on the population dynamics of the orange-tip butterfly *Anthocharis cardamines* (Dempster 1997 and personal communication), and, fortuitously, this included putting out pots containing the larval food plant, cuckooflower *Cardamine pratensis*. Muntjac ate considerable numbers of his plants, which was annoying for him, but his observations provided comparative information on grazing levels at that time. It was an unintentional bioassay and a forerunner of my own ivy trials. Jack complained that he felt that muntjac were watching and licking their lips every time he put out the plants. Overall, there was a significant rise in grazing levels, 1982–1993. Grazing on cuckooflower was less than 7% between 1982 and 1985, but then rose to 14% in 1988; after then, it varied between 10% and 13% apart from a dip to 7% in 1991 (Cooke 1994). These observations were consistent with increasing grazing pressure during the 1980s, but with a decrease in 1991 following the die-off of muntjac in the early months of that year (Section 8.2.2).

By the early 1990s, it was noticeable that coppice plots and rides in Monks Wood were less flower-rich in spring and summer (Wells 1994). In particular, the carpets of primrose, bluebells and wood anemone which had been so obvious in the two years following coppicing had ceased to be a feature of the wood. Similar losses were reported in woods where the oxlip was a speciality (e.g. Rackham 1975;

Figure 15.2   Grazed lords-and-ladies in Monks Wood in 1996.

Tabor 2002). Various reasons were proposed, including lack of or the wrong sort of management, changes in climate and pollution, but these authors all agreed that deer were implicated in the process. Not only did deer eat the inflorescences, but they also browsed the coppice regrowth, letting in light and promoting growth of more vigorous species that out-competed the vernal herbs. The oxlip is a very particular species, relishing wetness but requiring a moderate amount of shade – too much shade and it performs poorly, too little and it becomes out-competed.

Lords-and-ladies *Arum maculatum* is a toxic spring plant. Native mammals have only rarely been recorded grazing it (Diaz and Burton 1996), but slugs and snails will eat the plant (Lynne Farrell, personal communication). In 1995, Anita Diaz visited Monks Wood to study the reproductive ecology of this species, but when she realised at least half of the plants were being grazed she decided to investigate the effects of muntjac instead (Figure 15.2). Similar levels of grazing were found in Riddy Wood in Cambridgeshire and Wigney Wood in Bedfordshire (Diaz and Burton 1996). Oliver (2013) later reported that muntjac relish lords-and-ladies, being especially partial to the sheaths (which are termed spathes). The inflorescence of lords-and-ladies is an elaborate trap to enlist the services of specific pollinating insects, and the trap mechanism does not work if the plant is grazed. By comparing the most affected sites with others where there were lower or negligible levels of grazing, Diaz and Burton (1996) discovered that the ratio of seedlings to mature plants was much lower, demonstrating that reproduction and demography were affected. Grazing often occurred as the new inflorescences emerged from the leaf sheaths in early spring, and it was conjectured that the plant's chemical defences may be less at that time of year and that later in the season alternative food resources were more available. These effects at Monks Wood were seen when muntjac density was very high and food resources very low. In 2015, I studied grazing levels on inflorescences in Monks Wood to compare with the results from 1995 when the average rate was 62% (Diaz and Burton 1996). I found an average rate of 41%, so although a decrease occurred over the intervening 20 years, it was relatively slight. Inflorescences were apparently being actively sought out as some were well hidden in some better-lit areas besides rides. Because I had occasionally seen leaf grazing or plants dug up and left on the soil surface in nearby Raveley Wood, I assessed grazing rate there too – it was 43%. Neither of these woods had a particularly high deer density in 2015, which suggested that such levels of grazing may be widespread. Lords-and-ladies did, however, still seem abundant in both woods.

Dog's mercury and lords-and-ladies are by no means the only poisonous species taken by muntjac. Oliver (2013) mentioned lesser celandine, stinking iris *Iris foetidissima*, and woody species such as cherry laurel *Prunus laurocerasus* and holly *Ilex aquifolium*, and discussed how the deer managed to survive. Having a battery of specific detoxifying enzymes or bacteria was thought unlikely because they expose themselves to just too many different poisons. Their defence may be that they eat sufficient other vegetation to dilute the poisons to tolerable levels, and perhaps tolerance is built up as they take in these toxic foods.

Grazing by deer has also been implicated in the loss of wood anemones from Rockingham and Hatfield forests (Rackham 2006) and Shadwell Wood in Essex (Tabor 2006). Mårell et al. (2009) studied grazing by roe deer on wood anemones in a deciduous forest in France. Grazing could be serious with up to 80% of flower shoots taken locally by roe with smaller losses to small mammals and invertebrates. It was suggested that deer grazing 'may play a crucial role in the population dynamics of wood anemone'. By the early 1990s, anemones no longer flowered in Monks Wood. Wells (1994) drew attention to the loss of abundant populations of anemones in recently coppiced areas. However, I had not realised they had gone completely until they began growing in 1996 in one of my study exclosures erected in 1993. During the late 1990s, these were the only anemones recorded in the wood. By 2003, they were in five of the eight exclosures, but in none of the eight unfenced control plots. While undertaking his ornithological work, Richard Broughton noticed two wood anemones flowering along a ride in 2005, seven years after stalking began; and the following year, he reported scattered flowers in several places. I saw my first decent drift of anemones in 2010. Since then they have continued to spread and can again be seen flowering in profusion in some parts of the wood (Figure 18.3).

Lesser celandine is a species that is likely to be more overlooked than even wood anemone. It is grazed by muntjac and probably decreased in Monks Wood, but no one has any hard information. Looking back through my notes on primroses, I unearthed a reference to seeing 14 flowering celandines during 50 minutes spent searching for primroses inside the south-west fence on 4 April 2003. That was clearly an unusual occurrence at the time, but now the celandine is abundant in several areas of the wood. Both anemones and celandines have thin stalks, and grazing will only be noticed if it is looked for specifically. Lack of information on grazing on such species does not mean they escaped the attention of muntjac.

Enchanter's nightshade is yet another species that declined unnoticed in Monks Wood, being found in 27% of woodland plots in the early 1970s but in only 4% in 2006 (Brunsendorf 2006). Deer grazing was again likely to be implicated as the species declined in Wytham Woods, Oxfordshire, prior to deer control starting (Kirby and Morecroft 2010).

15.2.1 *Difficulties in studying grazing impacts*
Being able to prove that grazing has a definite effect can be far from straightforward. In 1999, Oliver Rackham succinctly summed up the main difficulty with working

on long-term ecological issues. He stated that by the time long-term problems have been noticed, it is often too late to begin making observations. He lamented the observations he would have started in the 1960s had he been forewarned of what problems were to become apparent by the 1990s.

Even if studies are undertaken, they will not necessarily prove whether muntjac have caused a problem – or, conversely, indirectly led to a plant population increasing in size. The rare ancient woodland indicator, crested cow-wheat *Melampyrum cristatum*, grows in a small area on the southern edge of Monks Wood, where it has been managed and monitored for many years (Wells 1994; Hughes 2005). Numbers were in the low hundreds from 1968 until the late 1970s, but then a hedge was cut down resulting in massive disturbance to the ground. The population numbered in the thousands in 1980 and 1981, apparently because the seed bank had been disturbed and germination was stimulated by reduced shade. The hedge was cut back and the ground disturbed by raking each winter, but the population declined to only 11 in 1993. That year I checked to see whether the plants were grazed but found no evidence (Cooke 1994). Numbers remained low until 1997, in part perhaps because management was carried out too late in the winter to stimulate germination. When management was brought forward to late autumn, numbers recovered until, by 2003, they were too numerous and tangled to count. In 2014, in response to a query from Peter Stroh of the Botanical Society of Britain and Ireland about whether muntjac grazed crested cow-wheat, I decided to check again and found just 0.4% of spikes to be grazed. I found two locations where plants were not grazed despite being intertwined with bramble whose leaves had been browsed by muntjac (Figure 15.3). While I had no evidence of significant levels of grazing in two summers separated by 21 years, the situations I had studied were: (1) when the deer population was extremely high but the cow-wheat very rare; and (2) when the cow-wheat was abundant but grazing pressure was low. So I do not know what happened when, in the mid-1980s, the muntjac density was high and the

Figure 15.3   Untouched crested cow-wheat, whereas the leading stem and a side shoot of the bramble have been browsed, Monks Wood, 2014.

cow-wheat was abundant but starting to decline. While variations in management can explain most of the changes in the plant's population, I cannot totally rule out the involvement of grazing by muntjac in the decline in the late 1980s.

Deer grazing on herbs is not simply a case of causing an annoying reduction in the number of flowers a visitor might see – if sufficiently severe, it can lead by a variety of processes to plants becoming smaller, reproduction failing and range being more restricted. So can grazing wipe out a species in a wood? The potential certainly exists but, even in a wood as well studied as Monks Wood, it can be impossible to prove cause and effect, especially if it is an event that happened in the past. The eventual impact of muntjac was totally unexpected during the 1980s, and by the time that problems were starting to be appreciated in the early 1990s, some species had already disappeared. Thus herb-Paris *Paris quadrifolia* declined from about 100 plants in 1972 and 1973 to zero in 1988, and has not been seen since (Wells 1994). I have seen herb-Paris grazed by deer elsewhere in the south of England, and its disappearance from Monks Wood occurred just after the muntjac population peaked. The two events may be connected but there is no way of proving it now. Oliver Rackham (2006) suspected the involvement of deer grazing in declines of this species in other woods, and Keith Kirby has told me that herb-Paris has increased in abundance in Wytham Woods, Oxfordshire, since fallow deer and muntjac have been controlled.

Climbing corydalis *Ceratocapnos claviculata* is a scarce plant of heaths and ancient woods on acid soils (Mabey 1996). It is a rare ancient woodland indicator, being described by Wells (2003) as being plentiful in 1997 in one area of Holme Fen, but nowhere else locally. Since then it has spread at Holme Fen and I have monitored its increase into and along several transects. The deer did not graze the corydalis and initially I thought it was probably responding to loss of bramble through muntjac browsing; however, as time progressed, bramble and ferns recovered but the corydalis continued to spread. It was noticeable that the areas being colonised by corydalis were those parts comparatively less browsed by muntjac. This was an example of a species of ground flora dramatically increasing in abundance in a site affected by muntjac, but it was not clear whether or how the two events were connected.

15.2.3   *Plants that appear to have benefited indirectly*
Seemingly clearer examples of plants benefiting from not being grazed are spurge-laurel and ground ivy *Glechoma hederacea* (Cooke et al. 1995; Crampton et al. 1998; van Gaasbeek et al. 2000; Kirby 2005). Ground ivy is toxic to stock (Cooper and Johnson 1984) and has a characteristic and strongly pungent smell, presumably due to the volatile oils that it contains (Oliver 2013). It increased from the 1960s or 1970s to the 1990s in Monks Wood and also in Wytham Woods, a site similarly affected by deer. In Monks Wood, muntjac tended to avoid areas of woodland where the ground was dominated by ground ivy (Cooke and Farrell 2001b; Figure 15.4). The difference was most marked in July and August when 3–4 times fewer deer were counted in plots where ground ivy was dominant.

Species of vegetation may benefit and increase if they are unpalatable to deer or if their tolerance of grazing affords them an advantage over grazing-sensitive species. Such species may increase further from soil enrichment by deer dung.

Many of the species to benefit are grasses and sedges, with the following being reported to increase in Monks Wood into the 1990s: wood small-reed *Calamagrostis epigejos*, wood false-brome *Brachypodium sylvaticum*, tufted hair-grass *Deschampsia cespitosa*, rough meadow grass *Poa trivialis*, pendulous sedge and pale sedge *Carex pallescens*.

Wells (1994) first drew attention to the fact that rides in Monks Wood had become much grassier in the previous ten years, and less floriferous in early and midsummer. *B. sylvaticum* had increased considerably, particularly where rides had been shaded by overhanging vegetation. Moreover, this species became much more abundant within woodland blocks by the mid-1990s (Cooke et al. 1995; Crampton et al. 1998); Kirby (2005) recorded a similar change in Wytham Woods. Some of the evidence for the involvement of muntjac in this process came from exclosures in Monks Wood erected by the University of East Anglia in 1978 in an area then dominated by dog's mercury (Cooke 2006). By 1994, the zone around the exclosures had become dominated by grasses, including *B. sylvaticum*, but this species was absent from inside. Observations were repeated in 2005, by which time several bramble-dominated exclosures had been destroyed by falling branches. The species increased significantly in these destroyed exclosures and became even more abundant in the control plots, but remained absent from the still-functioning exclosures. The fallen branches had not appreciably opened the canopy over the destroyed exclosures so a change in the amount of shade was unlikely to have stimulated grass growth. Willi and Sparks (2003) reported that, in 2002, Monks Wood had considerably greater cover (23%) of *B. sylvaticum* than 19 other local woods (range 0–9%), which would seem to discount a significant effect from a more broad-brush factor such as air pollution or climate change. Moreover, Monks Wood's large size made it less likely to suffer from agricultural eutrophication. The evidence points to deer activity driving the increase in this grass species in Monks Wood and elsewhere, such as in Wytham and Roudsea Woods (Kirby 2001a; 2001b). Brunsendorf (2006) confirmed that it grew much more extensively in Monks Wood in 2006 than in 1973.

Figure 15.4   A woodland floor dominated by ground ivy.

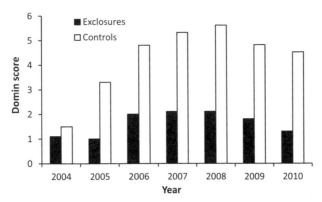

Figure 15.5   Average Domin scores for pendulous sedge in eight exclosures (black bars) and their respective control plots (white bars) erected in cleared areas in Monks Wood in 2004. Differences occurred despite stalking starting in 1998.

In 2004, new exclosures were erected in Monks Wood in the cleared edges of woodland blocks. *B. sylvaticum* was initially the most abundant species of ground cover, but decreased over the next six years inside the exclosures and their control plots. In the control plots, coppice regrowth was affected by browsing – and bramble and pendulous sedge increased. *B. sylvaticum* may have been squeezed out by competition. Within the exclosures, it failed perhaps due to the smothering effect of bramble and the coppice canopy, despite being shade tolerant, similar to the reasons for its absence from the 1978 exclosures, as described above.

The pendulous sedge story in Monks Wood is similar, although this is a species that performs best in unshaded situations (Rackham 2003). It increased markedly between 1966 and 1996 (Crampton et al. 1998), especially in damp areas where coppice regrowth was severely browsed by muntjac, so preventing canopy development (Cooke and Farrell 2001b). In control plots outside exclosures erected in coppice coupes in 1993, it had increased significantly by 2003 (Cooke 2006). In the exclosures erected in 2004 in cleared areas, pendulous sedge changed in abundance relatively little during the following six years under the coppice canopy (Figure 15.5). However, in control plots, it steadily increased up until 2008 because muntjac browsing inhibited canopy formation. By 2008, it was significantly more abundant outside the exclosures (e.g. Figure 18.1). The subsequent decrease was probably due to the proliferation of bramble. So, in such situations, the indirect influence of muntjac browsing was apparent for longer on this species than in the case of *B. sylvaticum*. In 2002, Monks Wood had 9% cover of pendulous sedge whereas 19 other local woods varied between none and 5% (Willi and Sparks 2003). Spread of pendulous sedge in deer-damaged habitat elsewhere in East Anglia has seriously affected the oxlip (Tabor 2005).

This account has focused on the direct and indirect effects of grazing. There are other ways in which muntjac may affect ground flora, such as by trampling or by feeding in one place and dunging in another (Putman 1994; 1996; 1998a; Kirby 2001a), but such processes are likely to have relatively minor effects.

CHAPTER 16

Impacts on specific ground flora and recovery following deer management

16.1 Introduction

Information is now examined for some of the other plant species for which data sets are available over periods of years. Special emphasis is afforded to the bluebell because this is the species studied for longest in Monks Wood. I have shared a great interest in deer impacts in conservation woods with the late Ray Tabor of the Essex Wildlife Trust. In 2006, he published an article on plant species that might be used as indicators both to monitor and to predict damage. Among the species of relevance to grazing damage, he highlighted bluebell, which he pointed out was easy to monitor in a quantitative fashion as both leaves and flowers were eaten and plant size varies. Other species recommended for study were common spotted orchid, dog's mercury and oxlip – all three feature in the following account. He also included species that increased indirectly as a result of not being grazed: ground ivy, *B. sylvaticum* and pendulous sedge, which have all been discussed in the previous chapter. Recording such species in varying degrees of detail helps to determine the level of impact in a wood and will aid prediction of future trends depending on what may happen to muntjac density.

16.2 Bluebells

16.2.1 Introduction
The bluebell is threatened globally and more than half of the world's population occurs in Britain (Plantlife 2004). Peace and Gilmour (1949) found that picking bluebells did not affect inflorescences in subsequent years providing trampling on

leaves was minimal. Plant vigour was reduced if leaves were removed or damaged; and in subsequent years bluebells became smaller and some failed to flower. So, while grazing on flowers can immediately reduce their number, grazing the leaves may lead to smaller bluebells and fewer inflorescences in the future. Damage early in the season has the greatest effect (Blackman and Rutter 1954).

Bluebell foliage contains toxins but is grazed by cattle and sheep (Knight 1964; Grime et al. 1988), and I have seen small, grazed bluebells that had been severely eaten by rabbits. Small mammals also appear to graze leaves and inflorescences to some extent (Cooke 1997), as do gastropods (Lynne Farrell, personal communication). It has been known for many years that muntjac eat bluebells (Jackson et al. 1977), and in my local woods they start feeding on newly emerged bluebell leaves each February if little other green food is available (Figure 16.1a).

The first reference to herbivores, including deer, visibly affecting bluebells seems to be Colin Tubbs writing in 1986 about impact in the New Forest. He said the 'shimmering blue carpets … are confined in the Forest to stock-free parts of Inclosures'. He related the case of one Inclosure where the boundary fence was partially relocated in the late 1960s, thereby exposing some of the bluebells to

Figure 16.1 Examples of grazing: (a) by muntjac on bluebell inflorescences and leaves; (b) by muntjac on dog's mercury; (c) by muntjac on an oxlip inflorescence; (d) by small mammals on inflorescences of cowslip *Primula veris*.

the Forest's grazers, including fallow deer. Within two years the density of plants had been reduced by more than 30%, the density of inflorescences by 50–75% and individual plants had become smaller. In contrast, bluebells remained vigorously flowering in the area that was still protected.

There are more recent examples of fallow deer affecting bluebells. For instance, Short Wood in Northamptonshire was seriously impacted by fallow deer by 2002. Once the problem was corrected in 2003, I found the bluebells recovered in just two years. Also, such grazing was suggested as the reason for the relative rarity of bluebells in Hatfield Forest in Essex (Rackham 2006).

16.2.2   Bluebells in Monks Wood

Following acquisition of Monks Wood by the Nature Conservancy, Sale and Archibald (1957) undertook an initial botanical survey. The bluebell was abundant on the eastern side of the wood and in the north-west corner. However, when Dick Steele mapped the wood's ground vegetation (Steele and Welch 1973), it remained abundant in the west side but not in the east. This change predated the main colonisation of the wood by muntjac. The western side of the wood still contained the principal aggregations of bluebells in the 1990s.

In 1993, my study on bluebells began with guidance from Lynne Farrell, and focused on recording in ten fixed 0.5 m quadrats along each of two 20 m transects randomly located in the south-west corner of the wood (Figure 12.4a). One of these transects was initially within an electric fence. A third transect midway between the other two was included from 1994, and all three were recorded each spring until 2014. Each transect was in a different sub-compartment of the wood. The south-west corner of the wood was enclosed by an 11 ha fence in the autumn of 1999. However, the fencing per se made little difference to deer density and grazing pressure because deer often had access through holes – the stalking that began in 1998 was more important for recovery of ground flora in that area (Cooke 2006). Bluebells were also recorded in four transects outside the fence along the western edge in 1995, 2000 and 2002–2005.

Average numbers of intact and grazed inflorescences are shown in Figure 16.2 for each spring for bluebells in the south-west corner. Prior to stalking starting,

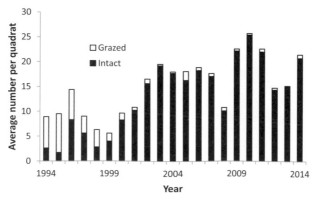

Figure 16.2   Bluebells in the south-west of Monks Wood: average numbers of intact and grazed inflorescences per 0.5 m quadrat. Stalking began in autumn 1998.

Figure 16.3   One of the bluebell transects in Monks Wood: (a) in May 1995 when there were few intact inflorescences; (b) in May 2007 after recovery.

numbers were low and grazing was severe. Bluebells will have been exposed to increased amounts of grazing in the 1980s as the muntjac population built up and peaked, and impact may have continued to increase as alternative food became less available (Section 15.2). After stalking started, grazing levels decreased and stayed low, while numbers of flowering bluebells increased. Visitors to the wood noticed the improvement in bluebells as early as 2001, and by 2003 bluebell density was clearly better in these stands. There was more than a tenfold increase in intact inflorescences between 1995 and 2003, and numbers were even higher in some later years (Figure 16.3).

Some inflorescences showed a different type of grazing damage, believed to be inflicted by small mammals. In such instances, inflorescences had been 'felled' by a small bite low down on the stem, with the flowers of the bitten inflorescence often being dismembered (see Figure 16.1d for an example of similar damage on cowslips). This contrasted with deer bites in which only the bitten stem remained. When grazing levels were low during the last five years of the study, 2010–2014, exactly 50% of recorded grazing was of this type. Grazing levels on leaves are shown in Figure 16.4; they began to fall as soon as stalking started, were low by 2004 and even lower by 2010. Rabbits were never seen in the part of the wood where transects were established, and roe deer failed to access the permanent fence until after the study finished.

Leaf length was used as an indicator of plant vigour. Leaves were found to lengthen at peak season by more than 1 cm per week so it was important to record bluebells at the same stage of development each year. It was also necessary to avoid automatically assuming bluebells with short leaves had been exposed previously to

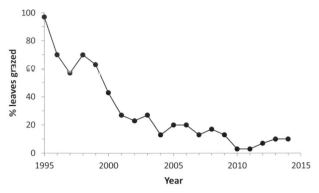

Figure 16.4   Bluebells in the south-west of Monks Wood: average percentage of grazed leaves. Stalking began in autumn 1998.

heavy grazing; some other form of abuse such as trampling, waterlogging or other adverse environmental conditions may produce the same effect (Rackham 1975). Leaf length did, however, increase during the Monks Wood study, with the shortest leaves in 1996 (average 16 cm) and the longest in 2005 (24 cm) associated with the reduction in deer density.

Results from four unfenced transects along the western edge of the wood were broadly similar. Percentages of leaves and inflorescences grazed decreased in 2000 (bluebells were not studied in 1999) and total inflorescences were lowest that year. Leaf grazing fell to 15% in 2005, and inflorescence grazing decreased to 2% in 2003. Numbers of intact inflorescences were lowest in 2000 (average 12 per 0.5 m quadrat) and highest in 2003 (29 per quadrat). Average leaf length varied from 17 cm in 1995 to 24 cm in 2004. Recording stopped along the western edge in 2005.

Sparks et al. (2005) reported that bluebells in Monks Wood in 1999 had shorter inflorescences and fewer flowers per inflorescence than in other local woods. Because of this work, I started measuring inflorescence height and number of 'bells' per stem in 2005 and, although our methods were different, found that the average number of bells had increased by 26% since 1999, but plant height had only increased by 7%. These measurements were repeated up until 2014, but neither changed significantly overall, 2005–2014, suggesting that recovery was essentially complete by 2005. For a summary of these effects on bluebells see Section 18.3.

These observations, though, were only part of the story – what happened to distribution in the wood? In 2002, their distribution in the western half of the wood was compared with Dick Steele's map from the early 1970s (Steele and Welch 1973; Cooke 2006). Bluebells did occur in the east of the wood in 2002, but nowhere were they dominant. While much of the bluebell's distribution in the west remained unchanged over the intervening 30 years, there were some modifications. Most notable was the way that the range contracted towards the western and north-western edges. There had been little change overall in the south-west corner of the wood, but elsewhere losses had occurred. The year 2002, however, saw stands of bluebells in a few localities where they had flowered only sparsely since the mid-1990s. Had the survey in 2002 been undertaken at the end of the 1990s, a greater reduction in range would have been indicated.

In 2005, bluebells were thinly scattered along the eastern edge of the wood. By 2010, there were good stands along that edge, and by 2015 bluebells were again magnificent in many of the areas in the east of the wood – where they had been recorded by Sale and Archibald (1957), but not by Steele and Welch (1973). Muntjac were not in the wood in the 1960s and had nothing to do with the earlier change in bluebell distribution. Crampton et al. (1998) found that the bluebell occurred in quadrats with similar frequency in 1996 as in the mid-1960s, so it must have had a quite restricted distribution in the 1960s too. But now that muntjac are controlled, bluebells are probably dominant in spring over a greater area of the wood than at any time in the last 60 years.

16.2.3   Relationships between bluebell grazing and deer density

Putman et al. (2011a) have argued that to assist management, critical threshold densities of deer should be established if possible, below which impacts are broadly tolerable. One difficulty with obtaining such information rests with determining deer density, especially where muntjac are concerned. Other issues are that different observers will have a range of views on what is 'broadly tolerable' and relationships between damage and density may vary between different sites. Nevertheless, the data set for bluebells and muntjac density at Monks Wood is probably the best and most relevant available today, and is worth examining.

Grazing on leaves leads to smaller bluebells and fewer flowers in the future, so impact on numbers of intact flowers is related to density a year or more before as well as to current density. However, grazing levels on inflorescences and leaves in any particular year are related more simply to deer density that year. Figure 16.5 shows for bluebells at Monks Wood how grazing levels on inflorescences and leaves changed with estimated muntjac density. These data were collected while bluebells recovered from severe grazing impacts. Because of differences in the availability of alternative food, the graph might look different had the study focused on the worsening situation starting in the early 1970s and going on into the early 1990s. We know that bluebells recovered well by 2003, five years after stalking started, but

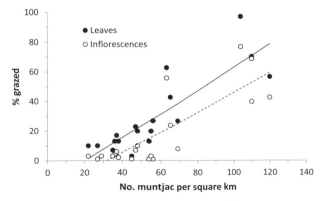

Figure 16.5   Bluebells at Monks Wood: the relationship between percentage of leaves grazed (black dots, solid line) and inflorescences grazed (white dots, broken line) and estimated muntjac density, 1995–2014 (updating and adapting Cooke 2006).

we have no information how quickly bluebells became damaged initially in the 1980s or early 1990s.

The broken line in Figure 16.5 indicates that there should be little or no grazing on inflorescences below a muntjac density of about 30 per km². There is, therefore, a threshold density at which grazing starts, and a higher threshold at which it becomes unacceptable. Just because a little grazing is noticed, it does not follow that it is unacceptable and requires urgent management. The situation becomes unacceptable for me at grazing levels of about 15% on inflorescences and 30% on leaves, which both correspond to a deer density in the range 50–60 per km². At these levels of impact, even relatively superficial monitoring should detect that there are grazing issues that warrant more careful consideration and monitoring. And as regards whether urgent action is needed, it should be remembered that, if there is an increasing deer population, the situation will probably worsen if nothing is done.

Deer scores of five and above in Monks Wood were usually associated with levels of grazing that were at least halfway to exceeding the thresholds of acceptability. This provides a means of determining the implications of muntjac grazing in bluebell woods over an area of countryside. In Cambridgeshire, the Wildlife Trust manages 19 woods that contain muntjac, but not fallow deer. Nine of these woods had maximum deer scores of at least five. I recorded the performance of bluebells in six of these nine: grazing thresholds were exceeded in three (Littless, Brampton and Raveley woods) and problems at all three were ameliorated by culling. In none were bluebells affected to the same degree as in Monks Wood. In the three other woods – Hayley, Lady's and Savages woods – the bluebells were relatively lightly grazed. At Lady's Wood, bluebells survived well at this small site without culling. Muntjac grazing on bluebells does not appear to have been such a widespread and serious problem as is sometimes stated or implied in the media.

The most affected of the Wildlife Trust's sites were Brampton and Littless woods. Brampton Wood is the third-largest ancient wood in the county (after Bedford Purlieus and Monks Wood), and was the first to be colonised by muntjac. In 1995, 8% of inflorescences and 47% of leaves were grazed (Cooke 1997; 2005). Despite stalking starting in 2000, grazing levels were 13% and 42% respectively in 2004. Since then, stalking has continued and the situation has improved. It is no longer possible to record in exactly the same locations because some have been scrubbed over and lost their dense stands, while alternative areas have been opened up and now host carpets of bluebells. Grazing rates had fallen by 2015 to 3% on inflorescences and 19% on leaves.

Littless Wood is in the nature reserve beside Grafham Water. It is part of a continuous block of woodland that extends to about 50 ha. Muntjac have been culled outside the reserve, but not before a damaging density occurred in the wood. An exclosure of 1–2 ha was erected in 2001 and this protected bluebells and other species of ground flora. However, muntjac developed a habit of grazing close to the fence which virtually eliminated bluebells and dog's mercury from within a few metres of the fence (Figure 16.6). Further afield in the wood, impact on bluebells in 2005 was patchy, varying from low to severe. In one transect well away from the fence, 90% of inflorescences and 75% of leaves were grazed, and surviving inflorescences were very small. Shooting outside the wood eventually appeared to lead to

Figure 16.6   An exclosure erected to protect ground flora was successful, but muntjac eliminated bluebells and other species from just outside the fence, Grafham Water nature reserve, 2004.

a recovery, at least in terms of grazing on inflorescences, which declined steadily from an average of 28% overall in 2005 to 3% in 2010.

It is probably not necessary for a woodland manager ever to have to worry about determining muntjac density or even monitoring grazing levels precisely, especially if he or she is in charge of a wood that does not have other grazing or browsing issues, such as damage to coppice. Simply seeing leaves heavily grazed over substantial areas and bitten inflorescences becoming easy to find might be enough to precipitate action such as stalking – but in order to 'see' the manager will need to take the time to look. If confirmation is needed by determining the actual levels of grazing, this can be quickly achieved as described in Section 12.6.1.

16.3   Dog's mercury

Dog's mercury rarely seems to attract much attention, although as Oliver Rackham (2003) said: 'It is a most important plant in a negative way; it determines where other plants will not grow.' Dog's mercury is tolerant of shade, although not of waterlogging. It is an ancient woodland indicator, but a poor ground flora may result where it occurs abundantly under a shady canopy.

Dick Steele mapped vegetation in Monks Wood in the early 1970s (Steele and Welch 1973), and dog's mercury was sufficiently abundant to cover about 34% of the wood. It seemed to have changed little in area since the late 1950s, judging from the description of the wood when it was first acquired by the original Nature Conservancy (Sale and Archibald 1957). However, the very dense population of muntjac in Monks Wood in the late 1980s and early 1990s had a considerable effect on dog's mercury, despite the fact that it is poisonous to stock and is said to show signs of extensive grazing only rarely (Cooper and Johnson 1984; Grime et al. 1988). Muntjac were known to eat dog's mercury (Jackson et al. 1977), but the extent of its decline in the wood was not appreciated until Cooke et al. (1995) pointed out that, by 1994, it had been reduced to covering only about 1% of the wood (1–2 ha). We also found that outside small experimental exclosures erected in the wood in 1978, dog's mercury stem height was reduced and plants were growing less densely than inside the exclosures. Similarly, I noted that Norma Chapman's captive muntjac had eliminated dog's mercury from their paddocks by the 1990s, although it continued to thrive outside. In addition, deer grazing was blamed for a scarcity of dog's mercury in the New Forest (Putman et al. 1989) and in Wytham Woods, Oxfordshire (Kirby 2005).

Muntjac will readily eat dog's mercury from the summer through to winter (Figure 16.1b). Levels of grazing on stems in the biggest stand left in Monks Wood reached 76% in early November 1994 and the stand contracted further by 12% between 1993 and 1998 (Cooke 2006). In April and May 1999, Sparks et al. (2005) used a quadrat method to compare cover and height of dog's mercury in the south of Monks Wood with its performance in eight other local woods. Cover was less than 1% in Monks, but average cover in the other woods was 43% with a range for individual woods of 3–69%. Average stem height in the other woods was 28 cm (range 22–32 cm), but in Monks Wood it was 5 cm.

In order to monitor the effects of stalking, I mapped the main dog's mercury stand in Monks Wood in late summer each year, 1998–2005, and grazing frequency and height of ungrazed stems were recorded, 1998–2008 (Cooke 2006). Stalking began in autumn 1998. Most of the dog's mercury was inside the deer fence erected in the south-west corner of the wood in 1999; indeed, its presence was one of the reasons for the fence being erected in that location. Despite the presence of the fence, deer density was similar inside and outside, and the response of dog's mercury did not differ markedly inside and out. Because of this observation, information from inside and outside the fence has been combined in Figure 16.7. The average percentage of stems grazed generally fell from 29% in 1998 to 5% in 2005 and to 2% in 2008 (although since then more serious grazing has been noted occasionally in autumn). Grazing levels fell in tandem with those on bluebell inflorescences (Cooke 2006), showing what was good for one species was good for the other. Intact stem height and stand area both increased up until 2005 in response to the reduction in grazing.

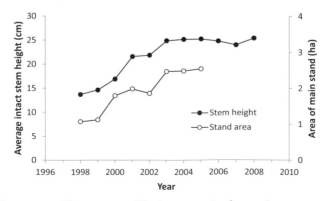

Figure 16.7   The recovery of dog's mercury in the south-west corner of Monks Wood, 1998–2008, after stalking began in 1998.

Small patches of dog's mercury appeared in other places in the wood. This trend continued over the next ten years, but by 2015 the species was still much less abundant than it had been in the 1970s. Colonisation foci ranged in size from single plants to patches up to several metres across, and most were in areas where the plant was known to have occurred in the early 1970s (Steele and Welch 1973). Grime et al. (1988) stated that it did not form a persistent seed bank and showed poor mobility. The most likely explanation for the new patches, sometimes many hundred metres from the surviving stands, must surely be recolonisation from a seed source. However, a record for dog's mercury in the Wilderness area (Walker 2005) appeared to demonstrate a degree of mobility. The Wilderness is a former arable field lying immediately to the south of the reserve that was cultivated between 1850 and 1960. It was then left to develop as secondary woodland.

Although there has been a recovery in size and a partial recovery in distribution with the reduction in muntjac grazing, the plant has failed to recover its former dominance. Many of the wood's compartments from which it has been lost are 'non-intervention' and are no wetter than they were, so neither management nor waterlogging will have been responsible for its current relative scarcity. Similarly, increasing shade cannot be implicated as the species is shade tolerant. During the period when muntjac density was high and dog's mercury was lost, the wood floor became dominated by grasses and sedges. Dog's mercury is now competing with these species and it may never fully recover its former distribution.

Elsewhere, my recording of dog's mercury has been intermittent or has only spanned a few consecutive years. Brampton and Aversley woods are two of the larger local woods where dog's mercury suffered heavy grazing by muntjac and stem height was reduced. In Brampton Wood, localised levels of grazing reached 82% in late 1994 when average stem height was 19 cm. Stalking started in 2000. Stems were still short at 20 cm in 2001, when the level of grazing was 10%, but had recovered to 27 cm by 2006 when 3% of stems were grazed. Aversley Wood was showing high muntjac impact when stalking started in the winter of 2000/1. In 2001, the level of grazing was 15% and average stem height was 21 cm; by 2006, the level of grazing had decreased to 7% and average stem height had improved to 29 cm. There was no evidence of distribution being affected in either wood, but no one specifically looked for changes.

The case of Short Wood in Northamptonshire has already been mentioned with regard to bluebell recovery. There, fallow deer were unintentionally kept in by fencing in 1998 and were not driven out until January 2003. By 2002, dog's mercury could only be found where protected by fallen branches. The wood's residual population of muntjac did not appear to graze the dog's mercury significantly, and its average height quickly recovered from 15 cm in 2003 to 25 cm in 2005.

Muntjac and fallow deer caused problems to conservation interests in Shadwell Wood in Essex (Tabor 2006). Dog's mercury was reduced in size and its population declined. In contrast to the situation in Short Wood, it was thought that muntjac were primarily responsible. Fencing the wood to exclude the deer resulted in plant size recovering within three years.

I studied potential grazing effects on dog's mercury at Marston Thrift in Bedfordshire and Raveley Wood in Cambridgeshire from 1998 until 2001. There was concern about muntjac impacts in both woods and stalking started in Marston

Thrift in 1998 and in Raveley Wood in 2000. Average annual grazing levels varied up to 8% in both woods but stems were not unusually short and there were no consistent changes in stem height, so monitoring was discontinued. This illustrates that even in situations where there are perceived threats from muntjac, dog's mercury is not necessarily measurably affected. However, if grazing rates exceed 10% and average height is less than 20 cm, then management action should be considered. Long-term survey is required to demonstrate a decline in abundance in a wood, but a single visit can provide information on possible loss of plant vigour and its cause, and small plants may be indicative of a reduction in abundance.

16.4   Oxlips, primroses and cowslips

16.4.1   Oxlips

The oxlip was only formally recognised as a separate species, rather than a hybrid of the primrose and cowslip, in the 1840s (Mabey 1996). It survives today in adjoining parts of northern Essex, western Suffolk and southern Cambridgeshire. There has been a long and concerted effort in Essex to improve the fortunes of oxlips, which was led by Ray Tabor. A survey of oxlips in the county in 2002 discovered that while the distribution remained broadly similar to that in 1974, the number of plants had declined significantly, with the population in Hempstead Wood decreasing from perhaps more than 2 million in 1974 to 30,000 plants in 2002 (Tabor 2002). Declines were blamed mainly on the activity of deer and on increasing shade caused by neglect or planting with conifers. The survey looked at grazing damage and found that overall more than 30% of inflorescences were grazed. In nearly one-third of Essex oxlip woods, over 50% of inflorescences were lost to grazing.

In Essex, roe and red deer were also present in some sites in addition to fallow and muntjac, but the last two species were more numerous and widespread, and Tabor (1999) devised an ingenious method for distinguishing the relative impacts of their grazing. A chestnut paling fence 1.2 ha in size was erected in Shadwell Wood in 1992. After a few months, muntjac gained access as the pales warped, but fallow deer were kept out. Permanent recording quadrats were set up inside and outside the fence, some being protected with chicken wire and some not. This arrangement meant oxlips in any quadrats with chicken wire were totally protected against deer; while oxlips in open quadrats inside the fence were protected against fallow deer but not muntjac; and in open quadrats outside the fence they were not protected at all. Recording over the three years 1996–1998 showed that oxlips just exposed to muntjac had high levels of grazing on inflorescences and low numbers of individual flowers per stem and per basal rosette; stem height was also reduced. Fallow deer ate whole plants rather than just the inflorescences and their extra impact decreased the size of leaves, proportion of rosettes flowering and number of inflorescences per rosette.

The Essex Wildlife Trust 'scored' 46 woods for deer activity and damage, with deer score being for all species present rather than for individual species. As deer score increased, so did the level of grazing damage (Tabor 2002; 2004). A deer score of three corresponded to roughly 20% of oxlips being grazed. Using these figures, the Wildlife Trust set targets that would protect its oxlips: deer scores should be less than three and grazing levels on oxlips should be less than 20% (Tabor 2011). These targets seem ambitious, but Tabor (2011) provided data for West Wood, where the cull rate was 0.7 deer per ha per annum over a period of five years. The deer score fell from six in the first year to three in the fifth, while grazing on oxlips decreased from 32% to 22%.

Hayley Wood in Cambridgeshire was set up as a reserve in 1962 especially to protect the oxlip, which was studied in great detail by Oliver Rackham (1975; 1999; 2003; 2006). Unfortunately, the oxlip declined in the reserve apart from where it was protected from grazing by fallow deer. By 1974, fallow deer were grazing most of the plants in the wood that were outside small experimental exclosures. Oxlips also suffered considerably in the heat and the drought of 1976 and in later hot summers. In 1980 and 1987, the Wildlife Trust constructed larger exclosures which fenced about 40% of the wood in total. Still the oxlip failed to prosper, and the Trust eventually fenced fallow deer out of 80% of the wood in 2002 (Figure 13.5).

Oxlips made a 'notable recovery' in Hayley Wood after 2002 in terms of size and number of plants (Rackham 2006 and personal communication): 104,000 in 2002, 139,000 in 2006 and about 220,000 in 2008. Estimates were derived by counting within transects and extrapolating to give approximate totals for each of the main sections of the reserve. Although earlier estimates were about 4 million in 1948 and probably less than 2 million in the early 1970s, it was considered that numbers may eventually fully recover (Rackham 2006).

Oliver Rackham did not measure grazing levels on inflorescences, so I began recording inside the fence in 2004. This was two years after the large fence had been erected and kept out fallow deer from most of the wood, but muntjac remained inside the fence and stalking started in 2004 to reduce their numbers. Single transects were set up in two coppiced areas and in a woodland block. In the first two areas, oxlips grew vigorously in response to the coppicing, but they were smaller with fewer inflorescences in the woodland. I recorded the percentage of stems where the whole of the flower head had been removed leaving part of the stem (Figure 16.1c). The average grazing figure exceeded 60% for each spring 2004–2010. Visitors, however, seemed happy with the floral display, perhaps because grazing levels in freshly coppiced plots tended to be lower at about 20%. In 2015, I recorded oxlips in the woodland transect, where they were still small and grazing levels were unchanged.

The long-term fate of oxlips in 11 woods was described by Rackham (2003). In addition to Hayley Wood, three other Cambridgeshire woods where muntjac have had a significant presence were covered – Gamlingay, Longstowe and Bourn – and in each wood there were declines in oxlips up to 2002. However, in Gamlingay Wood, Peter Walker and I measured a grazing level of 29% in 2003 and he has not recorded any appreciable change in oxlips in study plots since 1998, although declines due to other factors have been noted elsewhere in the wood. At nearby Waresley and Gransden woods, we recorded a higher grazing level of 78% in 2003, but again Peter did not see a change in number of oxlips in study plots.

If oxlip populations are stable or have started to recover, as inside the Hayley fence, it might be argued that it does not matter if grazing levels are so high. Muntjac have left enough flowers at Hayley Wood to produce adequate numbers of seedlings (Rackham 2006), and visitors during the flowering season are well pleased with what they see. But visitors generally do not go into the wood's quieter places, and are not aware of the massive declines that have occurred in the past.

16.4.2 Primroses and cowslips

In Monks Wood, Wells (1994) observed that the masses of primroses, so often a feature in the years following coppicing, had more or less disappeared. Van Gaasbeek et al. (2000) noted that, while primroses were widespread in the wood in the early 1980s, the species was virtually restricted to the south-east and south-west corners by 1999. Compared with other local woods, primrose density in Monks Wood was described as moderate, and these authors speculated that muntjac grazing was likely to be the major cause. However, Rackham (1999; 2003) considered that the primrose decline in other woods in Cambridgeshire was mainly due to the recent trend towards longer, hotter summers.

Although the primrose is adapted to live in grazed situations and is long-lived, it may not be able to sustain an exceptionally high level of flower grazing. My early work in Monks Wood on damage to primroses in two transects in 1993 showed local grazing levels on primrose flowers of more than 70% (Cooke 1994; 2006). However, additional losses at seed capsule stage were comparatively low because surviving capsules tended to be overgrown by other vegetation and so had some protection from vertebrate grazers. Pollinated flowers could still produce a seed capsule if the petals had been grazed but the base of the calyx left. Other species, such as small mammals, could apparently graze parts of the primrose flower, but my grazing records quoted here are based on the loss of whole flowers. Rabbits will take whole flowers, but were unlikely to be active in those parts of Monks Wood that were studied.

In contrast to flowers, grazing on leaves in the two transects in 1993 was comparatively low at 10–12% in terms of overall leaf loss. Seventy per cent of clumps had leaf grazing, but for most it only affected a small number of leaves. Some slug and snail damage was also recorded. As with oxlips, high levels of grazing on leaves might be expected to lead to smaller primroses in future years, but I decided that these rates of leaf grazing were relatively unimportant and focused in later years just on primrose flowers. Observations in 2015 showed that only 7% of clumps had any leaf grazing of the type for which muntjac deer were probably responsible, with the whole end of the leaf apparently nipped off, and it never affected more than two leaves per clump. In contrast, nearly 20% of clumps had signs of gastropod damage.

Monitoring the impact of stalking took the form of walking routes along rides and through woodland blocks and recording numbers of primrose clumps within 5 m. Where two or more plants had overlapping leaves, they were treated as a single clump. A clump needed to have at least five flowers or bitten stalks to be included, and those that been crushed by trampling or vehicles were excluded. Data in Table 16.1 are from two routes each of 2.4 km that were surveyed on five occasions during 2003–2014.

Grazing reached a minimum level in 2010, 12 years after stalking began. The highest number of primroses counted was in 2014, but otherwise there was no sign of an increase in response to a reduction in muntjac density. Van Gaasbeek et al. (2000) walked the perimeter of Monks Wood and every ride, a distance of 21 km, and counted 1,033 plants, equivalent to 49 plants per km. This cannot be directly compared with my figure of 22 flowering clumps per km in 2014 (Table 16.1), but it does suggest that primroses were no less abundant in 1999 than in 2014. On the

Table 16.1   Primroses in Monks Wood: numbers of primrose clumps and levels of grazing
on flowers along two routes with a total length of 4.8 km. Stalking began in 1998.

Year	No. of clumps	% grazing on flowers
2003	74	46
2005	50	46
2008	72	15
2010	52	4
2014	107	7

other hand, if primroses generally have been declining because of hot dry summers, then the reduction of grazing in Monks Wood by stalking may have been countered by weather conditions. Had the climate not become warmer recently, some increase in primroses might have been apparent.

In 2003, grazing on primrose flowers was also recorded in several other woods in Cambridgeshire and Bedfordshire with similar deer activity scores (Cooke 2006). Grazing levels ranged from 4% in Buff Wood and 5% in Wistow Wood to 50–60% in Brampton Wood, Raveley Wood and Waresley and Gransden woods, so the figure of 46% for Monks Wood was not unusually high.

Matt Hamilton and Aidan Matthews of the Wildlife Trust recorded grazing on unprotected primrose inflorescences in Littless Wood beside Grafham Water. Muntjac control outside the wood resulted in grazing levels falling from 70–80% during 2006–2008 to 20–30% during 2009–2010.

Cowslips were regarded as 'local' in Monks Wood by Steele (1973). They have been grazed in the past, both in the wood and on the grass fields to the south (Figure 1.6) where experiments on creating flower-rich meadows had taken place previously. Cowslips were also recorded on my primrose walks from 2008, but none of a total of 102 inflorescences was grazed. Additional observations were made in the spring of 2015, which was evidently a good one for cowslips in Monks Wood. I located 62 clumps along 450 m of the wood's southern edge, and not one of their 164 stems was grazed. Out on the experimental fields, there were several thousand inflorescences; the vast majority had no sign of deer grazing despite high levels of roe deer activity. However, about 2% of inflorescences had been bitten off and most were left untouched (Figure 16.1d). This was similar to damage reported by Rackham (1999) on oxlips in Hayley Wood. As with the 'felling' of bluebell inflorescences, small mammals were probably responsible (Section 16.2.2).

16.5   Orchids

Some of the rarer orchid species in Monks Wood have been regularly monitored and there have been concerns about the effects of deer grazing. Several species declined or disappeared in the 1980s when muntjac became particularly abundant.

In one of the fields within the wood, between 14 and 70 inflorescences of southern marsh-orchid *Dactylorhiza praetermissa* were counted during 1980–1987, but since then there have only been sporadic reports of small numbers (Wells

1994; Hughes 2005). Rainfall was considered to be the main factor affecting its abundance, but Wells (1994) also wondered whether grazing might be involved. This orchid now occurs on the grass fields beyond the south of the wood (Figure 1.6), where it sometimes hybridises with the common spotted orchid *Dactylorhiza fuchsii* (Roger Orbell, personal communication)

The bird's nest orchid *Neottia nidus-avis* was last recorded in 1984, but can be overlooked and it is possible that plants have been grazed by deer (Hughes 2005).

In recent years, the violet helleborine has been present in the wood as a small, discrete population. Numbers of plants or inflorescences ranged from 6 to 19 between 1978 and 1989, from 0 to 4 between 1990 and 2000, and have exceeded 10 each year since then (Wells 1994; Hughes 2005; Cooke 2006; Barry Dickerson, Peter Walker and Roger Orbell, personal communications). Deer grazing and drought may have contributed to the poor performance during 1990–2000. Known specimens have been caged against deer and rabbits for many years, but the increasing use of cages since 2000 (Figure 12.1) and decreasing grazing pressure have probably led to the increase in flowering. In 2009, 30 spikes were counted – some helleborines were left unprotected and grazing did not appear to be serious. There were 132 in 2013 with about 10% grazed. In 2015, most helleborines were enclosed within new fencing, but some were deliberately unprotected – no grazing was seen. In that year, 47 plants and 90 spikes were counted. In 2016, numbers were low in the wood and elsewhere in a dry season. This is an example of how small-scale protection can conserve a population of a rare species.

Wells and Cox (1991) undertook a ten-year demographic study, 1979–1989, of a population of bee orchids *Ophrys apifera* growing on one of the fields to the south of the wood. They concluded that grazing by rabbits, and later by muntjac, was more severe than grazing by slugs and snails, and could pose a serious threat to the population. It was perhaps fortunate that muntjac were most active on the fields during late winter, and not during the summer months when the orchids were most vulnerable.

The greater butterfly orchid *Platanthera chlorantha* was described in the early 1970s as scattered, sometimes occurring in large colonies after clearance (Steele 1973). Twenty years later, when I began looking for grazing damage, I came across several butterfly orchids that were individually caged, suggesting concern about grazing at that time. Roger Orbell has informed me that, in recent years, about ten have flowered with no sign of grazing.

In contrast to these rarer species, common spotted orchid and early-purple orchid have not been regularly monitored in the wood. I have studied grazing levels on common spotted orchids irregularly since 1993 and have made a few casual observations on early-purple orchids.

Common spotted orchids were widespread in the wood in the early 1970s (Steele 1973). In 1993, I studied populations in one of the open fields in the wood and along a major ride (Cooke 1994). In the field site, 50 intact inflorescences were selected in early June 1993 and monitored through to seeding, by which time 15 (30%) survived intact, the remainder having been grazed. Out of 116 common spotted orchids found along an 80 m length of the ride, only 5 (4%) survived to seeding. This was then the best ride-side location for this species in the wood, and the study demonstrated that casual observation could be misleading. During the first part of the flowering season, the number of intact inflorescences changed little because grazed ones were replaced. The median time from finding an orchid to it being grazed was only four days. Other vertebrate grazers, such as rabbits and brown hares, may have been involved, but muntjac deer were much more numerous in these locations and will almost certainly have been the main grazers. Very rarely, inflorescences were found dismembered, probably by small mammals. Picking by humans can be discounted as a major influence because most of the stalk usually remained attached to the basal leaves.

During the following 22 years, orchids were counted in the ride-side plot on eight occasions during peak flowering. In 1994, inflorescences numbered 23 with 51% being grazed. During 2003, 2007 and 2008, roughly 80 orchids were counted each summer with grazing levels of 19–30%, but in 2005 there were 234 and only 6% were grazed. So, while grazing damage had been reduced overall by 2003–2008, it varied from year to year. By this time, lesser damage to flower heads was more apparent and was likely to have been caused by slugs or snails. In 2009, 2010 and 2015, there was little grazing and numbers ranged from 172 to 253. In some recent summers, there have also been notable displays of common spotted orchids along other main rides of the wood. In 2015, more than 100 spikes were counted on one ride along a stretch of only 20 m. Orchids were flowering in many more places in the wood compared with the 1990s and inflorescences were surviving through to the seed stage.

Although there has been an underlying improvement brought about by culling deer, Wells (1994) and Hughes (2005) pointed out that orchid populations are susceptible to drought and to management changes. The period 1994–1997 was particularly dry (Hughes 2005), and this may have contributed to numbers being low in the 1990s. With these common spotted orchids, however, numbers were low in 1994 despite 1993 being a wet year. Also numbers were high in 2005 following dry conditions during 2004 and the early part of 2005, so the fluctuations in numbers did not seem to be closely associated with variations in rainfall.

Like the common spotted orchid, the early-purple orchid was described as 'widespread' in the wood by Steele (1973). Wells (1973) provided more details, saying it was 'frequent on well-trodden rides, especially at the south edge of the wood'. By the 1990s, however, the early-purple orchid was rare – a few inflorescences could usually be found in early May along the western edge. In 1994, however, early-purple orchids began to appear in one of my experimental exclosures that had been erected the previous year. There were ten inflorescences in 1995, eight in 1996 and nine in 1997. During this period, this 4 × 4 m exclosure held about half of the early-purple orchids that I saw in the wood. It was often used for demonstrating to visitors the apparent impact of deer on ground flora. By 2003, orchids had been shaded out in the exclosure but, by then, those along the western edge were thriving. A search in 2005 revealed 71 early-purple orchid inflorescences. This was probably the greatest concentration of early-purple orchids flowering in the wood for at least 15 years. The population has continued to increase, with little or no grazing in recent years, numbering over 400 in 2017 (Roger Orbell, personal communication). There is therefore circumstantial evidence that early-purple orchids in Monks Wood were affected by deer grazing, and that the situation improved after the reduction in deer density.

Table 16.2   Percentage grazing on early-purple orchids in Raveley Wood in relation to muntjac deer score in 20 years, 1996–1998 and 2000–2016 (data from Cooke and Baker 2017).

Muntjac deer score	Number of years	Percentage inflorescences grazed	
		Average	Range
3–4	9	4	0–8
5	7	10	0–27
6–7	4	34	3–64

The evidence from Monks Wood points towards muntjac grazing in the 1980s and 1990s contributing to a general reduction in numbers of both rare and common orchid species. Direct information on grazing at that time seems to be limited to just three species that were monitored in more detail: violet helleborine, bee orchid and common spotted orchid. However, visitors, including experienced botanists, will have been looking for inflorescences, rather than grazed stalks, which are easily overlooked. Thus, the lack of information on grazing on other species is not surprising. With the subsequent decrease in the muntjac population, there has been resurgence in several species.

At Raveley Wood, Martin Baker has monitored a population of early-purple orchids since the mid-1990s. Regular control of regrowth on elm suckers that had previously shaded the orchids has allowed the population to recover (Cooke and Baker 2017). Muntjac built up in the wood during the 1990s and deer scores were highest during 1998–2003 with a peak in 2001, when moderate impact of various types occurred (stalking began in the wood in 2000). Some grazing of inflorescences was reported and, overall, there was a significant relationship between muntjac deer score and the percentage of orchids grazed. In years when deer scores were low, then grazing was low (Table 16.2), but in years when deer score was higher, grazing was variable and sometimes very high. Some of this variability may have been due to the turnover of the few muntjac based in this small wood – and their opportunities and preferences. There has sometimes been a large population of rabbits in the wood, but these occasions did not coincide with high levels of grazing.

Knowledge of muntjac scores in Wildlife Trust woods in Cambridgeshire revealed that, out of 19 woods that did not have fallow deer, eight had muntjac scores in the range 5–7, and one, Littless Wood, had a higher score. Most of these woods are known to have early-purple orchids, and grazing from muntjac is likely to have been widespread, but rarely serious. Several instances of grazing were observed in Littless Wood and plants were subsequently caged (Matt Hamilton, personal communication).

16.6   Wood ferns

Ferns have dominated the ground floor over much of Holme Fen NNR. Bracken is probably the commonest species, but wood ferns (*Dryopteris* species) have also dominated large areas. The most abundant of the wood ferns is narrow buckler-fern *D. carthusiana*. This species was found in the ungrazed fenced plot in a long-term study in the New Forest but not in the grazed control plot (Putman et al. 1989), and Kirby (2001a) reported that most British ferns tend to be more common in ungrazed situations. In this account, the wood fern community is referred to simply as 'ferns'.

The only grazing seen on bracken at Holme Fen has been on new unfurling fronds in spring, and has been of negligible significance. The first substantial grazing on other ferns was recorded when I restarted deer surveillance at the reserve in the autumn of 2003 after a break of five years. At that time, an area of badly grazed ferns had damage similar to that seen on coppice regrowth and on field beans being grown outside Monks Wood (Figures 2.7b and 11.1). Peripheral fronds were directly grazed, but taller more central fronds were broken and bent down prior to being grazed. This area was revisited in September 2004 to set up and record damage in study plots. Signs of muntjac were obvious, including paths, tunnels through the ferns and fresh dung. Some tussocks were dead, some had small leaves but no mature fronds, while the remainder tended to have few fronds with varying levels of grazing.

In order to monitor fern height more broadly in the reserve, observations were made, starting in 2006, in five transects set up in 2005 to monitor bramble. The average of the maximum fern height in each plot increased yearly from 60 cm in 2006 up to 105 cm in 2013; height was slightly lower in 2014 suggesting that maximum overall height may have been attained in response to culling reducing muntjac density. This bald statement does, however, mask considerable differences between the five transects. Information from the most and least affected transects, which were about 2 km apart, is shown in Figure 16.8. The woodland block with the most affected ferns displayed signs of severe grazing and browsing (Figure 16.9a) with dead fern tussocks. In the transect in that block, fern height recovered throughout the monitoring period (Figures 16.8 and 16.9b). In contrast, height where ferns were least affected changed relatively little during monitoring.

Nuttle et al. (2014) reported that high densities of white-tailed deer in Pennsylvanian forests during 1979–1990 led to increases in fern cover, which could still be seen in 2010, despite lower deer densities after 1990. These observations

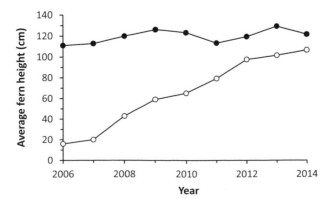

Figure 16.8   Holme Fen fern height in two transects that had previously been exposed to different levels of grazing impact: affected (white dots), relatively unaffected (black dots). Stalking began in 2005.

Figure 16.9   The start of a transect at Holme Fen where wood ferns
had been severely impacted by the time stalking began in 2005:
(a) May 2006; (b) May 2014. Note also the recovery of bramble.

were attributed to the deer avoiding ferns. If the muntjac at Holme Fen were less
attracted to ferns than to bramble, the fern community could have flourished as
the muntjac population built up in the 1990s and concentrated on the bramble.
However, where the bramble died back in some parts of the reserve, the muntjac
may have begun to graze the ferns severely. Then, with stalking, ferns will have
recovered more quickly as the bramble took a while to get the upper hand. By
2014, ferns may have been at their most abundant. This conjecture is supported
by the recent resurgence of muntjac and colonisation by roe deer appearing to
be associated with greater effects on bramble than wood ferns (Section 13.3.1). The
relative dominance of ferns and bramble in the future may depend on trends in
deer density, and also shade, because tree cover was removed from part of the
reserve from 2017 to encourage the spread of heather *Calluna vulgaris*.

CHAPTER 17

Indirect effects of muntjac on animals

17.1 Introduction

Indirect impacts on vegetation due to species' unpalatability or tolerance to grazing have already been discussed (Section 15.2.3). This chapter considers indirect effects on fauna due to a variety of mechanisms and examines whether there is evidence of recovery from impacts when deer are controlled. Generally, indirect effects have been less studied than direct effects, and there is relatively little information specifically on those caused by muntjac. Therefore, some reference is made to information on other deer species, including North American studies. The term 'cascade' is often used to describe a chain of events such as when the actions of deer lacking sufficient natural predation affect different trophic levels in an ecosystem, or when deer are controlled and a series of often unforeseen changes occurs (e.g. McShea et al. 1997; Rooney 2001). In view of the paucity of information and the unexpected nature of some interactions, I have occasionally indulged in exploring consequences and connections at Monks Wood. The approach is simple, though, and complex studies would be needed to prove cause and effect.

17.2 Muntjac

Among the species that muntjac browsing can affect are the deer themselves. Intraspecific competition increases as numbers approach the carrying capacity of a wood. It has been suggested that the population at Lady's Wood in Cambridgeshire declined after 1998 because muntjac had impacted the wood and decreased its carrying capacity (Section 13.1.2; Figure 13.6). At Monks Wood, there were die-offs in 1991 and especially in 1994 because deer density was too high for the impacted wood to support (Section 8.2.2). The body weight of dead adult males in 1994 was reduced by 40% compared with animals from elsewhere in Britain during the months of April–May; and that of dead adult females was reduced by 45% (Cooke et al. 1996). Body weight in culled deer recovered substantially by 1995 but weights were still about 10% below those from elsewhere in Britain, and remained at a similar level thereafter. The conclusions were that adult muntjac in Monks Wood were naturally lighter than those from elsewhere and any abnormal loss in weight in 1994 had been rectified by 1995.

17.3   Other species of deer

Displacement of water deer and roe deer by colonising muntjac has already been discussed (Sections 9.2 and 9.3). What happened to sightings of water deer when muntjac populations were controlled at sites from which the former species had been displaced is shown in Figure 17.1. At Monks Wood, water deer had become a rarity by 1998 when stalking of muntjac began. It took about ten years before some recovery was observed, and even then sightings tended to be restricted to occasional individuals moving through farm fields outside the wood. The situation at Holme Fen was similar with sightings recovering from none per winter up to a regular scattering of records beginning after about seven years of stalking. Bearing in mind the basic requirements of water deer as regards food and shelter, these partial recoveries may owe as much to increased emigration from Woodwalton Fen as they do to habitat recoveries within the other two reserves. There was also displacement of water deer from the south of Woodwalton Fen (Figure 9.1); stalking of muntjac began there in 2011. With data so far only available up to 2018, there is some evidence of increasing numbers of water deer: average sightings per visit in the south were on average about 20% higher in the seven winters after stalking began compared with numbers in the previous seven winters.

17.4   Red fox

The fox is most abundant where habitats are diverse with a wide variety of food and cover (Baker and Harris 2008). In conservation woodland in lowland England, its diet would be expected to comprise mainly lagomorphs, small mammals, fawns of the smaller deer species, ground-nesting birds, invertebrates and fruit. I have

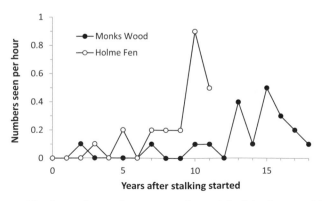

Figure 17.1   Numbers of water deer seen per hour at dusk in the years following the start of muntjac stalking: Monks Wood (black dots) and Holme Fen (open dots). Stalking began in 1998 at Monks Wood and in 2005 at Holme Fen.

Table 17.1   Information on foxes and muntjac in Monks Wood, 1986–2013.
Foxes were recorded as average number seen per hour of surveillance and
as a ratio of sightings, foxes to muntjac (data from Cooke 2014b).

Time period	Deer stalking	Foxes	Foxes seen per hour (ratio foxes:muntjac)	Muntjac
1986–1998	No	Rare	0.023 (1:780)	Population very large despite die-offs in 1991 and 1994
1999–2007	Yes	Rare	0.013 (1:340)	Population decline despite high recruitment
2008–2013	Yes	Resident and breeding	0.11 (1:19)	Decline continuing with lower recruitment

counted foxes seen during dusk surveillance, and for many years the frequency of fox sightings was much lower in Monks Wood than in Holme Fen or Woodwalton Fen. The reason for its rarity in the wood was not known, but may have been related to the intensity of fox control on adjacent keepered land (Cooke 2014b). Foxes have not been controlled inside any of these reserves. From the paucity of reported sightings, foxes appear to have been rare in Monks Wood during the 1970s and 1980s and their rarity may have been a factor in the muntjac's very successful colonisation (Cooke 2006; 2014b). Fox sightings were extremely low from 1986, when surveillance began, up to 1998 just before stalking started in the wood (Table 17.1). During this period, the muntjac-impacted wood was probably a poorer habitat for foxes as cover was much diminished and some of their preferred prey, such as small mammals, may have been rarer. With low numbers of this top predator, the muntjac population attained exceptional densities.

As stalking reduced the muntjac population in Monks Wood, so the fabric of the wood recovered. By about 2006, bramble thickets were starting to form, coppice regrowth was noticeably better and ash regeneration was beginning along ride edges. Foxes remained rare in the wood up until 2007, but were then seen with increasing frequency (Figure 17.2). A plausible explanation is that fox numbers increased in the countryside with the partial relaxation of control and by 2008

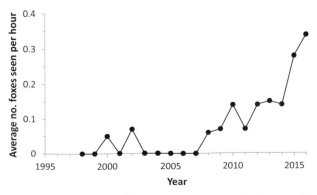

Figure 17.2   Numbers of foxes per hour seen at dusk at Monks
Wood, 1998–2016, after muntjac stalking began in 1998.

the recovering wood became a more suitable habitat. In 2016, the ratio of foxes to muntjac on camera traps set on animal paths was close to 1:1. Since fox numbers have increased in the wood, there is evidence from both the recruitment index and from stalking records that there are fewer young muntjac in the wood. Thus the average recruitment index was 37% during 2000–2007 and 19% during 2008–2013, while the average percentage of shot deer aged up to 35 weeks dropped from 34% to 11% (Cooke 2014b). So stalking appears to have aided a recovery in the fox population, which in turn is now aiding muntjac control.

The fox is the most significant natural predator of the muntjac fawns in this country, but like the muntjac it is subject to widespread attempts to control its numbers. Muntjac have made a success of colonising this country despite our resident foxes. In one study population in Suffolk, about 50% of fawns failed to live to two months and it was believed that most were taken by foxes (Chapman and Harris 1996). Charles Smith-Jones (2004) stated that, in rural areas managed for game or other interests, fox control allowed better fawn survival and led to increases in muntjac populations. There are several studies showing the importance of fox predation on young roe deer in Scandinavia, including at a national scale (Jarmeno and Liberg 2005). The study in Monks Wood indicated that a flourishing fox population may be a management benefit in woodland reserves where muntjac impact on conservation interests. People who are controlling both muntjac and foxes might consider how they can manage the balance between the two species to better effect.

17.5   Brown hare

Brown hares eat grasses, herbs and arable crops (Jennings 2008). They prefer short vegetation for feeding at night, but need areas with some cover for resting and feeding during the day. Woodland provides hares with shelter and feeding opportunities. A study in the King's Forest indicated considerable spatial overlap between hares and muntjac, but the former favoured grassier habitat (Wray 1995). It is possible therefore that muntjac activity in and around a site such as Monks Wood might have affected the suitability of habitat for brown hares in a number of ways. In a study in 1993/4 looking at the species' preferred habitats, hares were seen in the wood, but their highest density was on the adjacent experimental grass fields which were cut for hay in the summer (Cooke 2006). They were at lower densities on the arable fields but, because these were more extensive, absolute numbers may have been high in spring and early autumn.

Hares have been counted on dusk surveillance walks in Monks Wood since 1995 and a marked decline has occurred despite the reduction in muntjac numbers (Figure 17.3). It seems that other factors must have been responsible.

Hares are now much rarer on the grass fields to the south of the wood, probably because regular summer cutting has ceased (Figure 1.6). Locations of all hares seen on surveillance walks have been mapped since 2006. From then until 2016, only 6% were seen on the grass compared with 57% on the arable and 37% inside the reserve. The red fox is the main predator of adult hares and leverets, and may limit population growth and density (Jennings 2008). Hare sightings declined significantly before foxes began to increase in 2008, but it is likely that the increase in

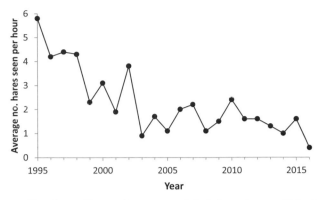

Figure 17.3   Numbers of brown hares seen at dusk in and around Monks Wood,
1995–2016 (updating Cooke 2006). Stalking of muntjac began in 1998.

foxes has led to greater mortality or avoidance of the wood by hares. The percentage of sightings inside the reserve decreased throughout the period 2006–2016, and was inversely correlated with the number of foxes seen per hour. During 2015–2016, when foxes were particularly numerous, no hares were recorded inside the reserve. So stalking may have indirectly led to more foxes and fewer hares. In addition, Richard Broughton has told me that he believes that increasing numbers of dogs being exercised in the wood have disturbed and occasionally killed hares.

17.6   Small mammals

Flowerdew and Ellwood (2001) comprehensively reviewed the possible impacts of deer populations on small mammals in lowland woods in Britain. Impact on habitat can affect food, cover and interspecific competition. Also deer will compete for food with the small mammals. As impacts become more severe, species could be lost until only wood mice *Apodemus sylvaticus* remain. These authors reported on a long-term study on wood mice and bank voles in Wytham Woods, where fallow deer and muntjac established dense populations during the 1980s and 1990s. Numbers of bank voles *Myodes glareolus* trapped decreased from the 1960s to the 1980s, but numbers of wood mice did not change significantly. At the same time, there was loss of cover in the woods attributable to increasing canopy cover and to deer browsing (Kirby et al. 1996; Kirby and Thomas 1999; Morecroft et al. 2001). Bank vole densities were five times higher in 2002 in deer exclosures erected in the Wytham Woods in 1998 than in open woodland (Buesching et al. 2010), whereas wood mice were evenly distributed. Bank voles are more dependent on a dense understorey to protect them from predators than the more alert and agile wood mice, so are more sensitive to loss of cover.

In the New Forest in Hampshire, an exclosure without deer and an enclosure with a fallow deer density of about 1 deer per ha were established in 1963 (Putman et al. 1989). Species composition and vegetation structure were markedly different by 1977. In particular in the grazed plot with deer, there was a conspicuous lack of vegetation in the height range 80–170 cm, that is below the browse line. Trapping of small mammals in 1983 and 1984 revealed wood mice, yellow-necked mice

A. flavicollis, bank voles, common shrews *Sorex araneus* and pygmy shrews *S. minutus* in the ungrazed plot, but only reduced numbers of wood mice in the grazed plot.

In 1971, rough estimates of the principal populations of small mammals living in Monks Wood's 157 ha, based on trapping representative areas, were 3,000 common shrews, 5,000 wood mice and 10,000 bank voles (Tanton 1969; Mellanby 1973). No direct quantitative information was collected on small mammals in Monks Wood when the muntjac population was at its peak, but five of the small exclosures erected in 1978 were still intact in 1994, and grazing of small mammals on bluebells was particularly evident inside three of them, suggesting they were havens compared with the rest of the wood. From the impact on the wood's vegetation that was apparent by the early 1990s, one would suspect that sensitive species had declined. My camera trapping videos since 2010 have indicated wood mice to be abundant but other species of small mammals to be very rare.

The hazel dormouse *Muscardinus avellanarius* is viewed as an indicator of good woodland habitat supporting a broad range of species (Bright et al. 2006). It has received much conservation attention in recent decades after suffering widespread declines in England and Wales. If deer browsing destroys the understorey, it can reduce dormouse food in the form of flowers and fruit, as well as affecting nest sites and arboreal pathways. These authors mentioned pollarding hazel at 1.5 m, instead of coppicing, to protect habitat from the ravages of muntjac. As part of the Species Action Plan for dormice, Brampton Wood in Cambridgeshire was selected as a reintroduction site in 1993 and a breeding population was established (Mitchell-Jones 2005). Much of the abandoned hazel coppice remained excellent dormouse habitat, but concern was expressed about the impacts of muntjac and increasing shade on the shrub and herb layers of the wood. Muntjac had been culled in Brampton Wood since 2000, but improvements in habitat did not begin until 2005 (Sections 13.1.2 and 14.1.3).

An illustration of the far-reaching and unexpected impact of deer comes from North America. A locally large population of white-tailed deer in New York State reduced numbers of the white-footed mouse *Peromyscus leucopus* probably because the deer reduced vegetation cover and competed with the mice for acorns (Ostfeld et al. 1996; Rooney 2001). One consequence of this was that lower numbers of mice increased the chances of gypsy moths *Lymantria dispar*, a prey species, reaching pest proportions.

17.7   Birds

In a major collaborative investigation on trends in populations of birds breeding in British woods, observed declines were considered to be caused by a range of factors such as increase in woodland maturation leading to unfavourable structural and compositional changes, reduction in active management and possibly increased deer browsing (Amar et al. 2006).

Rob Fuller and other BTO staff have investigated the effects of deer on woodland birds for many years. For instance, unique experimental work at Bradfield Woods in Suffolk using coppice plots showed that excluding deer protected the shrub layer and led to higher breeding densities of dunnock *Prunella modularis*, garden warbler *Sylvia borin*, nightingale and long-tailed tit *Aegithalos caudatus* (Holt et al. 2011). Dunnock and nightingale both require a dense understorey for nesting and bare ground for foraging, features that are detrimentally affected by deer browsing. Territory mapping and radiotelemetry showed that male nightingales had a strong preference for the deer exclosures (Holt et al. 2010). Numbers of roe deer, muntjac and fallow deer had increased substantially in and around Bradfield Woods from the 1980s, and at the same time nightingale territories declined from 15–20 to 4–8 (Gill and Fuller 2007).

Information on possible changes in bird populations has also emanated from long-term studies or recording of woodland bird populations that were started before deer became a concern. Thus at Wytham Woods, there has been Common Birds Census territory mapping since 1972 (Perrins and Overall 2001; Perrins and Gosler 2010). Declines in blackcap *Sylvia atricapilla*, garden warbler and chiffchaff *Phylloscopus collybita* since the 1970s were blamed on browsing by fallow deer and muntjac, although it was acknowledged that closing of the canopy due to cessation of forestry thinning also contributed to reduced growth of the understorey. Chiffchaffs could still be studied in the 1990s in patches of brambles that persisted beside the main access track where human disturbance appeared to have lowered deer browsing. Willow warblers *P. trochilus* declined as well, but unlike the chiffchaff they declined on a regional basis in Britain, indicating that other factors were significant. Disappearance of the nightingale in the 1980s was attributed to impoverishment of the understorey and was described as 'a significant aesthetic loss to the human visitor' (Perrins and Gosler 2010). A general reduction of bramble was thought to be especially important, and this may also have played some part in the declines noted for blackbird *Turdus merula*, song thrush *T. philomelos*, dunnock and bullfinch *Pyrrhula pyrrhula*. Blackbirds had higher breeding success where woodland vegetation was denser and nest predation was lower.

An important point made by Perrins and Overall (2001) was that although Wytham Woods had been a nature reserve owned and managed by Oxford University since around 1950, the woodland environment had not remained stable. Myxomatosis arrived in 1956, virtually eradicating the rabbit population thereby reducing grazing and affecting various predators such as weasels *Mustela nivalis* and stoats. The sparrowhawk disappeared as a breeding species in 1959 as a result of pesticide poisoning, and reappeared from the mid-1970s. Dutch elm disease killed most of the elms in the 1970s. Knopper galls on oak trees affected the production of acorns, and young oaks and sycamores *Acer pseudoplatanus* were being affected by different fungal species. Deer were one of a series of 'ecologically destabilising events'.

Another very important consideration is that many other factors might affect numbers of breeding birds, especially those such as the warblers and the nightingale which are long-distance migrants. I collated information on numbers of singing male nightingales in Monks Wood from local bird reports, from English Nature records and from personal observations from 1976 until 2005, except for 1982–1987 when there was a gap in the records. Numbers peaked at 18 in 1981 and there was a steady decrease from the late 1980s down to zero in 2005 (Cooke 2006). Since then there have been only sporadic records, although Richard Broughton has informed me of singing from a regenerating area just outside the wood in recent years. Nightingale populations have also suffered long-term declines in some other local reserves (Cooke and Farrell 2001b; Bowley 2010) and again have been rare or absent in recent summers. Monks Wood's ornithologists have tended to put most of the blame on woodland maturation (Hinsley et al. 2005). Nevertheless, muntjac browsing has destroyed nightingale habitat in the wood and elsewhere; in addition, the spread of coarse grasses and sedges may have rendered habitat even less suitable, so deer were probably implicated.

Other species requiring low vegetation and said to be vulnerable to deer browsing (Fuller 2004) have also decreased in Monks Wood since the 1970s, such as dunnock, song thrush, willow warbler and bullfinch (Hinsley et al. 2005), four of the species that declined in Wytham Woods. As with the nightingale, woodland maturation and deer browsing are likely to have been two of the factors involved. April censuses undertaken by Ian Wyllie in Monks Wood in the 1980s and 1990s have been repeated by Shelley Hinsley and Paul Bellamy each April since 2006. Of the four species listed above, the dunnock displayed increased breeding numbers during 2014–2016 (Shelley Hinsley, personal communication), and Richard Broughton has spoken of the bullfinch increasing in numbers especially in nearby regeneration. In addition, the group of migrant species which nest in scrub and dense cover, such as whitethroat *Sylvia communis*, lesser whitethroat *S. curruca* and garden warbler, have all shown some recovery in recent years. So there is evidence of a recovery beginning, but it needed about 15 years of stalking before it occurred.

Garden warbler, chiffchaff and willow warbler were counted from 1998 in spring and summer along transects in Holme Fen and Woodwalton Fen (Bowley 2010). Chiffchaff numbers held up well between 1998 and 2009, but numbers of the other two species decreased, the declines being especially severe at Holme Fen. When this study on breeding birds began, there was low browsing impact on the woodland understorey in Woodwalton Fen, but this increased to a high level in 2010. In Holme Fen, however, there was considerable browsing impact by 2004. Stalking began in Holme Fen in 2005 and in Woodwalton Fen in 2011. So the greater effects on garden warbler and willow warbler at Holme Fen could be explained by browsing problems being experienced earlier at that site. By 2009, there had been relatively little recovery in bramble at Holme Fen (Figure 14.8). Alan Bowley told me that, by 2016, there were no convincing signs of recovery in breeding numbers of affected passerines at Holme Fen, 11 years after stalking began.

Ron Harold has provided me with information on a survey undertaken during 2011–2013 to compare breeding populations of bird species in the woodland at Woodwalton Fen with those present in 1987–1989 (Harold 1991). Many factors, other than the impact of muntjac on habitat, might have affected population size over this period of more than 20 years. However, several species believed to have been affected by deer elsewhere registered declines of at least 40%: dunnock, nightingale, song thrush, garden warbler, willow warbler and bullfinch. It is likely, therefore, that impact by deer was implicated in the decrease in populations of a number of species.

Impacts on birds in the United States were reviewed by McShea and Rappole (1997). In a Pennsylvanian forest, species richness and abundance were reduced for birds nesting in the understorey in enclosures containing managed densities above 8 per km^2 of white-tailed deer (deCalesta 1994). Birds nesting in the canopy or on the ground were unaffected. Chollet and Martin (2013) found a 'continent-wide link' in North America between the increase in deer populations in forests and decreases in songbirds dependent on the understorey. The problem for understorey species is evidently widespread.

17.8   Amphibians and reptiles

Consideration of whether deer browsing affects amphibians and reptiles seems to have attracted little or no interest from other researchers. With my personal concern for herpetofauna, however, some attention has been given to how the terrestrial habitat of these species may be modified. For example, great crested newts *Triturus cristatus* prefer a mosaic of different terrestrial habitats with dense ground cover, such as might be detrimentally affected by muntjac browsing and grazing. I had monitored changes in newt populations in two breeding ponds close to the edge of Monks Wood over the period 1985–2005, so was able to assess whether there was any evidence that populations had been affected indirectly by muntjac (Cooke 2006). No such evidence was found – indications of a population decline at one site were blamed on increased shading of the pond.

Neither was there any evidence of detrimental indirect effects on common frogs *Rana temporaria* and common toads *Bufo bufo* in Monks Wood. Both species were said to be relatively numerous up until 1960, but were absent or rare by the early 1970s because ponds and ditches, where they bred, had dried out (Prestt 1973). It was at this time that muntjac began to colonise the wood. In recent decades, both species have been encountered more commonly in the wood as ponds have been (re-)created and better maintained.

The common lizard *Zootoca vivipara* lives in a range of habitats in Britain including open woodland, preferring undisturbed areas with good exposure to the sun (Inns 2009). In recent years, I have encountered them most frequently in Monks Wood and Woodwalton Fen in plots of failed coppice in late summer (Figure 17.4). At that time of year, newborn lizards are especially obvious when basking. Browsing by deer at Monks Wood provided the lizards with suitable, open habitat for a longer period of years than would otherwise have been the case. This benefit was more short-lived at Woodwalton Fen, where reed quickly colonised failed coppice.

Figure 17.4   A young common lizard on a dead sallow stool at Woodwalton Fen, August 2011 – a vertebrate species that has benefited from deer browsing.

17.9   Invertebrates

The impact of deer browsing on invertebrates was reviewed by Stewart (2001). Lowland mixed deciduous woodland has structural complexity and species-rich vegetation that offers niches for a multitude of invertebrates. Most commonly, impacts are mediated via alterations to the structure, species composition and quality of the vegetation. Deer also compete with invertebrates for plant food, although, perhaps surprisingly, they remove a smaller proportion of the primary productivity of a wood than the herbivorous invertebrates. Some changes will be beneficial for invertebrates. For instance, if browsing helps create and maintain open areas in woodland, then sun-loving insects and those needing to feed on flowers may benefit. Invertebrates associated with dung could increase if deer densities are high, and in turn provide extra food for vertebrates such as birds and bats.

With regard to specific damaging impacts, Stewart (2001) explained that:

- Many of the plant species preferentially browsed or grazed by deer are important for invertebrates, such as bramble, blackthorn, hawthorn, elder and members of the Compositae.
- Some insect herbivores feed specifically on common species in the field layer such as violets (*Viola* species), enchanter's nightshade and dog's mercury. As has been discussed in previous chapters, these can be heavily grazed by muntjac and presumably by other deer.
- A high density of deer results in a browse line below which relatively unpalatable grasses, sedges and bracken may be the only species to flourish.
- Creation of a browse line will change the microclimate at the level of the field layer which may detrimentally affect species such as snails, which require damp and shady conditions.

Butterflies and moths are traditionally better recorded than other orders of insects. Feber et al. (2001) recounted the story of how the fortunes of butterflies in the New Forest have fluctuated over the last two centuries depending on the density of fallow deer.

The white admiral *Limenitis camilla* was studied in Monks Wood in the 1970s by Ernie Pollard (1979). This was just after muntjac started to colonise the wood but well before there was any concern over their ability to damage ecological interests. He found that eggs were laid on honeysuckle and average height from the ground could be less than 1 m. In 1993, he and I examined the height of egg-laying sites in relation to muntjac browse lines (Pollard and Cooke 1994). Egg-laying sites were higher than they had been in the 1970s and indeed were higher than in Kent, which the muntjac had not then reached. We considered that as some eggs were still laid below 1 m they were likely to eaten by muntjac along with the leaves. Ernie Pollard was one of the founders of the UK Butterfly Monitoring Scheme which started in the 1970s. Count data from the scheme revealed that although the white admiral had decreased in the wood, it had done no worse than elsewhere in its range.

Jack Dempster's work during the 1980s and 1990s on cuckooflower and orange-tip butterflies in Monks Wood has already been mentioned (Section 15.2). Not only did muntjac eat his study plants, but they also ate butterfly eggs and larvae on them. There was, however, little impact on the abundance of flowers or of orange-tip larvae in the wood because other factors were more important (Dempster 1997 and personal communication).

These two examples demonstrate how detailed studies can provide insights into subtle effects that do not necessarily translate into measurable effects on populations. Population changes have, though, been detected to which browsing and grazing by muntjac deer are likely to have contributed. The role of muntjac in the proliferation of coarse grasses in Monks Wood has been discussed (Section 15.2.3). Pollard et al. (1998) reported that three butterfly species with larvae that feed on coarse grasses – large skipper *Ochlodes sylvanus*, speckled wood *Pararge aegeria* and ringlet *Aphantopus hyperantus* – increased relative to populations in other sites in eastern England. Similarly, moths with grass-feeding larvae also increased in the wood (Greatorex-Davies et al. 2005).

Monks Wood's main claim to lepidopteran fame is that the first British black hairstreaks *Satyrium pruni* were discovered there in 1828 (Thomas and Lewington 2010). I shared a laboratory at Monks Wood Experimental Station with Jeremy Thomas in the early 1970s when he was undertaking research on black and brown hairstreaks. The brown hairstreak *Thecla betulae* died out in the wood in the mid-1970s (Greatorex-Davies et al. 2005). Muntjac were still rare in the wood at that time and this butterfly's local extinction seems part of a general disappearance from sites in eastern England, with loss of hedgerows being a factor (Thomas and Lewington 2010). Both of these hairstreaks lay their eggs on blackthorn, but the black hairstreak is more of a woodland species than its relative. As blackthorn can be heavily browsed by muntjac and other deer, it might be assumed that conflict would result. However, black hairstreaks fare best on banks of blackthorn in sunny, sheltered positions, so the maturity of the bushes may keep habitat relatively safe from browsing. Browsing would, however, have potential to affect regeneration of blackthorn in the longer term. Management of the blackthorn in Monks Wood has meant that this hairstreak colony is still buoyant. At the woodland reserve of Glapthorn Cow Pastures in Northamptonshire, patches of blackthorn have been fenced by the Wildlife Trust to eliminate browsing by fallow deer and muntjac and so help conserve black hairstreaks.

Lupins *Lupinus perennis* are native to America and are grazed by white-tailed deer. In New England, this has caused declines of the endangered Karner blue butterfly *Lycaeides melissa samulis*, which is a specialist feeder on lupins (Miller et al. 1992; Rooney 2001).

Historically in this country, many woodland butterflies have fared well in coppice areas. However, deer browsing has been a disincentive to undertaking coppice management, resulting in areas becoming overgrown, shady and unsuitable (Feber et al. 2001). Fencing conservation coppice has met with some success, but if growth is too rapid, the successional window for butterflies may shorten to a couple of years. It comes back once again to a little browsing being beneficial, a lack of browsing being slightly less good and too much browsing being disastrous.

Young leaves tend to be more nutritious for herbivorous insects than older leaves because they contain more nitrogen and less tannin and other unpalatable substances. In their studies at Bradfield Woods, Holt et al. (2011) found that insect density per unit of foliage in the coppice plots to which deer had access was similar to that inside exclosures, so there was no evidence of any chemical response in plants following deer browsing. Where deer had access, however, there was much less foliage for insects!

There is little information on non-lepidopteran species. Observations were made by Putman et al. (1989) on beetles in the fenced plots in the New Forest described above (Section 17.6). After 22 years, there were more ground beetles (Carabidae) in the ungrazed plot and more rove beetles (Staphylinidae) in the grazed plot where there was a density of roughly 1 fallow deer per ha. Differences reflected habitat preferences at both order and species level (Stewart 2001). Ants and spiders were more numerous in the grazed plot.

Colin Welch studied Monks Wood's beetle fauna for many years, and concluded that the dramatic reduction in herbs must have affected beetles, including some phytophagous species and others feeding on pollen and nectar (Welch 2005). He pointed out that the surrounding farmland offered few opportunities for recolonisation once a species was lost.

Leadbeater (2011) has suggested that woodland managers may have a greater effect on invertebrate biodiversity than deer by removing dead timber and thereby eliminating habitat required by the many species of 'decomposers' that far outnumber the few charismatic light-loving species safeguarded by culling. However, in the woods discussed in this part of the book, dead wood is viewed by managers as an important conservation resource.

The sentiments in the last two paragraphs illustrate the huge uncertainty and concern over what happens to invertebrates in particular when there are fundamental changes to habitat composition.

CHAPTER 18

Recoveries in Monks Wood since control of muntjac began

18.1 A chronological summary

Monks Wood is one of the best-studied woods in the country, with its vegetation and fauna being described in great detail in the early 1970s prior to muntjac becoming firmly established (Steele and Welch 1973). Ecological information was updated in 1993 (Massey and Welch 1994), by which time the wood had become severely impacted, and it was reviewed again in 2003 (Gardiner and Sparks 2005) when deer stalking had occurred in the wood for five years. I documented recoveries primarily due to stalking up to 2005 (Cooke 2006), and this new account moves the review on to 2017. During the first 19 years of stalking, roughly 1,000 muntjac were shot in and around the wood. There is little published information on rates of recovery of woodland from severe browsing by deer. This chapter brings together information on recoveries in Monks Wood covered more comprehensively in the preceding four chapters.

The questions to consider are: (1) when did the various conservation features of the wood change after stalking started in 1998; (2) to what extent have they returned to how they were before muntjac became established; (3) how might changes in impacts relate to deer density? My conclusion in 2005 was that partial recoveries had occurred for many of the features that had been affected but none had totally recovered (Cooke 2006). Reviewing progress again enables statements to be made about how some species have recovered faster than others and reinforces the views of investigators elsewhere that most recoveries are slow and/or never totally reversible. Table 18.1 provides a chronological summary of what happened, and is followed by discussion of different types of impact.

None of the affected features of the wood has so far been found to have recovered to exactly how it had been in the 1970s. It should be appreciated that the aim of management was not necessarily to return Monks Wood precisely to how it was

Table 18.1   Recoveries in Monks Wood recorded since the start of muntjac stalking inside the wood in 1998. Statements in italics relate to impact assessments for the whole wood (see Section 12.5.3 and Appendix 1).

Year	Observations
1998	*Impact on vegetation was assessed as severe.* Stalking began.
1999	Bluebells were used as an indicator of grazing: grazing levels fell, but inflorescences did not increase in number and remained small with few flowers.
2000	*Overall impact remained at very high, having decreased to this level in 1999.*
2001	Numbers of flowering bluebells increased noticeably. *Overall impact became high.*
2002	Bramble began to recover in height. Ivy was being used as a browsing indicator and results showed signs of a decrease in browsing.
2003	Protected ash seedlings were more numerous and grew taller than those that were unprotected, demonstrating that browsing was still having a marked effect.
2004	The main patch of dog's mercury in the wood was less heavily grazed and covered more than twice the area it did in 1998. However, woody regrowth grew better in exclosures erected in spring 2004 than in open control plots (Figure 18.1).
2005	Bluebells had recovered in the main stands. New patches of dog's mercury were appearing in different parts of the wood. There was a good show of early-purple orchids in the west of the wood with little grazing. First flowering by wood anemone was noted outside exclosures. Primroses, however, continued to be heavily grazed. Bramble thickets began to form. Browsing on hazel coppice regrowth was less severe, but still sufficient to cause loss of stems and some stool death. The wood's condition was assessed by English Nature as being 'Unfavourable Recovering'.
2006	Many species remained scarcer than in the 1970s, e.g. honeysuckle and enchanter's nightshade. Young ash taller than 20 cm were first seen along ride edges.
2007	*Impact on vegetation was assessed as moderate.*
2008	Blackthorn, hawthorn, privet and dogwood were recorded as increasing in woodland transects. Sightings of foxes began to increase.
2009	Grazing was low on common spotted orchids and numbers were higher.
2010	Little or no grazing was recorded on primroses and cowslips. Bluebells were spreading along the east edge. Dog's mercury patches increased in size and number. The first drifts of wood anemone were noted. Basal growth and suckers on ash suffered only a little browsing and increased in abundance.
2011	Further improvements were seen on ride-side coppice regrowth, although problems still occurred towards the rear of plots. Water deer began to be seen regularly again.
2012	Blackthorn thickets started to form – browsing on young stems was variable.
2013	Ride-side shrubs grew well after the wet weather in 2012/13. Ash saplings taller than 1.3 m began to appear. Unprotected hazel coppice regrowth had a good chance of growing acceptably. *Overall impact was assessed as low.*
2014	There were dramatic displays of spring flowers, including bluebells, anemones and lesser celandines, especially along the east edge. Common spotted orchids were once again common along rides, and ride edges were generally more colourful in summer. Breeding populations of dunnocks and various warbler species began to recover.

Year	Observations
2015	Understorey of bramble, blackthorn and privet was well developed in places. Grazing was no longer a problem for violet helleborines. Lords-and-ladies was still well grazed, but remained abundant. Monks Wood was assessed by Natural England as being in 'Favourable' condition.
2016	Saplings of oak, field maple, hazel and hawthorn were becoming more abundant in the woodland. Spreading thickets of bramble occurred in well-lit areas. Bluebells were more abundant than ever previously documented. Wood ferns were noticeably more widespread. Despite the general recovery, some species, such as dog's mercury and the nightingale, remained much rarer than in the early 1970s. Other species, particularly grasses, sedges and ground ivy, were more abundant.
2017	*Overall impact remained low.* However, ash die-back had taken a firm hold since first being seen in the wood in 2015.

before muntjac colonised (Kirby 2001a; Gardiner and Kirby in Cooke 2006), but rather to control muntjac to a level at which woodland processes and communities could recover. The fact that the wood was assessed by its senior manager, Chris Gardiner, as being in 'Favourable' condition in 2015 showed that deer control had worked – having been assessed about ten years before as 'Unfavourable Recovering' (Section 12.5.2). Nevertheless, muntjac deer were still having measurable impacts and continuing with stalking was viewed as essential, especially employing stalking strategically to tackle local pockets of higher deer density.

18.2   Browsing impacts

Bramble, hazel and ash were studied in detail over a long period in Monks Wood. When stalking began in 1998, bramble was badly impacted by muntjac; the few mature bushes that remained were showing die-back. Stunted, browsed seedlings were, however, abundant, and as the muntjac population was controlled, so these initially provided the means for recovery. Monitoring bramble height indicated that recovery in growth was just beginning in 2002. By 2005, thickets were beginning to form in favourable positions. Since then, bramble has continued to grow and spread, and thickets have become extensive in some well-lit areas. The early years of bramble recovery were associated with a muntjac density estimated to be about 50 per km². Even by 2016, though, bramble was still not as abundant as in the early 1970s. In particular, it was rarer in some of the damper blocks of coppice from which it was displaced when pendulous sedge invaded as an indirect consequence of muntjac activity.

The success of hazel coppice was measured as the number of regrowth stems attaining a height of at least 1 m when compared to an ideal of 25 stems per live stool. In the absence of stalking in the 1990s, any unfenced hazel had minimal survival. Steady improvement occurred after stalking began so that there was at least 76% success of regrowth stems during 2009–2011. Failure to reach the ideal was due to live stools not having sufficient stems per stool and to some stools dying – deer browsing contributed to both components of failure. Opportunities did not arise to record a sample of unprotected plots after 2011, but occasional and opportunistic observations indicated that freshly cut unprotected stools could still be at risk of significant browsing damage. However, roe deer appeared to be contributing their share, and, on balance, muntjac browsing on hazel coppice may have become acceptable more often than not in a conservation context by 2013. The muntjac density at that time was estimated to be 20–30 per km².

Young ash exceeding 20 cm in height began to survive along the edges of rides in 2006. These grew and quickly increased in number. They occurred in smaller

Figure 18.1   One of the small exclosures on the hillside in Monks Wood with its control plot in the foreground; it was erected in 2004 and the photograph was taken in April 2007. By that time the exclosure was filled with young elms, but muntjac density had been high enough to inhibit their establishment in the control plot.

Figure 18.2   Similar views of the interior of the same woodland compartment: (a) March 1994 with a lack of understorey; (b) March 2018 with an understorey composed mainly of bramble and privet.

numbers inside the woodland blocks where less light and higher levels of browsing slowed growth and made establishment of saplings less likely. In 2015, however, ash die-back was recorded for the first time and the future of ash in the wood is uncertain. At the same time that ash seedlings were developing, the recovery of other woody species was becoming more apparent. Blackthorn (in the form of suckers) and privet were the most obvious of these species. By 2015, the developing understorey made it more difficult to see into much of the woodland from the rides (Figures 13.8b and 18.2). During the early years of the recovery by these trees and shrubs, muntjac density was estimated to be about 40 per km².

18.3   Grazing impacts

The main stands of bluebells in the south-west of the wood were fenced from the autumn of 1999 (a year after stalking began) but, as the fence never functioned as a complete barrier, this had little or no extra effect on recovery. Deer densities inside the fence fell primarily because of stalking generally in the wood. The bluebell is a good example of the sequence of recoveries that may take place. First, there needs to be a reduction in deer activity – in this case, grazing on bluebell leaves and inflorescences. This then has the immediate result of more inflorescences surviving intact. This may not happen if a plant species is especially palatable, but did occur with bluebells. As grazing levels fall so inflorescence survival improves further, but in addition, as vigour returns to the bluebells, they grow bigger and a higher proportion flower. Thus a stand recovers its former glory. But this has little effect on what happens away from the main stands. Where viable seed survives, germination may occur and seedlings have a much better chance of establishment, being exposed to a reduced level of grazing. By 2005, bluebell size and numbers flowering in the south-west and west of the wood had fully recovered. By 2016, their stands were more widespread and extensive than had been recorded at any time in the past; recolonisation of the east of the wood had more than compensated for any losses in coppice areas elsewhere.

Bluebells are only obviously affected when grazing is especially severe and after some other plant species have been badly impacted. As bluebells are poisonous, it is likely to be a species that muntjac do not relish and only turn to when they must. When stalking occurs and muntjac density declines, so a surviving deer will have proportionally more food available and will be less likely to select bluebells. This may be why bluebell stands recover relatively rapidly.

Like the main stands of bluebells, those of dog's mercury were also inside the south-west fence. Late summer grazing levels fell between 1998 and 2008, and stem height recovered completely by about 2005. At the same time, the area of the main stand increased, but has shown no clear change since then. Elsewhere in the wood, this slow coloniser still occupied less than 10% of its former coverage by 2016.

Roughly half of primrose flowers were still being grazed in 2005, but grazing levels fell to a minimum figure of only 4% in 2010. Ray Tabor (2006) reported that primroses in his study woods in Essex were not grazed: perhaps the deer ignored them to concentrate on the more succulent oxlips. There were increased numbers of primroses along rides in Monks Wood in 2014, but it is unclear what part climate change has played in influencing abundance in this part of the country. Any recovery along Monks Wood's rides has, however, been overshadowed by the losses in newly cut coppice plots since the 1980s. Instead of being actively managed, the plots have been abandoned since the mid-1990s to become woodland. Although these patches have developed other interests, they lack the spectacular vernal displays of old.

Other species that were part of those displays were wood anemone and lesser celandine. The former had stopped flowering in the wood by the early 1990s and was not seen flowering outside exclosures until 2005, seven years after stalking began. There are now fine drifts in the east of the wood (Figure 18.3) but they are more or less absent from the old coppice areas. Quantitative information on celandines is largely lacking but they have recovered in number and area in recent years. Lords-and-ladies is a species that is still heavily grazed yet it remains common and widespread in the wood, including in the shady centres of woodland blocks. Because species such as lesser celandine and lords-and-ladies are of no special conservation interest, their distribution and abundance have not attracted much attention.

Herb-Paris may have become extinct in the wood as a result of deer grazing in the 1980s, but has failed to reappear since deer have been controlled. Various orchid species were seriously affected by grazing, and bird's nest orchid died out in the 1980s. Grazing levels decreased after stalking started and numbers of early-purple orchid, common spotted orchid and violet helleborine have all increased. However, it took until about 2014 before one could say that common spotted orchid was once again 'common' in ride-side swards and until 2015 before grazing was no longer a problem for the main group of helleborines.

Some species, notably some grasses, sedges and ground ivy, increased when muntjac became abundant because they were rarely or never eaten by the deer. These have then inhibited the recovery of impacted flora, such as dog's mercury. In some cases, these species have decreased in recent years. For instance, pendulous sedge stands in failed coppice may have fragmented to some degree as trees such as aspen have colonised and combined with the existing standards to produce a denser

Figure 18.3   A drift of wood anemones in Monks Wood, April 2015 –
ten years after they began flowering again in the wood outside exclosures.

canopy. The unpalatable species are, though, more abundant and widespread than they were before muntjac became established.

The level of grazing on plants will depend not just on deer density but on characteristics such as palatability and length of the season that a species is available for grazing. Thus grazing levels on presumably relatively unpalatable bluebells fell below 10% in 2001, when estimated muntjac density was about 70 per km^2. Such grazing levels were not regularly attained for more palatable common spotted orchids until 2009 or for primroses until 2010, when deer density had fallen further to 40 per km^2. Lords-and-ladies was different, being toxic but still being actively sought by the remaining muntjac in 2015 (deer density 20–30 per km^2), and dog's mercury was unusual in having an extended period each year over which it was eaten. So although summer grazing levels on dog's mercury fell to less than 10% by 2000, it was nevertheless heavily grazed in some autumns – waning toxic properties may have facilitated and encouraged this later grazing. The recovery of floral populations along the edges and into the centres of the woodland blocks has generally been a slow process. But, with the exception of dog's mercury, this had occurred by 2016.

By 2017, the woodland blocks probably most closely resembled how they appeared in the early 1970s. They were grassier in 2017 – sheets of spring flowers were present in places (Figure 16.3b) – but the species balance and where they were growing had changed to some extent. The rides were also grassier but the spring and summer flowers added colour (Figure 1.5). The biggest change was in many of the old coppice plots where those that been heavily browsed in the 1980s and early 1990s were reverting to forms of woodland (Figures 13.8 and 13.10).

18.4   Indirect impacts on other fauna

In 2006, I concluded that a range of animal species had decreased in the wood, many of which were of conservation importance (Cooke 2006). Fauna appeared generally less rich than it was two or three decades previously. The indirect effects of muntjac will not have been responsible for all of these changes. Nonetheless, modifications to the structure and species composition of the wood's vegetation, mediated via deer activity, will have contributed, probably to a significant degree.

During the late 1980s and early 1990s, water deer were frequently recorded in the wood, especially in the open fields in the north, but then sightings petered out. It required ten years of muntjac stalking before water deer began to reappear on spring surveillance visits and 13 years before they were seen regularly again. Roe deer have been colonising since the early 1990s, yet it was 2005 before they were eventually recorded on a surveillance walk; initial colonisation may have been inhibited to a degree by muntjac. Since then, roe sightings have increased rapidly and this species may now be deterring water deer from more fully utilising the wood.

For the first 20 years of spring surveillance, fox sightings were rare, but then increased from 2008. Relaxation of fox control from 2002 in the countryside outside the wood probably helped in this recovery. However, it is possible that recovery in small mammal prey in response to habitat improvement in the wood also played a part. Sightings of brown hares decreased from 1995 and the dearth of records inside the wood in later years may have been related to the increase in foxes.

Figure 18.4   Silver-washed fritillary in Monks Wood on
bramble inside the coppice fence, August 2016.

Breeding populations of a range of bird species that require dense low cover decreased in the wood in the 1990s, and browsing by muntjac was likely to be one of the contributing factors. There have been a few reassuring signs of recent increases in some passerine species, such as the dunnock. Many species have not recovered, but a number of other factors have affected our national or regional populations.

Recent reviews of possible recoveries in butterflies and moths in the wood are lacking. Populations of some species were reported to have been altered indirectly by muntjac feeding, or conversely not feeding, on plants that were important to them (Pollard et al. 1998; Greatorex-Davies et al. 2005). The moth trap is no longer in operation but the butterfly transect is still routinely recorded for the national Butterfly Monitoring Scheme. However, information for neither group has been analysed to see if changes have occurred as the fabric of the wood has recovered. I have seen or been informed about notable increases in recent years, such as for silver-washed fritillary *Argynnis paphia*, purple emperor *Apatura iris* and marbled white *Melanargia galathea*, but these are probably more due to broader changes at national or regional level or to introduction, than they are the indirect result of muntjac control. The UK Butterfly Monitoring Scheme clearly shows the silver-washed fritillary (Figure 18.4) has increased on a national scale (Fox et al. 2015), but it may have done better in the wood in the last few years than it would have done had muntjac remained at high density. Its larval food plant is the common dog-violet *Viola riviniana*, and in the past I have seen patches of violets heavily grazed by muntjac.

18.5   Impacts associated with different deer densities during the recovery

My earlier review of impacts in Monks Wood described the types of impact that were apparent at different deer densities (Cooke 2006). Here I extend that analysis, a process complicated by the recent increase in roe deer. Associations between density and impacts may vary depending on whether the situation is improving, worsening or has been stable for a number of years. The majority of conservation woodlands in Cambridgeshire, and probably elsewhere in the main range of the muntjac, were tentatively estimated to have had densities of 20–60 deer per km², but a small number of woods had higher densities.

18.5.1   At least 100 muntjac per km² (1 or more per ha)

Such deer densities existed in Monks Wood through the 1990s up until 1998. They were associated with unacceptable damage to coppice regrowth, lack of tree regeneration, loss of the shrub layer including bramble thickets, and modification of the ground layer with loss of floristic interest and an increase in grasses and sedges. A density of this magnitude occurred again inside the coppice exclosure by 2015; lack of paths and difficulty of access inside the exclosure meant that no stalking had taken place for many years and density increased – it only required six deer inside this 6 ha exclosure to reach such a density. In recorded localities inside the fence, bramble suffered die-back, young ash decreased in abundance and survivors were heavily browsed, dog's mercury disappeared, but ground ivy increased. It was a microcosm of what had happened to the whole wood in the 1990s.

18.5.2   About 60 muntjac per km² (0.6 per ha)

Between 1999 and 2005, deer density decreased from 60–70 per km² to about 50, with an average of 60 per km². Ground flora in Monks Wood partially recovered during this period. Such densities were, however, still associated with an absence of drifts of some spring flowers as well as noticeable effects on woody vegetation such as coppice regrowth, shrubs and seedling trees (Figures 15.5 and 18.1).

18.5.3   About 35 muntjac per km² (0.35 per ha)

Between 2005 and 2015, muntjac density declined from 50 to 20–30 per km² with an average of about 35 per km². This period was marked by further recoveries in ground flora with bramble thickets developing again and seedling ash surviving and growing. The understorey became dense in places and coppice regrowth was more likely to be acceptable. Recoveries were, however, still incomplete and further time is needed to determine whether various impacts are irreversible. Some impact in unprotected hazel coppice might occur at about 25 muntjac per km².

18.5.4   About 20 muntjac per km² (0.2 per ha)

This deer density was proposed for the coppice exclosure during 2000–2005. Here impact was slight, this being the only area of the wood where ash saplings in the height range 20–130 cm were found in 2005.

18.5.5   About 10 muntjac per km² (0.1 per ha)

Beginning in 1993, Ken Lakhani and I studied coppice plots that were unfenced or electrically fenced – the fences partially excluded deer. Dung counting in these plots indicated their relative deer densities. Fenced plots where there was little or no damage to regrowth had roughly one-tenth as much dung as the unfenced and severely browsed plots. Thus the average density inside these fences was probably in the region of 0.1 deer per ha. The wood at that time offered little alternative food and the muntjac appeared determined to feed on palatable regrowth (many died of starvation during the winter of 1993/4). Had Monks Wood not been impacted at the time and had a range of food been readily available, then a general density of 10 per km² would probably not have had any measurable effect. At similar densities elsewhere, however, it would still be prudent to be aware of the attractiveness of certain species such as oxlips (Tabor 1999; 2004). Muntjac can become a nuisance in situations such as gardens at densities as low as 2 per km² (Section 11.4).

18.6   Other potential factors affecting Monks Wood

Management of the muntjac had an immediate impact in the wood after stalking began in 1998, and conservation features progressively improved over the following 20 years. These recoveries helped to confirm that muntjac had a major impact on the wood in the late 1980s and 1990s. One should never forget, though, that other influences are at work on our woodlands. As Chris Gardiner and Keith Kirby noted in their tailpiece to my review of muntjac impacts in the wood (Cooke 2006), shifts in the plant community of the wood occurred before the arrival of muntjac, while structural changes since clear-felling in the 1920s were likely to account for some of the changes chronicled in recent decades.

Successional change is of course continually taking place over much of the wood, although measurements within long-term transects showed that canopy gaps decreased only slightly from 9.5% in 1996 to 8.1% in 2008 (Mountford and Peterken 1998; Tanentzap et al. 2012b). Succession following the clear-felling was already quite advanced when the muntjac began causing problems in the 1980s. Presumably ash die-back will mean that the wood will become more open for a while.

Terry Wells (1994) suggested that several species of grasses and sedges had spread because of the fertilising effect of high levels of nitrogen compounds in the atmosphere. But Tim Sparks and his colleagues (2005) found that Monks Wood had much higher cover of such species than other nearby woods, indicating a general factor such as air pollution was not responsible. They also pointed out that Monks Wood's relatively large size should afford protection from agricultural eutrophication.

Climate change is another factor to be considered as it might have both direct and indirect effects. Many plants, including herbs, shrubs and trees, may respond to changes in rainfall and temperature. Rackham (1999; 2003) blamed loss of primroses on longer, hotter summers. Flowering bluebells may face increasing shade because trees are coming into leaf earlier. Bluebells were recorded each spring at what I regarded as peak flowering date, with the vast majority having 'bells' but with some still in bud and a few going to seed. I found that bluebells in the wood were flowering earlier on average by one day every two to three years during the period 1994–2015; during this time, the range of peak flowering dates was 27 days, with the date varying in response to spring temperature. They may remain approximately in tune with leafing times in (even) earlier springs in the future, but changes in climate will continue to have a role in determining the future structure and composition of the wood.

There were of course other species of moderately sized, mammalian herbivores in the wood when muntjac were an especial problem in the 1990s, but none approached the numbers of muntjac, let alone the biomass. In 1993/4, sightings during 96 midday or dusk walks along an 8 km fixed route were: muntjac 2,333, water deer 13, roe deer 2, rabbit 559 and brown hare 384. In recent years, roe deer have increased considerably, while the lagomorphs have decreased. In the 1990s, rabbits were abundant in certain localities in the wood, but were only rarely found in the woodland or coppice areas where the worst browsing and grazing impacts occurred. Feeding by small mammals or invertebrates may still be comparatively important on certain species of vegetation. Currently about half of grazing on bluebell inflorescences is inflicted by small mammals, but before stalking started this was insignificant when compared with the destruction wreaked by muntjac.

18.7   Recoveries elsewhere

That recovery in Monks Wood has taken so long and some changes are likely to be irreversible does not come as a complete surprise. Information from other parts of the world has demonstrated that recoveries driven by culling are slow and not totally predictable. A long-running study of the effects of manipulating densities of white-tailed deer in the Allegheny forests of Pennsylvania dates back to the 1970s (Tilghman 1989; Horsley et al. 2003; Nuttle et al. 2014). The conclusion in the last of these articles was that 'elevated deer densities cause significant profound legacy effects on understorey vegetation persisting at least 20 years'. The authors predicted that simply reducing deer density does not guarantee recovery of the understorey – some unwanted features may need removing by other means.

Impacts of white-tailed deer were studied by Tanentzap et al. (2011) in deciduous forests in south-west Ontario. Declines in all tree size classes were recorded between 1981 and 1996; culling then reduced deer densities significantly until 2009, but recruitment of small trees remained limited. Deer control was acknowledged as being essential, but other forms of intervention were considered necessary to accelerate recovery in tree density. Similarly, although non-indigenous red deer had been controlled in Fiordland National Park in New Zealand since the 1960s resulting in densities of less than 1 per km^2, recruitment of palatable forest species remained affected (Tanentzap et al. 2009). Processes causing slow recovery and types of active management to reinvigorate recovery were discussed by Tabor (2006) and Tanentzap et al. (2012a).

CHAPTER 19

Deer impacts at Woodwalton Fen

19.1 Introduction

So far, consideration of environmental impacts has focused on muntjac, but this chapter deals with Woodwalton Fen NNR where large populations of both muntjac and water deer have occurred in recent years. The main question being addressed is what contribution does the water deer make to undesirable impacts on vegetation in this wetland nature reserve?

Extrapolating from an earlier relationship found between density and numbers seen per hour during winter dusk surveillance (Section 7.6.2) suggests that water deer densities in the reserve may have increased from typically 20–50 per km² during the early 1990s to more than 100 per km² in 2011/12, but then decreased again (Figure 19.1). Reasons for these changes have been discussed (Section 8.3).

Muntjac density has never been studied directly but signs and camera trap results suggested it rose to 50 or more per km² by 2010, so combined density of the two species in the reserve might have peaked at more than 150 per km², an astonishingly high (and unsustainable) density. It needs to be stressed that these densities are based on the size of the reserve, and the deer populations, especially that of water deer, range over a greater area for much of the year. In summer, however, deer are more confined to the reserve.

Concern has arisen that water deer might be responsible for impacts in nature reserves or elsewhere (White et al. 2004b; Dolman et al. 2010). The potential for problems might be anticipated to be greatest in Norfolk where significant concentrations occur over wide areas; however, reserve managers have reported few, if any, issues in the county. David Nobbs, the warden at Wheatfen Broad, wrote in 2002 that they made an attractive addition to the fauna and caused little damage despite reaching densities of 40 per km², and when I met him in 2015, he confirmed they remained a benign presence. On the Bure Marshes reserves, Rick Southwood told me in 2010 that they were not causing any problems, with grazing being 'virtually unnoticeable'. He added that it would be useful for him, as site manager, if 'they did

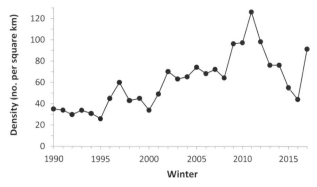

Figure 19.1   Estimated winter densities of water deer at Woodwalton Fen
NNR, 1990/1–2017/18, based on the size of the reserve. Deer considered
to be residents on adjacent farmland are not included.

a bit more'. Tim Strudwick, RSPB warden at Strumpshaw Fen on the north side of
the River Yare, had not noticed any problems from their grazing or browsing when
I visited in 2010, the local summer density on the reserve being about 20 per km².
The only issue raised was whether their paths in the reed beds allowed foxes easier
access to predate nests of rare birds, but he explained in 2016 that this concern had
not been realised.

In this chapter, consideration is restricted to the impacts of deer feeding
within Woodwalton Fen. Other effects, such as trampling rare flora, are likely
to be relatively insignificant, and potential impacts on surrounding farmland
are discussed briefly elsewhere (Section 11.2). Differences in the landscape at
Woodwalton Fen and on the Broads may have made higher densities and impacts
more likely at the former site. Broadland has great expanses of suitable habitat
for dispersal. In the past, Woodwalton Fen was an ecologically rich island in a sea
of arable fields, although since 2005 development of the Great Fen has probably
made the surrounding habitat more attractive to the west and north (Figure 10.1).
Emigration from Woodwalton Fen in other directions is rendered more difficult by
the Great Raveley Drain.

19.2   Early studies on impacts

Water deer must eat, and their biomass has at times exceeded one tonne per km² at
Woodwalton Fen, so the potential exists for them to cause impacts on native flora
and fauna. Until 2010, when camera trap studies began, relatively little was known
about what they were eating at the fen. On the vast majority of occasions when water
deer were seen feeding, they grazed at ground level. Examination by Lynne Farrell
and Tony Mitchell-Jones of stomach contents of six deer found dead during winter
and spring, 1977–1980, revealed that graze species predominated, with about half of
the diet being grasses and sedges (Section 2.9). However, a third of the fragments
were browse species with bramble leaves being the main component, so significant
browsing was evidently occurring when deer were hidden from view, either when
in cover or at night

The reserve is a patchwork of reed bed, mixed fen fields, grassland and sallow carr with blocks of birch or mixed woodland (e.g. Figures 1.7 and 4.3a). Up until the 1990s, there had been little concern about deer impacts on vegetation in the reserve, nor awareness of the potential for them to occur. Bramble was one of the species likely to show impacts first. Ron Harold, who was then reserve warden, wrote in an internal report in 1986 that bramble was the main constituent of the high field layer of the woodland, being present in dense, impenetrable patches. My initial studies on muntjac damage at Monks Wood in 1993 stimulated visits to the southern woodland at Woodwalton Fen in order to see if I could detect signs of muntjac impacts. During spring of 1994, I found places where both bramble and privet were heavily browsed, and most stems of bramble, privet, hawthorn and dog rose that I cut and left for the deer were quickly eaten. Nevertheless, scoring in 1995 indicated that overall levels of damage in that area of the reserve remained low. Around that time, Lynne Farrell and I concluded that deer were not perceived to be a problem on the reserve. This might have been because deer density was not particularly high or because the fenland habitat was robust (Cooke and Farrell 1995).

Deer impacts at Woodwalton Fen attracted little attention for another eight years. Then I took the opportunity to assess success of regrowth on coppice cut in the extreme south of the reserve in the winter of 2002/3. By the following autumn, birch was little browsed, hawthorn was heavily browsed and damage to sallow was intermediate. If this was a typical pattern of attractiveness, there would be concern over possible loss of hawthorn, with diversity in coppice species composition being seen as good for conservation. Densities of both species of deer were increasing steadily and perhaps worryingly, bearing in mind that problems from muntjac browsing were becoming apparent at nearby Holme Fen.

In 2008, I decided it would be prudent to score and assess deer impact in four areas of the reserve. Signs of the deer themselves were similar in all four areas, but impacts differed according to the nature of the habitat:

- In mixed woodland in the south of Woodwalton Fen, impacts were moderate.
- In open birch woodland and grassland in the south, they were low.
- In mixed fen and sallow carr in the north, they were slight.
- In reed bed habitat with some carr in the extreme north, they were also slight.

Although these results still did not raise major concerns, they did highlight four particular issues:

1. The mixed woodland had become more impacted and monitoring was needed.
2. It was also important to continue to monitor damage to woody vegetation in the north to see if problems developed, as, for instance, no form of protection was afforded to newly cut coppice. The north was viewed as being robust apart from the risk to coppice.
3. Although water deer had not to date ever been implicated in damaging conservation interests, they might be responsible for some of the browsing damage that was recorded.
4. Most of the concern centred on damage to woody vegetation, but it would be unwise to ignore totally the potential impact on ground flora. Comfrey *Symphytum officinale*, an abundant plant of the reserve, was conspicuously

well grazed by autumn 2008, so a targeted study was made. Grazing levels on comfrey plants in 11 compartments of the reserve varied between 7% and 88% with an average of 38%. Grazing on comfrey itself was not of concern, but here it was used to indicate what levels of grazing might occur more generally on palatable species.

To address the second and third issues in particular, browsing trials were undertaken in 2009 and 2010 in the most northerly 40 ha of the reserve, which was composed of reed bed, mixed fen and sallow carr. This area was selected because it was optimal water deer habitat and sightings of muntjac were relatively rare (Cooke 2009a). In an ivy trial in February 2009 (Section 12.6.3; Figure 12.6b), results were indicative of low browsing impact. Signs beside the groups of ivy suggested that water deer were responsible rather than muntjac. In spring 2010, stems were cut from trees in the reserve soon after they came into leaf, and their cut ends inserted into the ground in recently mown areas. The objective was to produce a worst-case scenario by maximising palatability and exposure. After three days, 60–70% of stems of hawthorn, birch and oak showed signs of browsing, as did 26% of sallow and 5% of alder. Alder stems were virtually untouched, but the remaining species were browsed, and the presumption was that water deer were responsible for the majority of any browsing. These results did nothing to allay growing concerns over browsing on hawthorn. Browsing on oak was not seen as a problem as this species is symptomatic of a drying site – wet alder woodland is a preferred habitat at the site. Birch has some value because it is short-lived and provides amounts of dead timber, but sallow is preferred because of the birds and invertebrates it supports.

Sallow and birch are especially abundant on the reserve, and regenerating trees of both species up to about 4 m in height were assessed for browsing damage. Of those surveyed, about 70% of both species had little or no browsing damage and their height was unaffected. Height was affected by browsing in most of the remainder. None of the sallow was dead, but 2% of the birch had died. As incipient sallow and birch can require manual removal from reed beds and other sensitive habitats, deer browsing is of some value on the reserve, but this may be more than offset by damage to regrowth stems in sallow coppice. It seems as if the deer cannot be useful in one area without being destructive in another. However, at Wheatfen Broad, David Nobbs told me that he does not coppice sallow carr, so water deer are, on balance, beneficial by browsing incipient sallow.

Sallow carr is still a conspicuous feature of Woodwalton Fen despite clearance programmes that have been undertaken for as long as I have been involved with deer surveillance. In the mid-1980s, coppice management was initiated to combat successional change to woodland that was occurring in some areas (Harold 1991). The original idea was to coppice on either 10- or 30-year rotations. Major clearance during 2002–2005 established distinct coppice areas which were managed as planned up until about 2011 (Alan Bowley, personal communication).

In summer 2010, a block of unprotected coppice cut the previous winter towards the northern end of the reserve was heavily and consistently browsed by deer. In view of the fact that deer populations in the reserve were higher than they had ever been, I decided to assess regrowth in blocks of sallow that had been coppiced in the north since 1998. Sixteen plots were found, of which four were deemed to have failed because of browsing. These four included the plot cut in 2009/10.

Table 19.1   Scores and impact stages in the mixed woodland in the south of Woodwalton Fen, June 2008 and April 2010–2018. Overall activity score is not a simple sum of the scores for the different species. Stalking of muntjac started in April 2011.

Year	Deer scores				Damage score	Impact stage
	Muntjac	Water deer	Roe	Overall		
2008	4	5	1	6	8	Moderate
2010	8	4	3	10	10	High
2011	8	3	1	8	10	High
2012	6	4	0	7	8	High
2013	7	4	0	7	9	High
2014	5	3	1	5	7	High
2015	6	4	3	7	8	High
2016	7	4	3	7	7	Moderate
2017	5	5	2	7	6	Moderate
2018	6	6	1	7	7	Moderate

When viewed in isolation, such a failure rate was not considered to be much of a problem by Alan Bowley, the Site Manager. One reason for this was that some areas were cut with the intention of eradicating sallow when it was growing where it was not wanted – usually such management was followed by stump treatment with herbicide. The background concerns, however, were that deer numbers were high and possibly still increasing and new coppice had been badly browsed.

As part of the work on issue 1 above, the worst damaged mixed woodland in the south was identified as a block of about 7 ha beside the western edge of the reserve. The east side of this block, an area of about 4 ha, had been scored in June 2008, and scoring was repeated in April 2010 (Table 19.1). Comparing results from the two years, there was an appreciable increase in both browsing damage and signs of muntjac. In 2010, evidence of a general browse line could be discerned throughout the area. This confirmed that deer management was required in order to improve regeneration and restore a shrub layer.

At the same time, grazing damage on conspicuous species such as yellow iris *Iris pseudacorus* and comfrey was becoming even more apparent – concern about issue 4 was increasing. For example, in May 2010, the average grazing level on the early leaves of iris was found to be 14% in fields of uncut fen vegetation. But where fields were cut during the previous winter, making the iris leaves more obvious, grazing increased fourfold to 56%.

So between 2008 and 2010, the situation visibly worsened. Significant impacts were apparent in the mixed woodland and there were more obvious effects on sallow coppice and flora. An additional concern was that browsing within mature sallow carr had reduced the abundance of species such as bramble. It seemed that water deer must be implicated in some of the browsing and grazing, but their exact contribution could only be guessed at. Scoring in 2010 (Table 19.1) indicated that muntjac were the principal contributors to woodland damage. But scoring for individual deer species, when two or more are present, is a crude and so far untested tool. An additional drawback is that when the peat soil is dry it fails to show slots and, when it is wet, slots might be too deep to be certain of their makers' identities; the fact that muntjac and water deer have browse lines of similar heights and bites of similar sizes did not help resolve the evidence. Nevertheless, it was strongly suspected that muntjac were the main culprits and a decision was made by Natural England to introduce muntjac stalking in 2011 to protect the woodland and its dependent fauna.

Just as it was felt that it was desirable to have better information on the deer species responsible for woodland impacts, so it was too with browsing on regrowth on sallow coppice and with grazing on flora. Alan Bowley and I decided that deploying camera traps could quantify relative amounts of browsing by the different species in mixed woodland, in mature sallow carr and in sallow coppice. This opened up an entirely new phase of work.

19.3   Browsing in mixed woodland and sallow carr

19.3.1   *Winter browsing*
Browsing studies to determine the relative contributions of different deer species commenced in winter between January 2011 and March 2012 (Cooke 2012b). Wildlife camera traps, set to take ten-second videos at nominal one-minute intervals, were situated in mixed woodland or sallow carr roughly 3 m from examples of target vegetation (Figure 19.2). Videos were used because photographs usually failed to prove unequivocally that browsing occurred (Figure 19.3a).

Bramble was considered to be a key species as it had evidently been affected by deer browsing. By January 2011, any surviving bramble bushes had few leaves remaining below the browse line, so bushes were constructed by cutting foliated stems from above the browse line and inserting these in the ground to make a bush about 60 cm in height. Naturally growing bramble bushes were studied in mixed woodland during late autumn and winter of 2011/12. Browsing on ivy was

Figure 19.2   A camera trap set to take videos of deer browsing on bramble.

Figure 19.3   Camera trap photographs. (a) A muntjac at a constructed 'bush' of ivy. Photographs usually fail to prove that browsing is occurring, rather than the deer just looking at or smelling the target vegetation. (b) A yearling doe water deer approaches a well-browsed sallow stool. (c) A brashed sallow stool being browsed by a buck water deer in summer 2012; this individual could be recognised by his torn left ear. (d) A water deer grazing on comfrey.

also studied. All ivy-clad trees in the reserve had browse lines and any ivy on fallen trunks or branches was quickly defoliated. Bushes roughly 60 cm high were constructed with cut ivy to provide information on browsing in the woodland and carr (Figure 19.3a). The studies focused on winter browsing because both of these species retain their leaves into winter and are important winter forage items. Moreover, both can suffer heavy browsing, which in the case of bramble may be associated with die-back of bushes.

For each video, records were kept about deer species involved and whether browsing or smelling vegetation occurred. During these two winters, a total of 426 videos of browsing on target vegetation were recorded in 356 camera days. Of these, muntjac featured in 98%, water deer in 1.4% and roe in 0.2%.

To provide a better indication of relative activity of the deer species in these areas, cameras were set on deer paths:

- In mixed woodland in 2010/11, 62% of records were of muntjac, 36% of water deer and 2% of roe.
- In mixed woodland in 2011/12, 73% were of muntjac, 26% of water deer and 1% of roe.
- In sallow carr in 2011/12, 22% were of muntjac and 78% of water deer.

So although there were water deer in the mixed woodland in the winter and their videos outnumbered those of muntjac in the sallow carr, they scarcely participated in the browsing in either habitat. This was a surprise as both bramble and ivy had been found in the rumen contents of water deer picked up dead during 1977–1980 (Section 2.9). These dead deer were examined before muntjac colonised the reserve and substantially reduced the abundance of these browse species. As muntjac increased over the intervening period of 30 years, so water deer may have had less opportunity to feed on these species. Muntjac were evidently actively seeking them out: this was especially true of ivy with 225 browsing videos in only 53 camera days. When they located a bush, they kept returning until it was defoliated. Water deer, on the other hand, showed relatively little interest in the bushes, and even when they found them the deer often sniffed, apparently suspiciously, rather than browsing. In videos of water deer, smelling was twice as frequent as browsing, whereas for muntjac, browsing videos were six times more numerous. It is possible that water deer were in part deterred by the smell of muntjac.

Videos also highlighted the fact that muntjac have a much more woody diet – they were sometimes videoed gnawing woody stems of ivy with their cheek teeth. Fully adult male water deer would be prevented by their tusks from feeding in this way, and females have not been filmed or seen browsing on woody stems either.

19.3.2   Muntjac and stalking

Culling of muntjac began in Woodwalton Fen in April 2011 in order to reduce browsing impact. As the mixed woodland was particularly badly affected, stalking started there, and 59 deer were reported to have been removed in the 12 months up to March 2012, equivalent to about 0.3 per ha per annum. Thereafter, fewer deer were shot and stalking was also undertaken elsewhere in the reserve. Up until April 2017, 93 had been shot, less than 0.1 per ha per annum overall. Neither water deer nor roe deer have been shot on the reserve.

The introduction of stalking in the mixed woodland in 2011 led to muntjac becoming more nocturnal (Table 12.1). A combination of camera work and surveillance revealed that, as deer were culled in the optimal habitat of the woodland, so their places were taken by other muntjac moving in from elsewhere in the reserve. Thus, in 2011/12, stalking affected density outside the woodland rather than within it. In subsequent winters, the muntjac presence decreased within the woodland too.

19.3.3   Bramble browsing through the year

From May 2013 until April 2014, cameras were positioned to video events at a single bramble bush in a clearing in the middle of the block of mixed woodland. The aim was to record occurrence of the deer and their browsing through the whole year, rather than just in the winter. Browsing was recorded steadily through 2013. By January 2014, there were few whole leaves remaining below the muntjac browse line, and by February only fragments of leaves were left below the browse line. New shoots were first seen in late March 2015 in an early spring, and browsing on these started in April. During the year, the bush spread sideways as several long lateral stems rooted.

The study began after two years of stalking. Muntjac were present more or less throughout the 12 months but were most numerous during January–April. Their browsing showed no seasonal peaks. Water deer were most apparent in August–September and particularly in November–December. There was likely to have been movement north during February–April to graze on the rides, open fen areas and farm fields. Most of their browsing on the bramble bush was done during August–December, although much of their attention in November–December was focused on grazing. There was no browsing January–April, when the bush had few leaves available. The presence of roe was largely restricted to the period December–February, the deer presumably entering the reserve woodland from the farmland to take advantage of its food and cover; they were able to reach bramble leaves above the browse line of the smaller species. Hemani et al. (2004) drew attention to the aggregation in winter of muntjac and roe on bramble in Thetford Forest.

Although water deer were more frequently videoed over the whole 12 months, muntjac were the more frequent browsers. Much of the browsing by water deer was recorded when the bush was actively growing. Roe deer contributed relatively little overall.

In midwinter 2013/14, cameras were also positioned on another bramble bush. This second bush suffered much browsing by muntjac and roe, but none by water deer. Overall contributions to winter browsing on the two bushes were: muntjac 63%, water deer 5% and roe 38%. These were very different from the figures of 96%, 3% and 1% respectively found for natural bramble bushes in midwinter 2011/12, indicating that roe deer rather than water deer took advantage of the reduction in muntjac numbers.

19.3.4   Changes in deer numbers and impacts in the woodland

Camera traps have been deployed since the winter of 2010/11 in the southern woodland to study various aspects of deer ecology, especially browsing and grazing on vegetation. The primary aim was not usually to monitor changes in deer numbers from winter to winter, but the data collected might tell us something about changes since stalking began in April 2011. It is appreciated that changes in numbers of videos may reflect behavioural alterations as well as changes in abundance. In Table 19.2, information is collated on numbers of each deer species caught on camera during December–February in four winters; no cameras were deployed in 2012/13. Muntjac data indicated that numbers were maintained between the first two winters covered, as discussed above; after that, numbers decreased considerably. In contrast, water deer records changed little during the four winters, despite the reduction in muntjac. This may be partially due to the highly territorial nature of water deer bucks in midwinter. However, dusk surveillance 2011–2018 showed some upturn in numbers more generally in the south of the reserve (Section 17.3). Roe deer were not as numerous in 2014/15 as in the previous winter, but were again

Table 19.2   Average number of deer that were recorded by camera traps in the mixed woodland during midwinter, 2010/11–2014/15. There were no observations in 2012/13. Deer rerecorded within five minutes were not counted.

Species	Deer videos or photos per ten camera days			
	2010/11	2011/12	2013/14	2014/15
Muntjac	15.9	16.8	7.9	7.8
Water deer	5.2	2.3	4.1	3.4
Roe	0.5	0.4	3.6	1.9

higher than previously. Despite the stalking, muntjac remained the species most frequently caught on camera in the woodland.

Since 2010, deer activity and damage scores have been determined each April for the core area of the mixed woodland block (Table 19.1). Deer scores are derived for each species but a single damage score is recorded for all deer. There was a reduction in muntjac deer score to 2014 with a corresponding change in overall damage score. Impact stage was assessed as 'high' until 2015, but improved to 'moderate' in 2016. Some of the previous impacts will take time to recover. For instance, browse lines on species such as hawthorn can persist even though browsing intensity has declined.

19.4   Browsing on sallow coppice regrowth

The initial study on browsing on sallow coppice regrowth involved camera traps being set up during May–July 2011 on target stools in five unprotected plots in the centre and north of the reserve (Cooke 2012b). These plots had all been cut in the previous winter. The management aim in three plots was to stimulate regrowth, but the goal was not clearly defined for the other two plots. However, for the purpose of this study, it was assumed that the aim in each case was to achieve good regrowth.

Browsing began in May and had started to decrease in intensity by the end of July. As usual, cameras were typically set to take ten-second videos at nominal one-minute intervals. In the whole study of 146 camera days, 99 browsing videos were recorded, with muntjac contributing 39% and water deer 61% (Figure 19.3b). So unlike results from the woodland and the mature sallow carr, water deer were the main browsers. The ratio of the total number of videos of each species (1.6 water deer to 1 muntjac) may provide a reasonable estimate of the relative abundance of the two species in coppice areas in summer. Both species browsed in a relaxed fashion on sallow regrowth, with videos tending to show deer browsing only briefly or ignoring stools as they walked past them, contrasting with the avid feeding of muntjac on ivy and bramble in winter.

The impact of the brief, but fairly frequent, episodes of browsing on the sallow was often to create and maintain densely packed new growth a few centimetres in height (Figure 19.3b). These fast-growing young stems seemed very palatable to the deer. In the absence of browsing for a week or two, stems grew, but might then have their tips removed by deer.

Figure 19.4   A plot of badly browsed sallow coppice being examined by the Site Manager, Alan Bowley.

At the end of the first growing season in 2011, all five plots were judged to have failed because of browsing (Figure 19.4). To determine whether recovery occurred during the second growing season, the plots were assessed again in late summer 2012. Browsing in 2012 was limited by dense growth of grasses and sedges, but many stools were dead and few seemed likely to survive. Browsing in 2011 was the ultimate cause of failure; being overgrown in 2012 was the proximate cause of death of stools. Evidently, if there is severe browsing in the first growing season because of high deer density and/or lack of protection, good protection is of no benefit in the second season if the stools and their regrowth become overgrown.

Whether stools might be protected by brash was investigated in 2012 in four newly cut plots. Three pairs of stools were selected in each plot. One stool in each pair was covered with cut sallow and the other was an unprotected control. Brashing reduced browsing impact. In every case for the 12 pairs, the brashed stool had higher regrowth, longer stems and less overall browsing. The brash also promoted growth of other vegetation so helping to protect the stool further. When it came to producing acceptable regrowth, the effects of brashing were more equivocal. Five of the 12 brashed stools did not have any regrowth above 1 m (all of the controls failed). Brashing protected regrowth against early-season browsing, but when stems grew higher than the brash, they were exposed and might be 'tipped' by deer. Results suggested, however, that there was a possibility that if an entire plot coppice was protected by brash then the overall deer presence would be reduced and acceptable growth may occur. In other words, if there were no easy pickings on unprotected stools, deer may forage less in those plots.

Camera traps were deployed again in 2012 to monitor browsing in these four plots. Muntjac deer were responsible for 24% of the browsing and water deer for 76%. The relative contribution from water deer was slightly higher than in 2011. Two individual deer proved particularly troublesome. A distinctively-marked and limping muntjac buck was responsible for all of the recorded browsing by that species in one plot. And a water deer buck with a torn left ear ranged over two of the plots and inflicted 74% of the total browsing by that species (Figure 19.4c). While it might be possible to shoot a troublesome muntjac, targeted shooting of individual water deer is not an option in summer as the close season extends from April to October.

Local winter flooding and staff changes meant no coppicing was undertaken for several years. Numbers of both water deer and muntjac declined in the reserve between the winters of 2011/12 and 2013/14, but, because there had been no coppicing, it was not known whether browsing pressure on regrowth had been relaxed. Therefore, in the spring of 2014, I undertook a small trial to test whether impact on regrowth might still be occurring. The aim was to set up a trial under conditions most likely to lead to browsing, so that if there was no significant impact, this would suggest that browsing was no longer a problem. A small patch of sallow was selected in an area with as high a density of water deer as anywhere on the reserve. In May 2014, leafing stems of sallow were cut in a patch a few metres across. Stems were cut at about 40 cm above ground and surrounding vegetation was cut and flattened. Any regrowth would then be invitingly above the surrounding vegetation at a convenient height for browsing. Cut stems of sallow were also regularly planted in the ground to encourage browsing. Camera work during June and July indicated that water deer were more active in this area than muntjac and were responsible for 92% of the browsing. The level of browsing was judged to be unacceptable with no regrowth above 1 m. Several water deer were involved in the browsing activity, but a buck with a distinctive broken tusk was responsible for most of it.

This trial indicated that it would be unwise to coppice sallow the following winter without giving the regrowth some protection (coppicing plans were eventually postponed). If it is known well in advance where and when coppicing might be undertaken, then a simple preliminary test of this type could be done to assess likely browsing impact. Cameras need only be used if information is required on which species or individuals inflict the worst damage. It is possible that deer populations in some sallow areas of the reserve might have been below the critical density associated with unacceptable impact. Similar quick and easy trials could be used to track change as well as predicting outcomes in specific localities.

19.5   Grazing on comfrey

Grazing on comfrey was also studied with camera traps (Figure 19.3d). One of the largest patches growing in the reserve was in the mixed woodland in the glade where bramble browsing was monitored from May 2013 until April 2014. Water deer had been videoed in the distance in early autumn 2013 grazing on this comfrey, so from May until December 2014 a camera was positioned on the patch in order to record deer presence and grazing. Water deer were the most abundant species on the patch and did virtually all of the grazing. Muntjac only contributed 4% to the grazing despite being present in most of the months. Roe deer were absent while the comfrey was palatable.

There was a little grazing by water deer in July and August, but most of the grazing occurred in September and October. By early October, the comfrey was severely grazed and starting to go brown and limp, which is normal at that time of year. Grazing continued until the last week of October, by which time the comfrey was no longer green. Grazing so late in the season probably has no consequence for the comfrey or for foraging insects (it provides good sources of nectar and pollen in summer). Comfrey remains an abundant and widespread plant on the reserve, but this study illustrated that the potential exists for water deer to be a threat to the survival of other, scarcer species of ground flora. It has been reported that a high density of water deer can pose a problem for growers of commercial flower crops (Childerley 2014).

19.6   Conclusions on the impact of water deer

Readers may be wondering why, in addition to introducing stalking of muntjac in 2011, Natural England did not simply fence the new sallow coppice, and perhaps shoot the water deer as well. Water deer were not shot because, at the time, there was no direct evidence that they were a problem and because of conservation concern for the species on a global scale. Coppice was not fenced because unprotected regrowth had usually survived in the past and the culling of muntjac should have provided extra protection.

The outcome of the studies was that water deer were, for the first time, demonstrated to cause unacceptable impact on unprotected coppice. They contributed to browsing in the woodland, but to a much lesser extent than muntjac. The less numerous roe deer also contributed to localised winter browsing. Some results indicated that roe deer benefited more than water deer from the reduction of muntjac numbers in the mixed woodland. Previously, Norma Chapman and her colleagues showed that muntjac could influence habitat use and local numbers of roe deer in the King's Forest in Breckland (Chapman et al. 1993).

At what density of water deer did browsing effects switch from being acceptable to unacceptable? Relevant information from the northern half of the reserve where water deer were the more abundant species included:

- The ivy trial in February 2009 pointed to low browsing impact during a winter when water deer density was estimated to be about 60 per km^2.
- The trials with tree stems in spring 2010, however, registered higher browsing impact, following a winter when density was estimated to be roughly 100 per km^2. And severe browsing on sallow coppice regrowth was noted that summer.
- Unacceptable impact occurred in the sallow coppice trial in May 2014, following a winter when density was estimated to be about 80 per km^2.

On that basis, unacceptable effects became much more likely at water deer densities of 80 and above per km^2. Such high densities seem to be rarely encountered. The state of the mixed woodland in the south of the reserve deteriorated between 2008 and 2010, coinciding with the increase in water deer density. However, muntjac sightings increased progressively and substantially for several winters after 2006/7 (Figure 8.1), and camera trap studies indicated they were causing most of the impact in the woodland. Some of the impact in the northern half of the reserve will also have been caused by muntjac but, because most plots were severely browsed, it is likely that the result would have been unacceptable even in the absence of muntjac. Other camera studies, not covered here, have confirmed that water deer readily browse on new summer growth on species such as blackthorn and hawthorn. Alder is evidently relatively unpalatable.

Whether significant browsing occurs on palatable species in the field will depend on local factors such as immediate deer density and even whether one particular individual has a liking for new woody growth. As was seen at Woodwalton Fen, a single deer can tip what might have been an acceptable result the other way. The question of protection for sallow coppice remains to be resolved. Currently, if the management purpose is to produce healthy regrowth, then protection would be recommended. Whether brashing all stems in a plot might be sufficient is still to be tested. Sallow in a fen site is likely to recolonise an area of failed coppice more quickly that had it been hazel in woodland.

There was a suggestion that water deer might be deterred from browsing bramble and ivy by the prior browsing of muntjac. If such an effect existed, then a change in behaviour might be expected after the muntjac population had been reduced, but this was only partially supported by later evidence. The year-round study (Section 19.3.3) demonstrated that water deer browsed at least as frequently as muntjac on new season growth from late spring until early autumn. No smelling was recorded throughout the 12 months, but there was no hard evidence that a reduction in the muntjac population led to increased browsing by water deer in midwinter.

The comfrey study demonstrated that water deer can be attracted to specific ground flora at certain times of year. It was also noticeable that browsing on bramble in the study glade dovetailed with grazing on comfrey, although observations were made in different years. Browsing on bramble peaked in August when grazing on comfrey was getting under way. Browsing then tailed off as comfrey grazing peaked in September and October. When the comfrey died down, the water deer turned their attention mostly to the grass sward. A brief examination of orchids in June 2017 did not detect any grazing on common spotted or early marsh orchids *Dactylorhiza incarnata* in fen fields frequented by water deer.

Landscape developments at the Great Fen should continue to benefit water deer, and the deer community will help to shape the development of habitats. The area will not be managed by the same philosophy that has been applied to its constituent NNRs. Alan Bowley (2013) said that the new philosophy was 'to allow more natural processes to operate in the absence of a defined end-point' and to 'experiment and record how land changes, rather than instigate management for a prescribed outcome'. It would, he continued, provide the opportunity to pursue in a more realistic fashion the concept of naturalness that should be a fundamental attribute of NNRs, as envisioned by Derek Ratcliffe (1977). The Wicken Fen Vision is a similar large-landscape proposal, for which Francine Hughes and her co-workers (2011) outlined how monitoring and evaluation would proceed in a paper which they titled 'a journey rather than a destination'. These authors appreciated that the roe deer in particular was a 'landscape species' capable of being a vector of habitat change. In the Great Fen, the deer community has already shown itself capable of causing significant change within both NNRs, and this will continue outside the old reserves if populations are not controlled. This is especially true if roe continue to colonise and red deer spread in from the east and become resident. How the laissez-faire approach will be reconciled with the current level of culling and with future deer impacts is not clear.

CHAPTER 20

Overview of management

20.1   Muntjac management at site level

These two species of deer are virtually at the opposite ends of the spectrum as regards impacts to conservation interests. Muntjac have had a considerable impact on conservation woodlands over much of lowland England, while water deer have, as far as we are aware, had only a minor impact in wetland reserves. An approach to management of muntjac might be viewed at different levels, such as site, county or national.

Some years ago, I made recommendations for tackling muntjac in conservation woodlands (Cooke 2004):

- Monitor the deer population and look for and monitor changes in features of conservation importance, including structure and natural processes, regeneration of young trees and coppice, ground flora and particular features of the wood in question.
- Determine objectively whether effects are unacceptable, whether muntjac deer are responsible, and what management may be required.
- Undertake management of vegetation and/or deer, and continue to monitor.
- Review progress regularly and continue with or amend management and monitoring as necessary.

Advice is now available within these pages and elsewhere on how to monitor, manage and make decisions. To the bullet points above, I would now add a certain amount of fine tuning. Thus, observe when during a year critical damage may occur and when control is most effective. Observe how, when and why muntjac move within the landscape and whether management needs modifying to take this into account. In recent times, this approach has come to be termed 'adaptive management'. Techniques and strategies need to evolve as situations change and knowledge increases; this is a statement that applies to all levels of deer management and to woodland management more generally. And an integrated approach, involving several management methods, is usually best, especially for a large wood where there may be multiple types of impact.

With a comparatively isolated site, it might be appropriate to focus solely on it, but in most cases consideration of management elsewhere in the local landscape is likely to be necessary. To give an example, the Woodland Trust has sanctioned stalking in Archer's Wood in Cambridgeshire since 2000, but the wood's recovery appears to have been inhibited by muntjac entering from Coppingford Wood, which it abuts at its southern tip. When I last assessed impact in Archer's Wood in 2015, it was high in the south but slight or low in the north.

20.2   Monitoring at site level

The monitoring methods that I have used locally have proved sufficiently sensitive and robust to detect and demonstrate impacts and recoveries. Thus at Holme Fen, recording bramble and fern height along transects quantitatively described the changes apparent in fixed-point photographs (e.g. Figures 16.8 and 16.9). The use of simple practical methods allowed a range of species and sites to be monitored with relatively little input.

Scoring signs was one of my main methods (Section 7.2.4). Deer score was based on four types of sign and damage on five, the frequency of each component sign being assessed as 0, 1, 2 or 3. Scores for each of these signs in Monks Wood are given in Appendix 2 for 1995–2017. The scoring visit in 1998 was made just as stalking in the wood was beginning. Each of the nine signs showed a highly significant decline over time.

The fact that vegetation recovered as muntjac populations were culled or kept out of sensitive areas was one piece of evidence that helped to confirm that muntjac were indeed at least partially responsible for the original impact. This was further confirmed where extent of impact was shown to be related to muntjac density. Earlier and other evidence that muntjac deer were responsible included:

- Impacts worsened as muntjac populations built up.
- Damage scores were higher in sites with high deer scores.
- Species of vulnerable vegetation survived inside exclosures but not outside.
- Whereas the converse was true for paddocks with captive muntjac, where the deer might show preferences for certain palatable species.
- Bioassays in woods with palatable species, particularly ivy, revealed the extent to which species might be taken.
- Camera traps demonstrated the ability of muntjac to defoliate certain species and browse woody stems.

Notwithstanding likely contributions from other factors, such as woodland maturation, air pollution, other herbivores and diseases, the mass of evidence points unequivocally to the fact that (1) muntjac can build up to sufficient densities to impact the structure and species composition of woodlands to a significant degree, and (2) management can permit (some) recovery to occur.

Although this chapter focuses on management, it provides an opportunity to ask how I might have improved surveillance and monitoring. I regret the following:

- Not starting counting the number of dung pellet groups seen per hour during scoring visits until 2005. Simple counts made annually over many years can be a useful guide to changes in deer density.

- Not recording a measure of cover of bramble inside monitoring plots. I focused on height attained by bramble, but as it grew, so it spread.
- Not taking more photographs prior to management being undertaken or becoming effective. I would then have had more options for fixed-point photography at later dates.

Those are activities that I might have thought of earlier. There are others in Monks Wood that I became aware of too late:

- I did not start counting the number of flowers on bluebell inflorescences until 2005, when Sparks et al. (2005) reported their observations – by which time recovery was complete.
- Had I known what was happening, I might have studied herb-Paris and crested cow-wheat in the mid-1980s when both were declining and the muntjac population had reached saturation point (Section 15.2.1).
- Similarly, I might have started routine observations of lesser celandines and other species along rides to monitor recovery.

This is also an opportunity to muse on long-term studies. George Peterken and Ed Mountford (2017) devoted a chapter in their book, *Woodland Development*, to a discussion of the merits and difficulties involved in the long-term monitoring of plots in woodland. Some of the particularly salient points made were:

- Studies become long term usually because of the efforts of individuals rather than organisations.
- Funding can be fickle. Although little may be needed, no organisation is likely to commit to open-ended funding.
- It is unrealistic to begin such a study without secure land tenure.

My long-term observations have rarely utilised permanent plots – the annual study of bluebells in Monks Wood from 1993 until 2014 is the longest I have worked with the same plots. Rather I have tended to record animals and plants in a routine quantitative manner, sometimes for more than 40 years. Many of these studies have, frankly, been superficial, but because they have been repeated for a number of years, they develop into more useful data sets. Their superficial nature has meant that many could be accommodated, especially if they needed recording at different times of year or not necessarily annually. The fact that I have lived and worked in the same area for about 50 years and most observations have been done in nearby secure nature reserves has meant that the years have ticked by and the data sets have grown in both number and length. New projects have continually suggested themselves and I have always found it easier to begin new projects than relinquish old ones. Probably the only benefit of being seriously ill occasionally is that it forces otherwise unpalatable decisions to be made.

Many projects could initially be better described as surveillance rather than monitoring, because measurements were not being made against a yardstick. Thus, having determined the level of grazing on bluebells in Monks Wood in 1993, I decided to continue to record in the same way each spring, simply to see what happened over an indeterminate period of years. Then when stalking began in autumn 1998, it became a monitoring exercise to see how observations changed compared to the baseline data of 1993–1998. Once bluebells had recovered from the effects of grazing, the rationale switched back to surveillance. Another point made by Peterken and Mountford (2017) was that it can be very difficult, impossible or even dishonest to specify a precise aim at the beginning of a long-term project. And many years after it has begun, it may have uses that would never have been envisaged at its inception. For instance, my bluebell observations have provided phenological data that demonstrate the species is flowering earlier in warmer spring weather.

20.3   Muntjac management in Cambridgeshire

20.3.1   *The threat from muntjac to conservation woodland*

How great is the threat from muntjac and how frequently is management needed? White et al. (2004b) reported that, at a landscape scale, models indicated that muntjac only appeared to exacerbate damage to woodland when present with other species of deer. The situation in Cambridgeshire provides an opportunity to describe what happens on the ground when only muntjac are present. For 19 woods managed by the Wildlife Trust that lacked significant populations of other species, four had slight impact, four had low impact, seven had moderate impact, three had high impact and one had very high impact. This sample contained eight woods of 10 ha or less. It did not include the two big NNRs of Monks Wood and Holme Fen, where impacts became severe, or the two largest holdings of the Woodland Trust (Aversley Wood and Archer's Wood), where high impacts developed. Some form of deer management specifically against muntjac was undertaken in more than half of the Wildlife Trust woods.

Muntjac will have attained sufficient densities to have caused observable impact in most woods in Cambridgeshire on unprotected coppice regrowth, tree and shrub regeneration and bramble growth. This in turn is likely to have affected some animal species dependent on particular forms of woodland structure and composition. Impacts on ground flora will have varied depending on factors such as palatability. Vulnerable species, such as oxlips, will probably have been heavily grazed in most woods where they occurred, whereas significant levels of grazing on bluebells and most other species were rarer. There will have been potential for similar effects elsewhere within the main range of the muntjac, although the presence of fallow and other larger deer species should have ameliorated the contribution from muntjac (Section 9.4).

20.3.2   *Approaches to management*

Impact from muntjac first became apparent locally in Cambridgeshire in the 1980s when coppice regrowth was found to be unacceptably browsed. Since then it has been the monitoring of impact, rather than the monitoring of deer numbers, that has driven management. A variety of barriers were used to try to prevent deer gaining access, including brash either laid over stools or inserted into the ground around individual stools. Alternatively, coppice plots were protected with more extensive barriers constructed from brash, paling, electrified strands, metal panels or from wire or plastic fencing materials. Fences which worked well were constructed from robust material that did not allow ready access under, through or over the fence. They tended to be less successful the larger they were because there was a greater likelihood of (1) deer being fenced in or being fenced out but determined to get back in again, and (2) damage from falling branches and trees.

Over time, it became clear that simple barriers were not necessarily enough on their own. Either inappropriate methods were being used or only small areas of woods were being protected. Gradually the reluctance to sanction culling was overcome until a tipping point was reached. My personal Damascene moment arrived in 1994 when I realised that unacceptable die-offs were an inevitable consequence of the muntjac population in Monks Wood being unsustainably large. For about 20 years, stalking has been used in many local conservation woods in addition to fencing. And this has resulted in reductions in muntjac density both locally in the relatively well-wooded polygon of 130 km² (Table 13.2) and more generally in woods in the former county of Huntingdon and Peterborough (Figure 13.7). The current position is broadly acceptable although there are some situations that give cause for concern.

It is important that such improvements should not be viewed as easy to achieve. Much effort has been directed at muntjac control in this area, and we have advantages over several other areas in East Anglia. Fallow and red deer are largely absent from much of western Cambridgeshire, and these species cause serious environmental impacts elsewhere. By virtue of its larger size, herding instincts and mobility, impacts from fallow deer are especially difficult to control, requiring a major effort from all concerned across a large tract of countryside. In sites where control is lacking, fallow may shelter there in daylight, and forage widely during hours of darkness. In contrast, the muntjac is small and tends to be solitary and territorial, although muntjac do not necessarily remain within woods throughout the day or the year.

We also have an advantage in western Cambridgeshire in that the few remaining interesting woods tend to be owned and/or managed by conservation organisations such as Natural England, the Wildlife Trust and the Woodland Trust, or by large private landowners. There has been a general commitment to improve biodiversity, aided by input from the Deer Initiative. Few landowners mean that liaison is simplified. Stalking has been undertaken by estate employees or by groups and individuals who appreciate that the primary aim is to reduce muntjac density in the woods to acceptable levels. For example, muntjac control in the three large local NNRs in recent years has been characterised by regular and frequent communication between Natural England, the stalkers and me. We have regularly met on the reserves to discuss and review the situation and to agree what is needed in the immediate future. Stalkers have been able to ask for management to be undertaken to facilitate deer control, such as cutting finger rides. For my part, I have provided a historic and holistic view of what was happening in a reserve. I have been able to warn of local pockets of high deer density or impact, which could often be quickly tackled by my colleagues. The main criterion for success of the management effort is that impacts decrease to acceptable levels and, by scoring and detailed observation of conservation features, I have provided what I hope has been relevant and up-to-date information. It is important that everyone involved in management has confidence in the dedication and competence of their partners.

Having a specialist on the ground to monitor and advise woodland managers and stalkers may be something of a luxury. The managers may be able to undertake monitoring of impacts themselves or be in a position to instruct someone else to do so. Meanwhile, stalkers are ideally placed to gather useful data of numbers and sexes of deer culled or seen per unit of effort that will help to understand how a population is changing. Many stalkers do this already, but all should be encouraged to do it irrespective of whether someone else is monitoring the deer.

20.3.3 Science and management

In 1993, Tim Clutton-Brock, who is widely known for his studies on the behavioural ecology of red deer, addressed the Annual General Meeting of the BDS on the subject of 'Deer and scientists'. He claimed that estate owners in the Scottish Highlands regarded scientists as 'only slightly less troublesome than midges', but went on to say that deer managers were fortunate that scientists had questioned traditional procedures and knowledge, such as the longevity of red deer stags.

Both muntjac and water deer are comparatively recent additions to our fauna, so there is little traditional knowledge attached to them that might need testing and if necessary correcting. Nevertheless, an issue of *Deer* magazine from 1989 carried an article from Norma Chapman entitled 'Muntjac biology for deer managers'. This was the summary of a talk given to the BDS's Services Branch the previous year. In it she explained (1) the specific identity of muntjac occurring in this country, (2) the fact that the males have a regular annual antler cycle, and (3) how to avoid leaving an orphaned fawn to starve to death when culling.

Since then, knowledge on other relevant aspects has gradually accumulated. When I started recording muntjac impact in the early 1990s, there was little information available on the subject and I had to teach myself how to recognise and interpret some of their signs, such as stem breakage. Resolving the veracity of 'traditional' knowledge has been limited to topics such as whether muntjac bucks use their tusks when fraying stems. My studies have been aimed at understanding better the natural history of the two species and providing information on recognising and monitoring damage in areas of conservation importance.

The point is often made that deer density and impact are not linearly related, so that reducing density does not necessarily lead to the recovery of affected vegetation. However, the knowledge base of this subject has not been especially extensive. Information in this book provides examples of different types of relationship. Impacts typically begin at certain threshold densities, but a single deer with unusual habits or tastes is capable of causing observable effects below an acknowledged threshold density. Threshold densities are different for different impacts. Above the threshold, although impacts are more likely to occur, they are not necessarily directly related to density. In some cases there is an upper density threshold beyond which the impact gets little or no worse. At Monks Wood, for instance, the graph of ivy taken in relation to estimated muntjac density (Figure 14.11) demonstrated that amounts eaten increased up to a density of about 70 deer per km², but there was only limited further increase at higher densities. Thus reducing deer density from, say, 120 to 80 per km² would be expected to cause little reduction in ivy browsed, but reducing density further to 40 per km² should result in considerable improvement. Yet, reducing deer density from 120 to 80 per km² would lessen grazing on bluebells (Figure 16.5). At Raveley Wood, early-purple orchids were comparatively lightly grazed at low deer scores, with variable levels at higher scores (Table 16.2). There, as deer score was reduced from a peak level, irregular improvement occurred until a threshold was reached and little or no grazing became the norm.

Looking at impacts across samples of woods, grazing on bluebells did not approach unacceptable levels (i.e. greater than 15%) until deer scores were moderately high (Sections 16.2.2 and 16.2.3). However, in Essex woods, a more worrying relationship was found between percentage of oxlip inflorescences grazed and combined score for all species of deer – fallow, roe and red were also present (Tabor 2002). Grazing could be as high as 60% at relatively low deer scores. Some woody vegetation is also vulnerable. Signs of browsing on bramble occurred even at the lowest deer scores (Figure 14.7). Tree seedlings are another feature particularly sensitive to deer browsing (Gill and Morgan 2010).

20.3.4   *Recoveries and new approaches to management*

As was seen at Monks Wood and elsewhere, reducing deer density can lead to recoveries in many of the affected species of vegetation, but this takes time and does not always result in full recoveries, in part because the nature of the wood has changed during the intervening years and seed sources may not have survived. But the goal should be to restore woodland processes to working order rather than to replicate some past state. The intervention of deer can be put in perspective by some event that changes the wood even more fundamentally. This is likely to happen with ash die-back. Currently, the development of the disease is being monitored in affected woods such as Monks Wood, and will be dealt with as the situation develops and management advice becomes available. Deer and other stressors, such as disease and climate change, may act in combination to produce greater effects.

Changes in composition and structure of vegetation in a wood mediated by deer browsing and grazing can have considerable indirect effects on animal life. Not surprisingly, recovery of animal populations can take many years as the vegetation has to recover first and there then has to be a local source of the animal species to recolonise. If a monitoring programme was being set up as stalking was beginning, then there are some obvious candidates to record, such as passerine birds that are dependent on bramble and scrub as breeding habitat. But sometimes even such species may need a long time to recover. On the other hand, some 'recoveries' can be unexpected.

Although fencing and culling have resulted in improvements in Monks Wood that I thought might be impossible 20 years ago, both of these forms of management have drawbacks. One of the big fences enclosed such a large area that it proved impossible to remove muntjac or to stop them gaining access. The second slightly smaller one worked well for several years but the dense habitat that developed meant no stalking occurred inside and muntjac density eventually reached a damaging level. We learnt that large fences several hectares in size should not be constructed without making adequate provision for deer control.

A more strategic approach to deer control is being adopted. Stalking tends to be more concentrated here in the larger woods, with a focus on specific times of year and localities known to attract higher concentrations of muntjac. Issues in smaller woods are more likely to be solvable by means of fencing certain sensitive areas as muntjac densities should be relatively low. Roe deer are so far only being culled in

Monks Wood and Holme Fen where they have been demonstrated to be a problem. Some people might suggest that the roe should also be controlled wherever there is muntjac stalking. However, there is a reluctance on the part of woodland managers to sanction culling in the absence of evidence of impact and because roe are a native species. In addition, it is likely that colonising roe exert some degree of competitive control over muntjac. Some water deer are being culled but nowhere is this being done for conservation reasons.

In addition to vegetation protection and deer control, more could be done to modify woodland management to reduce impact on sensitive vegetation (Mitchell-Jones and Kirby 1997; Putman 1998b; 2003). Malins and Oliver (2017) referred to the use of shooting and wire fencing to mitigate the impact of deer in forestry as 'metallic' management. They argued that the design and management of forests affect deer numbers and behaviour, so thoughtful practices could reduce the need for stalking and fencing. Ancient woodlands have not been recently designed but there could be opportunities for management to reduce impact. Thus, where there is to be ride-side clearance to diversify habitat, this could be in south-facing situations beside major rides to maximise levels of light and disturbance, and promote growth of woody vegetation. Stools in newly cut coppice are sometimes left unnecessarily exposed, when more careful cutting could have provided some protection. Modifying methods or subtly redesigning a management plan might accommodate a deer density that would otherwise need metallic management.

While on the subject of reducing shooting and fencing, views on acceptability of impact may be revisited. Evidence of grazing and browsing in a wood does not inevitably mean deer should be actively managed. A manager must decide whether the situation is (becoming) unacceptable and intervention is necessary. Currently, some managers may view the occurrence of an alien species, such as muntjac, as unacceptable in its own right. However, if it is subsequently shown that muntjac struggle to establish viable colonies in western and northern Britain, their presence there may be perceived differently. Similarly, a manager within the main range of the species in this country may decide that stable, moderate impact is acceptable after all and no management is necessary.

The concept of 'cultural carrying capacity' is becoming more important (Watson et al. 2012). This is the number of deer in an area that society will tolerate. Attempts may be needed to integrate all views as far as possible; transparency in decision-making can help in this respect. In time, perhaps development of contraceptives will mean stalking can be at least partially replaced and repellents may start to take over from fencing. But in the meantime, observing, monitoring and reviewing changes and options are ongoing.

Since 2000, muntjac numbers have been reduced in Cambridgeshire woods although the species has continued to colonise the Fens during that time. There has been no direct attempt to restrict the range of muntjac. It would have been difficult to try to prevent its north-eastward spread into fenland because it was spreading west from Norfolk anyway. So while both density in a key habitat and probably also overall numbers have been reduced, the range of the species has increased. At the moment, infilling the Fens at low density does not ring alarm bells, but the situation will continue to be monitored. To some extent, an increase in deer numbers nationally has been inferred from surveys indicating range expansion rather than from real data on deer numbers. Some caution is needed with this approach as in Cambridgeshire it could be argued that range expansion for muntjac across the Fens has not been accompanied by an increase in numbers. This is one reason why the BTO's abundance index is welcome for clarifying that real increases have occurred for national populations of some deer species, including the muntjac (Harris et al. 2017).

20.4   Muntjac management at national level

A key element of controlling muntjac is culling, but stalking was described by Rory Putman (1988) as a perpetual treadmill and an extremely expensive proposition to contemplate as a long-term solution. National Gamebag Census data collected by the Game & Wildlife Conservation Trust indicated that the total number of muntjac culled increased threefold between 1995 and 2009 (Aebischer et al. 2011). The increase was presumably driven by continuing colonisation and rising concern at the problems muntjac might cause.

Anywhere that stalking succeeds in reducing deer density, the population's productivity could increase through increased fecundity. However, information on whether this (always) occurs in muntjac populations is equivocal. John Gough (1999) related the differences between muntjac populations at Kineton in Warwickshire and Yardley Chase in Northamptonshire. During the early years of culling the dense population at the former site, 48% of deer were at least four years old and 45% of adult females were pregnant; in contrast, deer at Yardley Chase had been culled for several years and only 19% were at least four years old with 73% of females being pregnant. At Yardley Chase, fecundity was much higher and demography was tilted more towards younger animals. Moreover, during the mid- to late 1990s in Monks Wood, the recruitment index was particularly high (Figure 8.2), which was believed to be mainly in response to the earlier die-offs. On the other hand, in Wytham Woods in Oxfordshire, winter culling during 1998–2003 reduced the populations of both fallow deer and muntjac by about 90%, but pregnancy rates for both species were found to be unaffected by density (Buesching et al. 2010). Similarly, Kristin Waeber has explained to me that no density-dependent effect was found on muntjac fecundity during the study of muntjac sources and sinks in Thetford Forest during 2008–2010 (Waeber et al. 2013).

One problem that definitely occurs is that stalking becomes progressively more difficult as deer numbers fall and survivors are more wary and more difficult to shoot as cover increases. And a muntjac population can quickly recover if stalking stops or if there is a die-off, such as when, after roughly half of Monks Wood's muntjac died in 1994, sightings returned to their former level in three years. Rory Putman (2003) emphasised that management of the impact of deer cannot be solely addressed by control of populations. It is unrealistic to expect that stalking can effectively contain and control deer populations on a national scale. Nonetheless, the roe deer was virtually driven to extinction in England by man during the eighteenth century (Whitehead 1964; Yalden 1999).

The idea now is not to try to exterminate muntjac in this country but to manage them (and other deer) in such a way that their impacts are acceptable. As Chapman et al. (1994a) explained, extermination 'would not be possible or universally welcome. Many people like muntjac and enjoy their presence in the countryside.' There were a number of attempts during the 1990s to make recommendations on a national level with regard to information requirements and approaches for a management policy (Chapman et al. 1994a; 1994c; Staines 1995; Chapman and Harris 1998). These needs were identified as:

- Fencing specifications to prevent muntjac escaping from captivity and from negotiating fences intended to keep them out.
- Information on how to cull muntjac effectively and humanely.
- An investigation to determine whether it is possible to eliminate isolated populations.
- Co-ordination of management over large areas.
- Data on population dynamics of muntjac over the whole country.
- Population models to assist predictive management.
- Better quantification of impacts.
- Addition of the species on to Schedule 9 of the Wildlife and Countryside Act, 1981.

These recommendations were aimed at preventing or slowing down further spread and reducing numbers within the known range in Britain. The last recommendation was acted upon in 1997 in order to make it illegal to release muntjac into the wild except in certain circumstances, but has failed to prevent further spread (Section 5.2.2). Most managers are now aware of types of fencing to use although the minimum height of 1.5 m recommended by the Forestry Commission to exclude muntjac would probably not keep them out of a large fenced area for long, and certainly would not be expected to contain them. How to cull effectively and humanely is now covered by Smith-Jones (2004) and Downing (2014) among others, and at least one isolated population of muntjac has been eliminated by shooting (Smith-Jones 2017). Studies, including major investigations by ITE in the East Midlands (Staines et al. 1998) and by the University of East Anglia and the Forestry Commission in Thetford Forest (Waeber et al. 2013), have added considerably to our knowledge of basic ecology, population dynamics and levels of culling that might be necessary. Nevertheless, studies on more northerly populations are still lacking. The most significant advance over the last 20 years has probably been the organisation of management effort.

Organisations such as the BDS, the Deer Initiative and the British Association for Shooting and Conservation are committed to ensuring that deer are managed sustainably, humanely and safely. They are involved in training stalkers, awarding Deer Stalking Certificates, producing and promoting best practice, facilitating the setting up of DMGs, advising landowners and many other activities. There are regular meetings organised with a view to discussing current issues on deer management so as to communicate what is going on and to advance an entire range of topics, including deer control, sale of venison and research priorities (Figure 20.1). Such meetings vary in scale from those focusing on single sites up to the National Deer Management Conference which was held in 2017 at Leamington Spa in Warwickshire.

The concept of control at a landscape scale is now widely accepted. However, for deer to be controlled in an area, landowners need to be convinced that there are too many deer and stalkers need to be committed and understand what is needed – and there needs to be good communication and feedback between everyone involved. One of the issues is that people view deer from different perspectives. Deer may be variously seen as 'a pest, a game meat resource, a valuable sporting quarry or a valued part of our wild fauna' (Wilson 2003), resulting in some people wanting to reduce their numbers, other people wanting to cull selected individuals but not affect overall numbers, while yet more people do not wish to see them culled at all. In some areas in recent years there have been failures to control fallow deer in particular, because understanding and/or commitment were not as good as they might have been. But we have to keep trying to improve techniques and communication. In Monks Wood in the 1990s, I saw the extent to which woods might become degraded in a world where there was no deer control. It was not pleasant for woodland life in general or for muntjac in particular.

Despite a more co-ordinated approach to deer control in recent times, all of our species of deer are continuing to increase in number and the Forestry Commission and Natural England have recently granted a four-year contract to the Deer Initiative to improve the condition of native woodlands in England (Watson 2017). Five priority areas, each of about 1,000 km², have been selected based initially on the condition, importance and density of woodland SSSIs. Four of the priority areas have significant populations of muntjac: Wye Valley, Oxford, Suffolk and Rockingham – the last one includes part of Cambridgeshire. The objective is to aid Natural England's Biodiversity 2020 Strategy to have at least 50% of SSSIs in 'Favourable' condition and at least 95% in 'Favourable' or 'Unfavourable Recovering' condition by 2020. The work will encompass: meeting landowners, managers, agents, local stakeholders, volunteers and others; monitoring deer density, movement and habitat use; training volunteers who will undertake activity and impact surveys; erecting exclosures for monitoring and demonstration; undertaking collaborative culls; and organising local Deer Management Units for continuing management. In order

Figure 20.1   David Hooton (in the red hat) discusses coppice protection at Bradfield Woods in Suffolk with deer managers and advisors during one of the Deer Initiative's Deer Awareness Days, November 2017.

Figure 20.2   In harsh winter weather, newly independent muntjac are especially vulnerable to starvation. The hard winter of 2009/10 affected foraging behaviour in Woodwalton Fen but was not sufficiently severe to increase mortality levels (camera trap photograph).

to aid planning, collecting and sharing cull data within these priority areas and beyond, online tools were developed and launched in 2017 by the Deer Initiative and the Sylva Foundation. Also in 2017, the Deer Initiative appointed a research officer tasked with developing an adaptive package aimed at collaborative decision-making with regard to deer management.

Concentrating on controlling muntjac where they are having the worst impact makes good sense. Nevertheless, it would be prudent to try to prevent them from becoming established in important areas around the edges of the current main range, such as in parts of Devon or Kent.

Chapman et al. (1994a; 1994c) pointed out that one or two exceptionally hard winters could reduce muntjac populations appreciably; and then it would be important to have an agreed management approach to regulate their recovery. More than 20 years later, we are still waiting for that type of hard winter, but our climate has changed substantially and milder winters are the norm. The winters of 2009/10, 2012/13 and 2017/18 have been the harshest in recent years, but failed to produce any reports of significant die-offs (Figure 20.2). We need to continue taking other action rather than pinning our hopes on the weather.

20.5   Water deer

There is a need to be vigilant as regards water deer and to report instances of where they have caused problems or are suspected of doing so. With natural or coppice regeneration, a high-density population can evidently browse palatable species, such a sallow, to a detrimental degree. This should be preventable by using suitable protection for a sufficient period of time. Pollarding has been used successfully on the Claxton Estate in Norfolk, and could be used as an alternative form of management where providing dense woody cover close to the ground is not one of the aims of management. Of possibly greater concern is grazing of water deer on scarce flora. The potential exists for problems to occur, but this is a topic about which we need more information.

Similarly, we need to determine whether the activity of water deer on crop fields causes specific local impacts or might do so in the future. An option might be for an organisation such as the Deer Initiative to collect and collate relevant items of information on possible crop damage in order to decide whether a more detailed study was needed. It would be wise to remember how damage to crops has recently escalated in South Korea (Jung et al. 2016; Chun 2018).

In 2010, the water deer was added to Schedule 9 of the Wildlife and Countryside Act, 1981, but, as with the muntjac, this action has not resulted in preventing further spread. There has been no specific study of the impact of culling on population growth, but currently losses to stalking, motor traffic and other factors appear insufficient to stop the national population increasing. As with muntjac, an exceptionally hard winter with low temperatures, high precipitation and prolonged snow cover could deplete water deer numbers.

The phrase 'management of water deer' has a second meaning because, in some farmland situations in Bedfordshire and Norfolk, there is what amounts to animal husbandry to support (commercial) stalking. Practices include predator control, habitat manipulation and selective culling to avoid health problems from overstocking, as unusually high densities can result.

CHAPTER 21

The present and the future

21.1   The present situation

National risk assessments for the two species are described in Section 11.1. In the 100 or so years since muntjac were first released in England, they have spread, aided and abetted by man, through much of the country and have a more tenuous toehold in Wales, Scotland, Northern Ireland and the Republic. Any cost–benefit evaluation is inevitably subjective. As a conservationist who nevertheless likes and respects muntjac, I have to admit that, although it has brought benefits to these shores, these are insufficient to tip the balance in its favour. And I suspect that most people who are aware of the problems caused would agree with this assessment. Nentwig et al. (2018) scored the environmental and socio-economic impacts of 486 alien species of vertebrates, invertebrates and fungi in Europe, and Reeves' muntjac was the ninth worst, one place behind the sika. The water deer was not among the 100+ worst offenders. However, views change and it is not impossible that in future conservationists and policymakers will learn to develop a greater degree of tolerance in order for people to continue to enjoy seeing potentially problematic deer species in woodland reserves and elsewhere.

A cost–benefit analysis for water deer reveals some significant differences when compared with the muntjac. Although the species is regarded as causing few impacts, it has been shown to browse sallow coppice to an unacceptable degree at high density at Woodwalton Fen, and have possible local effects on crops or pasture on farmland adjacent to large wetland populations. Compared with the muntjac, however, these impacts are rarely seen and have cost little in monetary terms. Otherwise it has the same other drawbacks as the muntjac, but being less common it is neither involved in so many traffic accidents nor is it as likely to become implicated in transmitting disease. It also has a similar range of positive attributes. The most important of these is that our national population is becoming increasingly special on a global scale. I have only rarely tried to convince our relevant authorities of the need to conserve water deer – currently it is spreading despite lack of intentional protection

– but I have tried to persuade the conservation community to look on the species less negatively. Hopefully, this book may help. As regards the cost–benefit equation, the presence of water deer in this country is markedly positive, particularly when viewed on a world scale. And because people are now aware of problems that deer can cause, the water deer is being monitored closely, still with a degree of suspicion. Should problems arise, they are likely to be noticed, reported and tackled, unlike what happened with the muntjac in decades past.

Conservation action in this country has accidentally aided the expansion of water deer populations and protected the species. Wetland sites, such as fens and marshes, are regarded as key places for safeguarding by a range of conservation bodies, and water deer seem to have colonised many of these sites in eastern England, occasionally helped by a clandestine introduction. In other cases, they appear to have moved substantial distances before they have settled in reed beds and associated habitats. By one means or the other, they have been reported from Gibraltar Point in Lincolnshire, and south along the coasts of Norfolk and Suffolk. Inland wetlands have also been colonised, including a variety of protected Broadland sites in Norfolk, Redgrave and Lopham Fen and Lakenheath Fen in Suffolk, Woodwalton Fen and Ouse Fen in Cambridgeshire and Tring Reservoirs in Hertfordshire. A substantial proportion of the national population must occur inside nature reserves and these animals remain protected from hazards such as shooting and road traffic providing they stay in the reserves and do not inflict unacceptable damage on the vegetation. Muntjac enjoy comparatively less unintentional protection from conservation bodies. A lower proportion of the overall population lives inside reserves, such as woodlands, and many of the deer that do reside there are likely to be exposed to stalking.

21.2   Future spread and abundance

Acevedo et al. (2010) modelled maps showing how favourable each 10 km grid square in Britain was predicted to be for each of our deer species. They began with the distribution maps available in October 2007 on the NBN Gateway. An assumption was made that deer occurred where there was a record and were absent where the square was blank. A total of 26 environmental variables were selected that might affect deer distribution. These were of five types: spatial (longitude and latitude), climatic, topographic (altitude and slope), human disturbance (urban areas and distance to roads) and habitats (of eight basic types). The macro-ecological requirements of each species were then estimated from its recorded distribution. The process involved calculating the environmental 'favourability' of each 10 km square in Britain for each species. Plotting out maps based on these data might indicate the further potential spread.

The map for muntjac showed an area of maximum favourability south and east of a line between the Humber and Severn estuaries with the likelihood of encountering muntjac petering out north and west of a line between North Yorkshire and West Glamorgan. Similarly, presence of muntjac became progressively less likely in a south-westerly direction from Somerset and Dorset towards Cornwall. Referring to the latest BDS map (Figure 5.3), the distribution of muntjac appears more or less 'complete' apart from at the extremities of its coverage in Kent, north-east and south-west England, and south-east Wales. Time will tell whether this scenario is played out. The predicted map for water deer suggested a main distribution east of a line heading north from the Isle of Wight with the species becoming rare north of the Humber (Figure 21.1). If this turns out to be correct, then water deer have still to colonise much of the East Midlands as well as the south-east corner of England. The manner in which they have spread south-west through Buckinghamshire in recent years suggests that they may colonise Berkshire before the East Midlands or the extreme south-east. This exercise does, though, imply that they should not colonise any further north than the muntjac and their spread to the west will be much more restricted; this means that the water deer are likely to be exposed to significant competition from the muntjac wherever they occur in woodland sites. The technique also provided information to help determine how changes in different environmental variables might affect colonisation. Thus, one variable that was important for both species was the extent of seasonality measured as the standard deviation of monthly temperatures. On this basis, if our winters become colder and/or our summers hotter, both the muntjac and the water deer might benefit provided that abnormally hard winters do not occur.

Making predictions can be a difficult and unrewarding occupation. Charles Smith-Jones (2004) suggested that many areas of the Scottish Lowlands would be fully colonised by muntjac within the following decade. Thirteen years later, however, he noted (Smith-Jones 2017) that the species had only been reported from a few limited parts of Scotland where it was unlikely to persist. Its status in Scotland was summarised by Edwards and Kenyon (2013) as 'reported but not confirmed as being resident'. Scottish Natural Heritage (2016) stated that muntjac should be managed 'to prevent their establishment and spread', with at least one of the organisation's recent press releases recommending muntjac should be shot on sight. The responses of Natural Resources Wales (2011) to the presence of (apparently illegally released) muntjac and the potential introduction of water deer were similarly unenthusiastic; neither species would seem to have much of a future in Britain outside England, at least in the short term.

Figure 21.1   Crowle Moor NNR in northern Lincolnshire is just south of the River Humber. In time, will it be reached by water deer?

Both are still colonising new areas, but what is happening to their overall populations? Mammal recording during the BTO's Breeding Bird Survey has demonstrated that the national population index of the muntjac has more than doubled since 1995 and shows little sign of slowing down (Wright et al. 2014; Harris et al. 2017). The large number of 'new' squares reported for muntjac in the national BDS survey of 2016 (Smith-Jones 2017; Figure 5.3) supports a continuing increase. Where the species reaches its ecological and physiological limits in Britain, however, fecundity and survival may decline and a high proportion of records could be of isolated individuals rather than established populations.

The more limited evidence for water deer also points to a continuing increase in number in England. The population at Woodwalton Fen has fluctuated markedly in recent years, but looking generally at Cambridgeshire, the species has become more uniformly distributed (Hows et al. 2016) with new colonies beginning. Charles Smith-Jones (2017) referred to anecdotal reports of increasing numbers in East Anglia and Bedfordshire, and I have had similar reports from Norfolk, Suffolk, Bedfordshire and Buckinghamshire.

The future of both species depends on many factors, and it is difficult to guess, for instance, how weather patterns might change. In general, the climate will become warmer, but does that rule out the possibility of another winter as hard as 1962/3? Probably not, but what we currently view as exceptionally mild winters such as 2015/16 also seem to be detrimental for water deer. And what will be the effect of some of the huge national infrastructure projects? The installation of the new high speed railway, HS2, will be disruptive in the short term, but should eventually blend into the background as far as deer are concerned. Similarly, new transport links along the Cambridge–Milton Keynes–Oxford corridor will themselves not have a long-term impact on colonisation by water deer. However, associated housing development may well disturb or destroy farmland that would otherwise have been suitable. Large-scale tree planting of a Northern Forest proposed across England in a belt between Liverpool and Hull could provide muntjac with a firmer base to colonise further north. This is in addition to the new Heartwood, National and Heart of England forests which are all located centrally within the main range of the muntjac.

21.3   Future management and related studies

Management of muntjac populations to reduce impacts in woodlands currently relies primarily on culling and on the use of physical barriers, particularly wire fencing. The approach is based on protecting key areas for appropriate lengths of time with the most effective types of fencing, and having culling methods and strategies to facilitate management of populations to acceptable levels.

Stalking methods evolve to solve specific issues. Further consideration is probably needed on a regular basis to adjust both our interpretation of optimal/acceptable densities and the balance between stalking effort and deer density to facilitate the ease with which they can be culled. Stalkers can become less enthusiastic as deer numbers decrease and bramble cover increases. However, a detailed discussion of stalking methods is outside the remit of this book. The new project of the Deer Initiative to focus in particular on five priority areas in England to help

Natural England's Biodiversity 2020 Strategy (Watson 2017) is an example of both a co-ordinated approach on a national scale and the fact that stalking procedures are continuing to develop. I hope that stalkers, woodland managers and others can be encouraged to continue to contribute to monitoring. More extensive science-based monitoring of deer populations and impacts is needed to assess effectiveness (Watson et al. 2012).

There have been, and still are, broad-scale management and monitoring programmes, details of which are not in the public domain. It is important that information that is likely to be more generally applicable is available to deer and woodland managers in future. In this way, managers will be better placed to understand, for example, exactly what intervention is needed and how quickly and to what extent woodland recoveries might occur. Similarly, information gathered on how impacts can be ameliorated by other forms of management, such as leaving or providing alternative food, should be widely communicated.

Novel forms of management have been discussed, some for many years (e.g. immuno-contraception), others more recently (e.g. the introduction of lynx *Lynx lynx*). Progress on immuno-contraception seems slow and quite how small deer can be treated safely, regularly and reasonably easily and cheaply is difficult to imagine. While, if an introduction of lynx goes ahead, it is likely to be to an area where muntjac do not occur or where few would be predated; also it could have negative effects by dispersing deer and by reducing the density of foxes, which are known to be capable of having some control over muntjac populations. The message seems to be that there is unlikely to be any major new form of management in the near future.

Observations presented in these pages have sometimes scratched the surface of intriguing issues that deserve more detailed study. Thus, simple monitoring of muntjac in Lady's Wood has shown that they colonised and inflicted some damage to the fabric of the wood but species such as bluebells were not perceptibly affected. The deer population then stabilised at a lower density. This has implications for muntjac control by asking the question, 'Under what conditions is management unnecessary?' If stalking is not needed in some situations then this releases manpower to be deployed where it is definitely required. There are subsidiary questions which suggest scenarios for investigation, such as:

- Are certain small woods less likely to be affected by the muntjac they hold?
- What part does culling muntjac in adjacent woods or in the wider countryside play in such situations?
- What might happen in woods close to the edge of the current range of muntjac in this country, including how might roe deer compete with muntjac?

While there has been an emphasis on problems that high densities of deer can cause, more thought needs to be given to subtle conservation benefits, including such aspects as aiding dispersal of plant seeds through woods. Similarly, issues such as fencing leading to increased rates of succession need to be understood in order to make balanced management decisions. Moreover, better knowledge and consideration of confounding factors are necessary otherwise there is a risk of deer being wrongly blamed and managed.

The principal thrust of work on water deer should be to keep a watching brief to record and act on situations in which impacts may occur. In particular, is there evidence of deer from dense wetland populations raiding farmland to a damaging extent? Should deer be suspected of such impacts, evidence could be collected in the form of analysis of dung found on farm fields and of gut contents from shot deer. Information from managed and stalked farmland populations would help in this respect. Within wetland nature reserves, further work is required on whether water deer can affect populations of nationally scarce, but locally abundant, ground flora. Again, dung and gut analysis could be done. Related to such issues are questions such as how far do water deer travel on to farmland, especially at night, and how do muntjac range over areas comprising small woods and surrounding farmland? These could be subjects for radio tracking or satellite tagging, as could the ranging of muntjac in suburban areas. Differences in activity between the sexes and age groups needs more research. We should also be aware that there may be new situations where muntjac populations cause impacts.

It seems unlikely that there will be government support for the attempted eradication of either species. Olaf Booy of the National Wildlife Management Centre and his 29 collaborators (2017) proposed a scheme to prioritise such management of invasive non-native species. For each species, an evaluation of risk would be combined with an assessment of the feasibility of eradication. The water deer was one of the species chosen to test the approach. It was judged to present a low risk (categories were low, medium and high), while the feasibility of eradication was medium (categories for the likelihood of success were very low, low, medium, high and very high). A matrix plotting risk category against feasibility of eradication was used to determine priorities. The water deer was graded as being of low priority for eradication (grades were low, moderate, high, very high and highest). The muntjac was not included in this exercise, but I would guess that its risk would be categorised as high but the feasibility of eradication would be very low, so its priority for an eradication attempt would, like the water deer, also be low.

In 2018, the UK government signalled its intention to widen and strengthen laws governing the management of certain invasive species when Defra issued a consultation document. The list of species included the muntjac, but not the water deer.

Work discussed above mainly relates to management for conservation reasons. Many other studies and trials are being undertaken to ameliorate different problems, such as those aimed at reducing the number of traffic accidents involving deer. Similarly, there have been projects aimed at promoting the consumption of British venison, but it is larger species such as fallow deer that are more likely to be consumed in increasing numbers. While details of these types of project are excluded from this account, potential new work on basic natural history is considered in the next section. It is, though, debatable whether some topics raised here are more relevant to management or natural history.

21.4   Natural history research in the future

Little is known about dispersal of young deer, especially water deer. Fawns could be caught and tagged in areas such as isolated wetland reserves or agricultural land and the public asked to report sightings. If funds allow, some could be satellite-tagged. This exercise should provide novel information on potential for migration, habitat selection and survival.

Figure 21.2   Thorpe Marshes Nature Reserve, Norwich, is within the city limits.
It has evidently been reached by water deer moving through suitable habitat
beside the River Yare. The hum of traffic on the A47 dual carriageway can be heard
on the marshes, and some less fortunate deer have become road casualties.

As reported in Section 3.2.1, information is equivocal on how and why male water
deer mark vegetation with scent. This could be best investigated by means of camera
traps. Cameras could also be used to attempt to detect early signs of water deer
entering suburbia. Water deer occur, for instance, in Thorpe Marshes on the edge
of Norwich, taking advantage of the corridor afforded by suitable riparian habitat
along the River Yare (Figure 21.2), but will they penetrate further into the city? And
are there areas on the Broads where they may come into quiet village gardens?

Knowledge of growth of canine tusks in buck water deer is incomplete.
I have taken the data available from deer found dead and from my local stalker
as far as I can. Trying to interest other stalkers to measure tusk length has been
unsuccessful. But a co-ordinated effort among stalkers would reveal more about
growth during the species' first winter and also about when tusks become close-
rooted. Alternatively, a study on tusk development in captive individuals could be
productive, providing it was done on deer that grew and developed as well as those
in the wild. A research project on cementum layers in teeth is also worth considering. For water deer, the relationship between age and number of layers needs to
be established with material from animals of known age. There is an opportunity to
combine studies on tusk development and cementum layers in water deer.

The age and gender structure of populations of both species could be an interesting research topic. There would need to be thought given to reducing bias as
much as possible, both concerning the probabilities of seeing the age groups and
sexes and the selection of animals for culling. And more data on weights of both
species of deer shot in this country would be informative: to what degree and why
do average weights vary between sites and what weight is attained by the largest
individuals?

I have given pointers in this book to population stability and factors that affect
population size. It is clear that both species can colonise and remain in specific,

secure sites for many years. The water deer, however, has had many introductions in this country which have failed and has been lost from woodlands after muntjac have colonised. Long-term surveillance of water deer could provide more information on the future for this species and how it might be managed. The Great Fen in Cambridgeshire provides a situation in which regular and routine observation could help to understand better how water deer, muntjac, roe deer and perhaps red deer utilise a large area of wetland, woodland and agricultural fields and interact with one another and with human activities. Regular counts of deer through much of the open landscape in the area are already organised by the Wildlife Trust. A similar approach could be made in and around Wicken Fen; that area currently has muntjac, roe and red deer, but few water deer, although they may increase in the future. Results from the two areas could be compared and contrasted.

Critical work needed for both species involves determining densities in various habitats and in various parts of their ranges in this country. Such investigation should extend into the wider countryside. As more information becomes available, so calculations of regional or national population sizes will become more reliable.

21.5   The future of the species elsewhere

Muntjac have been decreasing in numbers in China and their range appears to have contracted because of hunting and habitat loss (Timmins and Chan 2016). They are now on the Chinese Red List (Jiang et al. 2015) and are on the radar of the IUCN, so the relevant conservation authorities are concerned. In the short term, however, it is likely that trends of the recent past will continue. The situation in Taiwan is said to be stable (Timmins and Chan 2016), apparently with species protection and extinction of predators offsetting habitat losses. Nevertheless, the IUCN authors have recommended that habitat and population levels need to be monitored in both China and Taiwan.

Due to escapes and releases, muntjac are well established in parts of Japan, and have a presence in Ireland and several countries on the European mainland. In none of these countries are they welcome, and steps have been taken to contain or eradicate them. Doubtless the struggle will continue between those people who wish to move and liberate them and those who wish to prevent such actions.

The population of muntjac in Britain is significant at a global scale. The size of the island of Taiwan is no more than half the area of the main range of muntjac in England, so our population must be much higher than that in Taiwan. The population in China was estimated to be 2–2½ million by Sheng (1992a). The current estimate for muntjac in British *woodland* of 128,000 (Mathews et al. 2018) represents at least 5% of the number in China 25 years ago, but since then the Chinese population has declined to an unknown extent. The size gap between the two populations should continue to narrow.

The last written estimate for the number of water deer in China is the total of less than 5,000 proposed by Min Chen in Fautley (2013). While there is no agreed figure for the total in Britain (Section 10.5), I would argue that there are now at least as many water deer in this country as in China. Much higher numbers of water deer occur on the Korean peninsula, but the situation in South Korea is volatile with up to one-third being killed each year (Chun 2018).

Water deer are certainly faring much better in this country than in China. Here, they are still expanding their main range and are probably becoming more numerous at the same time. We seem to be at the stage where a wandering water deer anywhere within the main range in eastern England is likely to be within reach of a breeding population. Moreover, in areas such as the Broads and large parts of the coast of Norfolk and Suffolk, their distribution is more or less continuous. This is a much healthier situation for the species compared with that in China, where populations appear to be in slow decline. The vast reserve at Yancheng may be 4,500 km², but it supports less than 400 individuals living in small isolated populations (Min Chen in Fautley 2013), down from 1,000 in the early 1990s (Zhang 1994). Min Chen has told me that she regards our system of small, reasonably well-protected nature reserves as more effective than the much larger reserves in China that allow development of different types.

There have been a number of recommendations to improve the situation for water deer in China. On behalf of the IUCN, Harris and Duckworth (2015) made the following specific recommendations:

- Increase the size of the reserve at Poyang Lake to include all of the species' range and introduce night patrols to combat poaching.
- Link isolated populations in the Yancheng Reserve with corridors of suitable habitat.

They also made more general recommendations to establish protected areas, strengthen measures to manage and safeguard the habitat and the deer, and educate people to reduce the threats of encroachment and poaching. With regard to Korea, these authors suggested that measures were required to prevent poaching and to provide substantial areas of secure habitat. More recent communications from South Korea, however, revealed a surprisingly large and persecuted population (Jung et al. 2016; Chun 2018). Further detailed and broader information is required from Korea, especially from the North.

Richard Fautley (2013) studied the genetics of water deer and found that each of the extant Chinese populations and the population in England were genetically different from one another. His results from China indicated a severe and recent decline, whereas those from England suggested a single source of introduced stock from a Chinese location where they no longer existed. He proposed that the English population was therefore 'a valuable conservation resource' and that the feasibility of translocating our animals back to China should be examined. Not only are they relatively numerous, but they are also genetically distinct and represent an extinct subpopulation.

To try to offset losses, there is already a captive-breeding and release programme at Shanghai using stock collected from the Zhoushan Islands (Chen et al. 2016); deer were released in 2010 and their movements monitored (He et al. 2016). In time, consideration might be given to whether deer from an English source could be used to found a new separate population or to augment the existing programme. In order to achieve such an objective, deer need to be caught and transported back to China. It is well known that water deer do not travel well (Cooke and Farrell 1998; Smith-Jones 2017), and research on this problem is needed before translocation back to China can be contemplated. Some animals were successfully translocated in the opposite direction in order to provide the origins of our population, but the cost in terms of animals lost remains unknown, and standards of animal welfare have improved considerably since that time. A translocation from Korea would seem more practical, but perhaps less acceptable genetically. At least the continued presence of this species globally is more secure now we have a viable population in this country.

At this point, right at the end of the narrative, I am reminded of Ken Thompson's 2014 book *Where Do Camels Belong?* This was about the ecology and impacts of invasive species and attitudes towards them. The title asked whether camelids belong in the Arab world where they occur now, in the Americas where they evolved and where many species still live, or even in Australia where there are huge numbers of feral dromedaries *Camelus dromedarius*? Thompson argued with good reason that scientists and conservationists tend to view every non-native species as a potential problem and every effect they cause as harmful. The water deer has been living here in the wild for about 70 years and so far its detrimental impacts have been minimal and relatively easily managed. It is continuing to spread and, should it cause new and more significant impacts, these will be quickly recognised and hopefully managed. The water deer is a species that is, on balance, beneficial in this country. It is flourishing here because we have stable blocks of quiet, suitable countryside. Many of our deer live in the Broads and coastal wetlands, where they are very much at home. Perhaps in time people will come to accept that they also belong here.

Appendix 1

This framework describes effects on woodland indicators at different stages of deer impact. Impact stages can be assessed for each indicator in a wood, and overall impact determined. Intermediate stages are named as slight, moderate and very high impact respectively. From Cooke (2009a) with minor changes.

Indicator	No impact	Low impact	High impact	Severe impact
Unprotected coppice regrowth	Nil	Stem loss < 50%	Stem loss > 50%	Severe stem loss + death of stools
Tree regeneration	Nil	Possible minor effects	Reduced for some species	Little/no regeneration of some species
Shrub layer	Nil	Possible minor effects	Density reduced in some species	(Virtually) eliminated
Browse lines	Nil	Nil or on occasional stems	Obvious on some species, signs of general one	General browse line throughout wood
Stem breakage	Nil	Nil to rare	Occasional	Frequent and conspicuous
Fraying	Nil	Rare	Occasional fresh fraying, much more old fraying	Fresh and old fraying frequent
Ivy browsed day 1	0–10%	11–30%	31–60%	61–100%
Ivy defoliated day 7	0–20%	21–60%	61–90%	91–100%
Ground flora	Nil	Possible minor effects	Flowering reduced in some species	Some species (virtually) eliminated
Grazing on bluebell leaves	0–5%	6–15%	16–40%	>40%
Grazing on bluebell inflorescences	0	1–5%	6–20%	>20%
Grazing on oxlips/ primroses	0–10%	11–30%	31–80%	>80%
Grazing on dog's mercury stems	0	1–5%	6–20%	>20%
Height of dog's mercury	>25 cm	>25 cm	21–25 cm	Up to 20 cm

Appendix 2

Scores for signs of muntjac (deer, dropping, slots and paths) and their damage (browsing, stem breakage, browse lines, fraying and grazing) in Monks Wood, 1995–2017, excluding 1997 when no scoring was undertaken. Each score may be 0, 1, 2 or 3. Stalking began in 1998.

Year	Deer signs				Damage signs				
	Deer	Droppings	Slots	Paths	Browsing	Stem breakage	Browse lines	Fraying	Grazing
1995	1	2	3	3	3	3	3	3	2
1996	2	3	2	3	3	3	3	3	2
1998	2	3	2	3	3	3	3	2	2
1999	1	2	1	2	2	2	2	2	2
2000	1	1	1	2	2	2	2	1	1
2001	1	2	2	2	2	1	2	1	1
2002	0	2	2	2	2	1	2	1	1
2003	1	1	1	1	2	1	2	1	1
2004	2	2	1	2	2	1	2	1	1
2005	1	2	1	2	2	1	2	1	1
2006	1	2	1	2	2	1	2	1	1
2007	1	1	1	2	2	2	2	1	1
2008	1	1	1	2	2	0	2	1	0
2009	0	1	1	1	2	0	2	1	1
2010	1	1	1	1	2	1	2	1	0
2011	1	1	1	1	2	0	2	1	1
2012	1	1	1	1	2	1	1	1	1
2013	0	1	1	1	1	1	1	0	1
2014	0	1	2	2	1	1	1	1	1
2015	0	1	1	2	1	1	1	1	1
2016	0	1	1	1	1	0	1	0	1
2017	0	0	1	1	1	0	1	0	1

References

Acevedo, P., Ward, A.I., Real, R. and Smith, G.C. (2010) Assessing biogeographical relationships of ecologically related species using favourability functions: a case study on British deer. *Diversity and Distributions* 16: 515–528.

Aebischer, N.J., Davey, P.D. and Kingdon, N.G. (2011) *National Gamebag Census: Mammal Trends to 2009.* Fordingbridge: Game & Wildlife Conservation Trust.

Aitchison, J. (1946) Hinged teeth in mammals: a study of the tusks of muntjacs (*Muntiacus*) and Chinese water deer (*Hydropotes inermis*). *Proceedings of the Zoological Society of London* 116: 329–338.

Amar, A., Hewson, C.M., Thewlis, R.M., Smith, K.W., Fuller, R.J., Lindsell, J.A., Conway, G., Butler, S. and MacDonald, M. (2006) *What's Happening to Our Woodland Birds?* Royal Society for the Protection of Birds Research Report 19, British Trust for Ornithology Research Report 169.

Anderson, D. (1973) Mammals in Bedfordshire 1946–1971. *Bedfordshire Naturalist* 26: 67–69.

Anderson, D. (1989) Mammals. Report of the Recorder. *Bedfordshire Naturalist* 43: 25–30.

Anderson, D. and Cham, S.A. (1988) Muntjac deer (*Muntiacus reevesi*) – the early years. *Bedfordshire Naturalist* 42: 14–18.

Anon. (2005) New population estimates for British mammals. *Deer* 13(4): 8.

Anon. (2008) Schedule 9 for CWD? *Deer* 14(6): 5.

Anon. (2010) Mortality reports sought for muntjac and CWD. *Deer* 15(5): 5.

Anon. (2013) How well can deer swim? *Deer* 16(9): 31.

Anon. (2014) Liver fluke survey 2014. *Deer* 17(1): 8.

Anon. (2017) Celtic connections. *Deer* 18(4): 10–12.

Appleton, G. (2017) Lend me your ears. *BBC Wildlife* 35(3): 30–34.

Arnold, H.R. (1984) *Distribution Maps of the Mammals of the British Isles.* Abbots Ripton: Institute of Terrestrial Ecology.

Arnold, H.R. (1993) *Atlas of Mammals in Britain.* London: HMSO.

Arnold, H.R. and Jefferies, D.J. (1985–2018) Mammal report for each year, 1984–2017. *Annual Reports of the Huntingdonshire Fauna & Flora Society*, volumes 37–70.

Bacon, L. (2005) *Cambridgeshire & Peterborough Provisional Mammal Atlas.* Cambourne: Cambridgeshire and Peterborough Biological Records Centre.

Baiwy, E., Schockert, V. and Branquart, E. (2013) *Risk Analysis of the Reeves' Muntjac Muntiacus reevesi.* Risk analysis report of non-native organisms in Belgium. Gembloux: Cellule Interdépartementale sur les Espèces Invasives.

Baker, P.A. and Harris, S. (2008) Fox *Vulpes vulpes*. In S. Harris and D.W. Yalden (eds) *Mammals of the British Isles*, 4th edn. Southampton: The Mammal Society. pp. 407–423.

Balgooyen, C.P. and Waller, D.M. (1995) The use of *Clintonia borealis* and other indicators to gauge impacts of white-tailed deer on plant communities in northern Wisconsin, USA. *Natural Areas Journal* 15: 308–318.

Banham, T. (2003) In the public eye. *Deer* 12(9): 18–20.

Barrette, C. (1977a) The social behaviour of captive muntjacs *Muntiacus reevesi* (Ogilby 1839). *Zeitschrift für Tierpsychologie* 43: 188–213.

Barrette, C. (1977b) Fighting behaviour of muntjac and the evolution of antlers. *Evolution* 31: 169–176.

Battersby, J. (ed.) and the Tracking Mammals Partnership (2005) *UK Mammals: Species Status and Population Trends*. Peterborough: JNCC and Tracking Mammals Partnership.

Blackman, G.E. and Rutter, A.J. (1954) *Endymion nonscriptus* (L.) Garcke. *Journal of Ecology* 42: 629–638.

Blakeley, D., Chapman, N., Claydon, K., Claydon, M., Harris, S. and Wakelam, J. (1997) Studying muntjac in the King's Forest, Suffolk. *Deer* 10: 156–161.

Booy, O., Mill, A.C., Roy, H.E., Hiley, A., Moore, N., Robertson, P., Baker, S., Brazier, M., Bue, M., Bullock, R., Campbell, S., Eyre, D., Foster, J., Hatton-Ellis, M., Long, J., Macadam, C., Morrison-Bell, C., Mumford, J., Newman, J., Parrott, D., Paine, R., Renals, T., Rodgers, E., Spencer, M., Stebbing, P., Sutton-Croft, M., Walker, K.J., Ward, A., Whittaker, S. and Wyn, G. (2017) Risk management to prioritise the eradication of new and emerging invasive non-native species. *Biological Invasions* 19: 2401–2417.

Booy, O., Wade, M. and Roy, H. (2015) *Field Guide to Invasive Plants & Animals in Britain*. London: Bloomsbury.

Bowley, A. (2010) Variations in population size of a number of breeding bird species at Holme Fen with relation to vegetation change. Unpublished thesis, Birkbeck College, University of London.

Bowley, A. (2013) The Great Fen – the challenges of creating a wild landscape in lowland England. *British Wildlife* 25: 95–102.

Bows, A.T. (1997) Deer predation on traditionally managed coppice woodlands and responses by management. *Deer* 10: 226–232.

Bray, D.W. (1980) Taking muntjac seriously. *Quarterly Journal of Forestry*, 74: 229–232.

Bright, P., Morris, P. and Mitchell-Jones, A. (2006) *The Dormouse Conservation Handbook*, 2nd edn. Peterborough: English Nature.

British Deer Society (2015) Deterring deer. Accessed at: http://www.bds.org.uk/index. php/advice-education/deterring-deer (3 November 2018).

Broughton, R.K., Hill, R.A., Bellamy, P.E. and Hinsley, S.A. (2011) Nest-sites, breeding failure, and causes of non-breeding in a population of British marsh tits *Poecile palustris*. *Bird Study* 58: 229–237.

Brunsendorf, A. (2006) Changes in the ground flora at Monks Wood NNR Cambridgeshire: A study of the impact of management and deer grazing on the ground vegetation. Unpublished M.Sc. thesis, University College London.

Buckingham, W.G. and Buckingham, B.M. (1985) Keeping muntjac in the garden. *Deer* 6: 202–204.

Buesching, C.D., Clarke, J.R., Ellwood, S.A., King, C., Newman, C. and Macdonald, D.W. (2010) The mammals of Wytham Woods. In P.S. Savill, C.M. Perrins, K.J. Kirby and N. Fisher (eds) *Wytham Woods: Oxford's Ecological Observatory*. Oxford: Oxford University Press. pp. 173–196.

Bullion, S. (2009) *The Mammals of Suffolk*. Ipswich: Suffolk Wildlife Trust.

Carden, R.F., Carlin, C.M., Marnell, F., McElholm, D., Hetherington, J. and Gammell, M.P. (2011) Distribution and range expansion of deer in Ireland. *Mammal Review* 41: 313–325.

Carne, P. (1964) Distribution records. *Deer News* 1(5): 15.

Carne, P. (1981) Britain's most enigmatic deer. *Shooting Times and Country Magazine*, 9–15 April 1981, 4.

Carne, P. (1999) What future for Chinese water deer? *Stalking Magazine*, September 1999: 18–21.

Carne, P. (2000) *Deer of Britain and Ireland*. Shrewsbury: Swan Hill Press.

Chaplin, R.E. (1977) *Deer*. Poole: Blandford Press.

Chapman, D.I. (1977) Deer of Essex. *Essex Naturalist* (New Series) 1: 3–50.

Chapman, D.I. and Chapman, N.G. (1982) The taxonomic status of feral muntjac deer (*Muntiacus reevesi*) in England. *Journal of Natural History* 16: 381–387.

Chapman, D.I. and Dansie, O. (1969) Unilateral implantation in muntjac deer. *Journal of Zoology, London* 159: 534–536.

Chapman, D.I., Chapman, N.G., Matthews, J.G. and Wurster-Hill, D.M. (1983) Chromosome studies of feral muntjac deer (*Muntiacus* sp.) in England. *Journal of Zoology, London* 201: 557–588.

Chapman, D.I., Chapman, N.G. and Colles, C.M. (1985) Tooth eruption in Reeves' muntjac (*Muntiacus reevesi*) and its use as a method of age estimation (Mammalia: Cervidae). *Journal of Zoology, London* 205: 205–221.

Chapman, N.G. (1989) Muntjac biology for deer managers. *Deer* 7: 414–415.

Chapman, N.G. (1992a) Reeves' muntjac (*Muntiacus reevesi*) in Britain. In H. Sheng (ed.) *The Deer in China*. Shanghai: East China Normal University. pp. 145–148.

Chapman, N.G. (1992b) Chinese water deer: more records. *Transactions of the Suffolk Naturalists' Society* 28: 1–2.

Chapman, N.G. (1993) Reproductive performance of captive Reeves' muntjac. In N. Ohtaishi and H. Sheng (eds) *Deer of China*. Amsterdam: Elsevier. pp. 199–203.

Chapman, N.G. (1995) Our neglected species. *Deer* 9: 360–362.

Chapman, N.G. (1996) Muntjac in the wars. *Deer* 9: 562–563.

Chapman, N.G. (1997a) Upper canine teeth of *Muntiacus* (Cervidae) with particular reference to *M. reevesi*. *Zeitschrift für Säugetierkunde Supplementum* 2, 62: 32–36.

Chapman, N.G. (1997b) Are your brambles eaten? *Deer* 10: 236.

Chapman, N.G. (2004) Faecal pellets of Reeves' muntjac, *Muntiacus reevesi*: defecation rate, decomposition period, size and weight. *European Journal of Wildlife Research* 50: 141–145.

Chapman, N.G. (2007) How old is that muntjac? *Deer* 14(2): 25–28.

Chapman, N.G. (2008) Reeves' muntjac *Muntiacus reevesi*. In S. Harris and D.W. Yalden (eds) *Mammals of the British Isles*, 4th edn. Southampton: The Mammal Society. pp. 564–571.

Chapman, N.G. and Harris, S. (1991) Evidence that the seasonal antler cycle of adult Reeves' muntjac (*Muntiacus reevesi*) is not associated with reproductive quiescence. *Journal of Reproduction and Fertility* 92: 361–369.

Chapman, N.G. and Harris, S. (1996) *Muntjac*. London: Mammal Society and Fordingbridge: British Deer Society.

Chapman, N.G. and Harris, S. (1998) Muntjac: where do we go from here? In C.R. Goldspink, S. King and R.J. Putman (eds) *Population Ecology, Management and Welfare of Deer*. Manchester: Manchester Metropolitan University. pp. 32–37.

Chapman, N.G., Claydon, K., Claydon, M. and Harris, S. (1985) Distribution and habitat selection by muntjac and other species of deer in a coniferous forest. *Acta Theriologica* 30: 287–303.

Chapman, N.G., Claydon, K., Claydon, M., Forde, P.G. and Harris, S. (1993) Sympatric populations of muntjac (*Muntiacus reevesi*) and roe deer (*Capreolus capreolus*): a comparative analysis of their ranging behaviour, social organization and activity. *Journal of Zoology, London* 229: 623–640.

Chapman, N.G., Harris, S. and Stanford, A. (1994a) Reeves' muntjac *Muntiacus reevesi* in Britain: their history, spread, habitat selection and the role of human intervention in accelerating their dispersal. *Mammal Review* 24: 113–160.

Chapman, N.G., Harris, A. and Harris, S. (1994b) What gardeners say about muntjac. *Deer* 9: 302–306.

Chapman, N.G., Claydon, K., Claydon, M. and Harris, S. (1994c) Muntjac in Britain: is there a need for a management strategy? *Deer* 9: 224–236.

Chapman, N.G., Furlong, M. and Harris, S. (1997a) Reproductive strategies and the influence of date of birth on growth and sexual development of an aseasonally breeding ungulate: Reeves' muntjac (*Muntiacus reevesi*). *Journal of Zoology, London* 241: 551–570.

Chapman, N.G., Claydon, K., Claydon, M., Forde, P.G., Harris, S. and Keeling, J. (1997b) History and habitat preferences of muntjac in the King's Forest, Suffolk. *Deer* 10: 289–294.

Chapman, N.G., Brown, W.A.B. and Rothery, P. (2005) Assessing the age of Reeves' muntjac (*Muntiacus reevesi*) by scoring wear of the mandibular molars. *Journal of Zoology, London* 267: 233–247.

Chen, M. (2012) Preliminary survey on distribution of Chinese water deer in Hanshan and Chuzhou area, Anhui Province, China. *Deer Specialist Group, Newsletter* 24: 13–19.

Chen, M., Zhang, E., Yang, N., Peng, Y., Su, T., Teng, L. and Ma, F. (2009) Distribution and abundance of *Hydropotes inermis* in spring Yancheng coastal wetland, Jiangsu Province, China. *Wetland Science* 7: 1–4.

Chen, M., Liu, C., He, X., Pei, E., Yuan, X. and Zhang, E. (2016) The efforts to re-establish the Chinese water deer population in Shanghai, China. *Animal Production Science* 56: 941–945.

Chevrier, T., Saïd, S., Widmer, O., Hamard, J.-P., Saint-Andrieux, C. and Gaillard, J.-M. (2012) The oak browsing index correlates linearly with roe deer density: a new indicator for deer management? *European Journal of Wildlife Research* 58: 17–22.

Chiang, P. (2007) Ecology and conservation of Formosan clouded leopard, its prey, and other sympatric carnivores in southern Taiwan. Unpublished D.Phil. thesis, Virginia Polytechnic Institute and State University.

Childerley, P. (2014) Herd reduction. *Sporting Rifle* 107: 90–92.

Chollet, S. and Martin, J.-L. (2013) Declining woodland birds in North America: should we blame Bambi? *Diversity and Distributions* 19: 481–483.

Chun, K. (2018) 171,000 deaths … How did it become a 'protected animal'? (in Korean). Accessed at: https://news.joins.com/article/22304459 (3 November 2018).

Ciuti, S., Muhly, T.B., Paton, D.G., McDevitt, A.D., Musiani, M. and Boyce, M.S. (2012) Human selection in elk behavioural traits in a landscape of fear. *Proceedings of the Royal Society B* 279: 4407–4416.

Clark, J.S. (1974) Mammal report. *Annual Report of the Huntingdonshire Fauna & Flora Society* 26: 65–72.

Clark, M. (1974) Deer distribution survey 1967–72. *Deer* 3: 279–282.

Clark, M. (1981) *Mammal Watching*. London: Severn House.

Clark, M. (2001) *Mammals, Amphibians and Reptiles of Hertfordshire*. Watford: Training Publications.

Claydon, K., Claydon, M. and Harris, S. (1986) Estimating the number of muntjac deer (*Muntiacus reevesi*) in a commercial coniferous forest. *Bulletin of the British Ecological Society* 17: 185–189.

Clutton-Brock, T. (1993) Deer and scientists. *Deer* 9: 17–18.

Coles, C. (1997) *Gardens and Deer: A Guide to Damage Limitation*. Shrewsbury: Swan Hill Press.

Collini, G. (2004) Calling muntjac. *Deer* 13(3): 10–12.

Cooke, A.S. (1994) Colonisation by muntjac deer *Muntiacus reevesi* and their impact on vegetation. In M.E. Massey and R.C. Welch (eds) *Monks Wood National Nature Reserve: The Experience of 40 Years 1953–93*. Peterborough: English Nature. pp. 45–61.

Cooke, A.S. (1995) Muntjac damage in woodland. *Enact* 3(3): 12–14.

Cooke, A.S. (1997) Effects of grazing by muntjac (*Muntiacus reevesi*) on bluebells (*Hyacinthoides non-scripta*) and a field technique for assessing feeding activity. *Journal of Zoology, London* 242: 365–369.

Cooke, A.S. (1998a) Survival and regrowth performance of coppiced ash (*Fraxinus excelsior*) in relation to browsing damage by muntjac deer (*Muntiacus reevesi*). *Quarterly Journal of Forestry* 92: 286–290.

Cooke, A.S. (1998b) Some aspects of muntjac behaviour. *Deer* 10: 464–466.

Cooke, A.S. (1998c) Colonisation of Holme Fen National Nature Reserve by Chinese water deer and muntjac, 1976–1997. *Deer* 10: 414–416.

Cooke, A.S. (1999a) *Muntiacus reevesi* (Ogilby, 1839). In A.J. Mitchell-Jones, G. Amori, W. Bogdanowicz, B. Kryštufek, P.J.H. Reijnders, F. Spitzenberger, M. Stubbe, J.B.M. Thissen, V. Vohralík and J. Zima (eds) *The Atlas of European Mammals*. London: Poyser. pp. 382–383.

Cooke, A.S. (1999b) *Hydropotes inermis* Swinhoe, 1870. In A.J. Mitchell-Jones, G. Amori, W. Bogdanowicz, B. Kryštufek, P.J.H. Reijnders, F. Spitzenberger, M. Stubbe, J.B.M. Thissen, V. Vohralík and J. Zima (eds) *The Atlas of European Mammals*. London: Poyser. pp. 398–399.

Cooke, A.S. (1999c) The effects of the Easter floods on the deer at Woodwalton Fen. *Annual Report of the Huntingdonshire Fauna & Flora Society* 51: 44–45.

Cooke, A.S. (2001) Information on muntjac from studying ivy. *Deer* 11: 498–500.

Cooke, A.S. (2004) Muntjac and conservation woodlands. In C.P. Quine, R.F. Shore and R.C. Trout (eds) *Managing Woodlands and Their Mammals: Proceedings of a Joint Mammal Society/Forestry Commission Symposium.* Edinburgh: Forestry Commission. pp. 65–69.

Cooke, A.S. (2005) Muntjac deer. In T. Collins, B. Dickerson, P.E.G. Walker and T.C.E. Wells (eds) *Brampton Wood: A Natural History.* N.p.: Huntingdonshire Fauna and Flora Society. pp. 83–85.

Cooke, A.S. (2006) *Monitoring Muntjac Deer* Muntiacus reevesi *and Their Impacts in Monks Wood National Nature Reserve.* English Nature Research Report 681. Peterborough: English Nature.

Cooke, A.S. (2007) Deer and damage scoring for woodland monitoring. *Deer* 14(5): 17–20.

Cooke, A.S. (2008) Muntjac activity in woodland reserves of the Wildlife Trust for Cambridgeshire. Unpublished report for the Wildlife Trust.

Cooke, A.S. (2009a) Chinese water deer *Hydropotes inermis* in Britain. *International Urban Ecology Review* 4: 32–43.

Cooke, A.S. (2009b) Chinese water deer on farmland. *Deer* 15(2): 14–17.

Cooke, A.S. (2009c) Classifying the impact of deer in woodland. *Deer* 14(10): 35–38.

Cooke, A.S. (2010) Chinese water deer on the Huntingdonshire fens. *Annual Report of the Huntingdonshire Fauna & Flora Society* 62: 52–57.

Cooke, A.S. (2011) Deer in Huntingdonshire and the Soke of Peterborough. *Annual Report of the Huntingdonshire Fauna & Flora Society* 63: 51–61.

Cooke, A.S. (2012a) Chinese puzzle. *Deer* 16(2): 10–15.

Cooke, A.S. (2012b) The relative contributions of Chinese water deer (*Hydropotes inermis*) and muntjac (*Muntiacus reevesi*) to browsing at Woodwalton Fen National Nature Reserve. *Nature in Cambridgeshire* 54: 21–26.

Cooke, A.S. (2013a) A sinister story from the Fen. *Deer* 16(9): 24–25.

Cooke, A.S. (2013b) Muntjac deer in Cambridgeshire. *Nature in Cambridgeshire* 55: 3–21.

Cooke, A.S. (2013c) What you see and what you don't. *Deer* 16(6): 17–20.

Cooke, A.S. (2014a) Muntjac in the woodlands of western Cambridgeshire, 1994–2013. *Deer* 16(10): 25–27.

Cooke, A.S. (2014b) Outfoxing the muntjac. *Deer* 17(2): 30–35.

Cooke, A.S. and Baker, M. (2017) Early-purple orchids and muntjac deer in a Cambridgeshire wood. *Nature in Cambridgeshire* 59: 57–63.

Cooke, A.S. and Farrell, L. (1981) The ecology of Chinese water deer (*Hydropotes inermis*) at Woodwalton Fen National Nature Reserve. Unpublished report for the Nature Conservancy Council, Huntingdon.

Cooke, A. and Farrell, L. (1983) *Chinese Water Deer.* Warminster: British Deer Society.

Cooke, A.S. and Farrell, L. (1987) The utilisation of neighbouring farmland by Chinese water deer (*Hydropotes inermis*) at Woodwalton Fen National Nature Reserve. *Annual Report of the Huntingdonshire Fauna & Flora Society* 39: 28–38.

Cooke, A.S. and Farrell, L. (1995) Establishment and impact of muntjac (*Muntiacus reevesi*) on two National Nature Reserves. In B.A. Mayle (ed.) *Muntjac Deer: Their Biology, Impact and Management in Britain.* Farnham: Forestry Commission and Trentham: British Deer Society. pp. 48–62.

Cooke, A.S. and Farrell, L. (1998) *Chinese Water Deer*. London: Mammal Society and Fordingbridge: British Deer Society.

Cooke, A.S. and Farrell, L. (2000) A long-term study of a population of Chinese water deer. *Deer* 11: 232–237.

Cooke, A.S. and Farrell, L. (2001a) Timing of the rut in a wild population of Chinese water deer. *Deer* 12: 22–25.

Cooke, A.S. and Farrell, L. (2001b) Impact of muntjac deer (*Muntiacus reevesi*) at Monks Wood National Nature Reserve, Cambridgeshire, eastern England. *Forestry* 74: 241–250.

Cooke, A.S. and Farrell, L. (2002) Colonisation of Woodwalton Fen by muntjac. *Deer* 12: 250–253.

Cooke, A.S. and Farrell, L. (2008) Chinese water deer *Hydropotes inermis*. In S. Harris and D.W. Yalden (eds) *Mammals of the British Isles*, 4th edn. Southampton: The Mammal Society. pp. 617–622.

Cooke, A.S. and Lakhani, K. (1996) Damage to coppice regrowth by muntjac deer *Muntiacus reevesi* and protection with electric fencing. *Biological Conservation* 75: 231–238.

Cooke, A.S., Farrell, L., Kirby, K.J. and Thomas, R.C. (1995) Changes in abundance and size of dog's mercury apparently associated with grazing by muntjac. *Deer* 9: 429–433.

Cooke, A.S., Green, P. and Chapman, N.G. (1996) Mortality in a feral population of muntjac *Muntiacus reevesi* in England. *Acta Theriologica* 41: 277–286.

Cooper, M.R. and Johnson, A.W. (1984) *Poisonous Plants in Britain and Their Effects on Animals and Man*. MAFF Reference Book 161. London: HMSO.

Corbet, G.B. (1971) Provisional distribution maps of British mammals. *Mammal Review* 1: 95–142.

Crampton, A.B., Stutter, O., Kirby, K.J. and Welch, R.C. (1998) Changes in the composition of Monks Wood National Nature Reserve (Cambridgeshire, UK) 1964–1996. *Arboricultural Journal* 22: 229–245.

Croft, S., Chauvenet, A.L.M. and Smith, G.C. (2017) A systematic approach to estimate the distribution and total abundance of British mammals. *PLoS ONE*, 12(6): e0176339. Accessed at: https://doi.org/10.1371/journal.pone.0176339 (3 November 2018).

Crowther, R.E. and Evans, J. (1984) *Coppice*. Forestry Commission Leaflet 83. London: HMSO.

Danilkin, A. (1995) *Behavioural Ecology of Siberian and European Roe Deer*. London: Chapman & Hall.

Dansie, O. (1977) Muntjac *Muntiacus reevesi*. In G.B. Corbet and H.N. Southern (eds) *The Handbook of British Mammals*, 2nd edn. Oxford: Blackwell. pp. 447–451.

Dansie, O. (1981) Are muntjac solitary? *Deer* 5: 254–255.

Dansie, O. (1983) *Muntjac*, 2nd edn. Warminster: British Deer Society.

de Nahlik, A.J. (1992) *Management of Deer and Their Habitat*. Gillingham: Wilson Hunt.

Dean, G. and Cooke, A.S. (2015) Chinese combat. *Deer* 17(4): 22–24.

deCalesta, D.S. (1994) Effects of white-tailed deer on songbirds within managed forests in Pennsylvania. *Journal of Wildlife Management* 58: 711–718.

Deer Initiative (2008) *Records and Surveys: Ageing by Teeth.* Accessed at: http://www. thedeerinitiative.co.uk/uploads/guides/110.pdf (3 November 2018).

Deer Initiative (2009) *Management: Cull Planning.* Accessed at: http://www. thedeerinitiative.co.uk/uploads/guides/115.pdf (3 November 2018).

Deer Initiative (2012) *Woodland Impact Survey.* Accessed at: http://www. thedeerinitiative.co.uk/uploads/guides/183.pdf (3 November 2018).

Deer Initiative (2018) About wild deer. Accessed at: http://www.thedeerinitiative. co.uk/about_wild_deer/ (3 November 2018).

Deer Initiative (n.d.) *Deer on Our Roads: Counting the Cost.* Accessed at: http://www. thedeerinitiative.co.uk/uploads/docs/26.pdf (3 November 2018).

Defra (2004) *The Sustainable Management of Wild Deer Populations in England: An Action Plan.* Wrexham: Deer Initiative.

Defra (2009) *Summary of Responses to the Consultation on the Review of the Wildlife and Countryside Act, 1981.* London: Defra.

Demidecki, M. and Demidecki, J. (2016) Chinese water deer and Reeves' muntjac at Wilstone Reservoir, Hertfordshire. *Country-Side* 34(4): 8–11.

Dempster, J.P. (1997) The role of larval food resources and adult movement in the population dynamics of the orange-tip butterfly (*Anthocaris cardamines*). *Oecologia* 111: 549–556.

Diaz, A. and Burton, R.J. (1996) Muntjac and lords & ladies. *Deer* 10: 14–19.

Dick, J.T.A. (2017) Research update from the emerald isle. *Deer* 18(3): 12–13.

Dick, J.T.A., Freeman, M., Proven, J. and Reid, N. (2010) First record of free-living Reeves' muntjac deer (*Muntiacus reevesi* Ogilby 1839) in Northern Ireland. *Irish Naturalists' Journal* 31: 151.

Dolman, P.M. (2011) *Woodlark and Nightjar Recreational Disturbance and Nest Predator Study 2008 and 2009.* Report to Breckland District Council. Norwich: University of East Anglia.

Dolman, P., Fuller, R., Gill, R., Hooton, D. and Tabor, R. (2010) Escalating ecological impacts of deer in lowland woodland. *British Wildlife* 21: 242–254.

Donnelly, J. (1995) *Population Estimation of Fallow and Muntjac Deer in Northamptonshire Using Faecal Pellet Group Counts.* Closure Report, Experiment Y9/216. Leicester: De Montford University.

Downing, G. (2010) *Woodland Stalking.* Shrewsbury: Quiller.

Downing, G. (2014) *Stalking Muntjac.* Shrewsbury: Quiller.

Dubost, G. (1971) Observations éthologiques sur le muntjak (*Muntiacus muntjak* Zimmermann 1780 et *M. reevesi* Ogilby 1839) en captivité et semi-liberté. *Zeitschrift für Tierpsychologie* 28: 387–427.

Dubost, G. (2016) Sexual dimorphism across 3 stages of development in polygynous Artiodactyls is not affected by maternal care. *Current Zoology* 62: 513–520.

Dubost, G., Charron, F., Courcoul, A. and Rodier, A. (2008) Population character-istics of a semi-free-ranging polytocous cervid, *Hydropotes inermis.* *Mammalia* 72: 333–343.

Dubost, G., Charron, F., Courcoul, A. and Rodier, A. (2011a) The Chinese water deer, *Hydropotes inermis* – a fast-growing and productive ruminant. *Mammalian Biology* 76: 190–195.

Dubost, G., Charron, F., Courcoul, A. and Rodier, A. (2011b) Social organization in the Chinese water deer *Hydropotes inermis.* *Acta Theriologica* 56: 189–198.

Edwards, T. and Kenyon, W. (2013) *Wild Deer in Scotland*. Scottish Parliament Information Centre Briefing 13/74. Edinburgh: Scottish Parliament Information Centre.

Eichhorn, M.P., Ryding, J., Smith, M.J., Gill, R.M.A., Siriwardena, G.M. and Fuller, R.J. (2017) Effects of deer on woodland structure revealed through terrestrial laser scanning. *Journal of Applied Ecology* 54: 1615–1626.

Ellwood, S. (2000) Using a dung clearance plot method for estimating fallow, roe and muntjac numbers in mixed deciduous woodland. *Deer* 11: 417–423.

European Union (2017) *Invasive Alien Species of Union Concern*. Luxembourg: Publications Office of the EU.

Fautley, R.G. (2013) The ecology and population genetics of introduced deer species. Unpublished D.Phil. thesis, Imperial College, London.

Fautley, R., Coulson, T. and Savolainen, V. (2012) A comparative analysis of the factors promoting deer invasion. *Biological Invasions* 14: 2271–2281.

Fawcett, J.K. (1997) *Roe Deer*. London: Mammal Society and Fordingbridge: British Deer Society.

Feber, R.E., Brereton, T.M., Warren, M.S. and Oates, M. (2001) The impacts of deer on woodland butterflies: the good, the bad and the complex. *Forestry* 74: 271–276.

Feer, F. (1982) Quelques observations éthologiques sur l'hydropote de Chine *Hydropotes inermis* (Swinhoe, 1870) en captivité. *Zeitschrift für Säugetierkunde* 47: 175–185.

Fitter, R.S.R. (1959) *The Ark in Our Midst*. London: Collins.

Flowerdew, J.R. and Ellwood, S.A. (2001) Impacts of woodland deer on small mammal ecology. *Forestry* 74: 277–288.

Forde, P. (1989) Comparative ecology of muntjac *Muntiacus reevesi* and roe deer *Capreolus capreolus* in a commercial coniferous forest. Unpublished Ph.D. thesis, University of Bristol.

Fordham, W.H. (1963) The muntjac deer in Cambridgeshire. *Nature in Cambridgeshire* 6: 42.

Fox, R., Brereton, T.M., Asher, J., August, T.A., Botham, M.S., Bourn, N.A.D., Cruickshanks, A.L., Bulman, C.R., Ellis, S., Harrower, C.A., Middlebrook, I., Noble, D.G., Powney, G.D., Randle, Z., Warren, M.S. and Roy, D.B. (2015) *The State of the UK's Butterflies 2015*. Wareham: Butterfly Conservation and the Centre for Ecology & Hydrology.

Freeman, M.S., Beatty, G.E., Dick, J.T.A., Reid, N. and Provan, J. (2016) The paradox of invasion: Reeves' muntjac deer invade the British Isles from a limited number of founding females. *Journal of Zoology* 298: 54–63.

Fuller, R.J. (2001) Responses of woodland birds to increasing numbers of deer: a review of evidence and mechanisms. *Forestry* 74: 289–298.

Fuller, R. (2004) Why are woodland birds declining? *BTO News* 253: 5–7.

Fuller, R.J. and Gill, R.M.A. (2001) Ecological impacts of increasing numbers of deer in British woodland. *Forestry* 74: 193–199.

Fuller, R.J. and Henderson, A.C.B. (1992) Distribution of breeding songbirds in Bradfield Woods, Suffolk, in relation to vegetation and coppice management. *Bird Study* 39: 73–88.

Fuller, R., Marshall, M., Eversham, B., Wilkinson, P. and Wright, K. (2016) The increasing importance of monitoring wildlife responses to habitat management. *British Wildlife* 27: 175–186.

Gardiner, C. (2005) Monks Wood NNR: a bibliography, 1993–2003. In C. Gardiner and T. Sparks (eds) *Ten Years of Change: Woodland Research at Monks Wood NNR, 1993–2003*. English Nature Research Report 613. Peterborough: English Nature. pp. 167–184.

Gardiner, C. and Sparks, T. (eds) (2005) *Ten Years of Change: Woodland Research at Monks Wood NNR, 1993–2003*. English Nature Research Report 613. Peterborough: English Nature.

Gilbert, C., Ropiquet, A. and Hassanin, A. (2006) Mitochondrial and nuclear phylogenies of Cervidae (Mammalia, Ruminantia): systematics, morphology, and biogeography. *Molecular Phylogenetics and Evolution* 40: 101–117.

Gill, R.M.A. (1992a) A review of damage by mammals in north temperate forests: 1. Deer. *Forestry* 65: 146–169.

Gill, R.M.A. (1992b) A review of damage by mammals in north temperate forests: 3. Impact on trees and forests. *Forestry* 65: 363–388.

Gill, R. (2004) Population increases, impacts and the need for management of deer in Britain. In C.P. Quine, R.F. Shore and R.C. Trout (eds) *Managing Woodlands and Their Mammals: Proceedings of a Joint Mammal Society/Forestry Commission Symposium*. Edinburgh: Forestry Commission. pp. 55–60.

Gill, R.M.A. and Fuller, R.J. (2007) The effects of deer browsing on woodland structure and songbirds in lowland Britain. *Ibis* 149 (Supplement 2): 119–127.

Gill, R.M.A. and Morgan, G. (2010) The effects of varying deer density on natural regeneration in woodlands in lowland Britain. *Forestry* 83: 53–63.

Gill, R.M.A., Thomas, M.L. and Stocker, D. (1997) The use of portal thermal imaging for estimating population density in forest habitats. *Journal of Applied Ecology* 34: 1273–1286.

Goldberg, E. and Watson, P. (2011) Landscape scale deer management to address biodiversity impacts in England. *Quarterly Journal of Forestry* 105: 29–34.

Goldsmith, J. (1972) Norfolk Mammal Report 1970. *Transactions of the Norfolk & Norwich Naturalists Society* 22: 201–219.

Gough, J. (1999) Observed differences in the muntjac deer populations of Kineton and Yardley Chase. *Deer* 11: 185–187.

Greatorex-Davies, N., Sparks, T. and Woiwod, I. (2005) Changes in the Lepidoptera of Monks Wood NNR (1974–2003). In C. Gardiner and T. Sparks (eds) *Ten Years of Change: Woodland Research at Monks Wood NNR, 1993–2003*. English Nature Research Report 613. Peterborough: English Nature. pp. 90–110.

Green, P. (2008) Deer on the pill. *Deer* 14(8): 12–15.

Green, P. (2012) Whatever happened to the bluetongue virus? *Deer* 16(4): 43.

Green, P. (2013) Deer – friend or foe to the farmer? *Deer* 16(9): 14–18.

Grime, J.P., Hodgson, J.G. and Hunt, R. (1988) *Comparative Plant Ecology*. London: Unwin Hyman.

Grue, H. and Jensen, B. (1979) Review of the formation of incremental lines in tooth cementum of terrestrial mammals. *Danish Review of Game Biology* 11: 1–48.

Guo, G. and Zhang, E. (2002) The distribution of the Chinese water deer (*Hydropotes inermis*) in Zhoushan Archipelago, Zhejiang Province, China. *Acta Theriologica Sinica* 22: 98–107.

Guo, G. and Zhang, E. (2005) Diet of the Chinese water deer (*Hydropotes inermis*) in Zhoushan Archipelago, China. *Acta Theriologica Sinica* 25: 122–130.

Hailstone, M. (2013) The great British deer survey. *Deer* 16(6): 12–15.

Hamrick, B., Strickland, B., Demarais, S., McKinley, W. and Griffin, B. (2013) *Conducting Camera Studies to Estimate Population Characteristics of White-Tailed Deer.* Mississippi State University Extension Service Publication 2788. Starkville, MS: Mississippi State University.

Harding, S.P. (1982) The muntjac project: a brief progress report. *Deer* 5: 450–451.

Harding, S.P. (1986) Aspects of the ecology and social organisation of the muntjac deer (*Muntiacus reevesi*). Unpublished D.Phil. thesis, University of Oxford.

Harold, R. (1991) *The Birds of Woodwalton Fen National Nature Reserve.* Peterborough: English Nature East Region.

Harris, R.A. (1981) Survey of the fallow deer population and its impact on the wildlife conservation objectives of Castor Hanglands NNR. Unpublished report for Nature Conservancy Council.

Harris, R.A. and Duff, K.R. (1970) *Wild Deer in Britain.* Newton Abbot: David & Charles.

Harris, R.B. and Duckworth, J.W. (2015) *Hydropotes inermis.* IUCN Red List of Threatened Species 2015: e.T10329A22163569. Accessed at: https://www.iucnredlist.org/species/10329/22163569 (3 November 2018).

Harris, S. and Forde, P. (1986) The annual diet of muntjac (*Muntiacus reevesi*) in King's Forest, Suffolk. *Bulletin of the British Ecological Society* 17: 19–22.

Harris, S. and Yalden, D.W. (eds) (2008) *Mammals of the British Isles*, 4th edn. Southampton: The Mammal Society.

Harris, S., Morris, P., Wray, S. and Yalden, D. (1995) *A Review of British Mammals: Population Estimates and Conservation Status of British Mammals Other Than Cetaceans.* Peterborough: JNCC.

Harris, S.J., Massimino, D., Gillings, S., Eaton, M.A., Noble, D.G., Balmer, D.E., Proctor, D. and Pearce-Higgins, J.W. (2017) *The Breeding Bird Survey 2016.* British Trust for Ornithology Research Report 700. Thetford: British Trust for Ornithology.

He, X., Chen, M. and Zhang, E. (2016) Home range of reintroduced Chinese water deer in Nanhui East Shoal Wildlife Sanctuary of Shanghai, China. *Animal Production Science* 56: 988–996.

Hemani, M.-R., Watkinson, A.R. and Dolman, P.M. (2004) Habitat selection by sympatric muntjac (*Muntiacus reevesi*) and roe deer (*Caprolus capreolus*) in a lowland commercial pine forest. *Forest Ecology and Management* 194: 49–60.

Hemani, M.-R., Watkinson, A.R. and Dolman, P.M. (2005) Population densities and habitat associations of introduced muntjac *Muntiacus reevesi* and native roe deer *Capreolus capreolus* in a lowland pine forest. *Forest Ecology and Management* 215: 224–238.

Hemani, M.-R., Watkinson, A.R., Gill, R.M.A. and Dolman, P.M. (2007) Estimating abundance of introduced Chinese muntjac *Muntiacus reevesi* and native roe deer

Capreolus capreolus using portable thermal imaging equipment. *Mammal Review* 37: 246–254.

Hewison, A.J.M. and Staines, B.W. (2008) European roe deer *Capreolus capreolus*. In S. Harris and D.W. Yalden (eds) *Mammals of the British Isles*, 4th edn. Southampton: The Mammal Society. pp. 605–617.

Hinsley, S.A., Bellamy, P.E. and Wyllie, I. (2005) The Monks Wood Avifauna. In C. Gardiner and T. Sparks (eds) *Ten Years of Change: Woodland Research at Monks Wood NNR, 1993–2003*. English Nature Research Report 613. Peterborough: English Nature. pp. 75–85.

Hofmann, R.R. (1985) Digestive physiology of the deer – their morphophysiological specialisation and adaptation. In P.F. Fennessy and K.R. Drew (eds) *Biology of Deer Production*. Wellington: Royal Society of New Zealand. pp. 393–407.

Hofmann, R.R., Kock, R.A., Ludwig, J. and Axmacher, H. (1988) Seasonal changes in rumen papillary development and body condition in free ranging Chinese water deer (*Hydropotes inermis*). *Journal of Zoology* 216: 103–117.

Hollander, H. (2015) Reeves' muntjac (*Muntiacus reevesi*) and sika deer (*Cervus nippon*) in the Netherlands. *Lutra* 58: 45–50.

Holt, C.A., Fuller, R.J. and Dolman, P.M. (2010) Experimental evidence that deer browsing reduces habitat suitability for breeding common nightingales *Luscinia megarhynchos*. *Ibis* 152: 335–346.

Holt, C.A., Fuller, R.J. and Dolman, P.M. (2011) Breeding and post-breeding responses of woodland birds to modification of habitat structure by deer. *Biological Conservation* 144: 2151–2162.

Hooper, M.D. (1973) History. In R.C. Steele and R.C. Welch (eds) *Monks Wood: A Nature Reserve Record*. Huntingdon: Nature Conservancy. pp. 22–35.

Horsley, S.B., Stout, S.L. and deCalesta, D.S. (2003) White-tailed deer impact on the vegetation dynamics of a northern hardwood forest. *Ecological Applications* 13: 98–118.

Hows, M., Pilbeam, P., Conlan, H. and Featherstone, R. (2016) *Cambridgeshire Mammal Atlas*. N.p.: Cambridgeshire Mammal Group.

Hughes, D. (2005) Rare plants in Monks Wood NNR 1993–2003. In C. Gardiner and T. Sparks (eds) *Ten Years of Change: Woodland Research at Monks Wood NNR, 1993–2003*. English Nature Research Report 613. Peterborough: English Nature. pp. 128–132.

Hughes, F.M.R., Stroh, P.A., Adams, W.M., Kirby, K.J., Mountford, J.O. and Warrington, S. (2011) Monitoring and evaluating large-scale, 'open-ended' habitat creation projects: a journey rather than a destination. *Journal for Nature Conservation* 19: 245–253.

Humphries, J. and Dutton, J. (2015) Camera trapping of muntjac in the Forest of Dean. *Mammal News* 172: 6–7.

Inns, H. (2009) *Britain's Reptiles and Amphibians*. Old Basing: WILDGuides.

Jackson, J.E., Chapman, D.I. and Dansie, O. (1977) A note on the food of muntjac deer. *Journal of Zoology, London* 183: 546–548.

Jarmeno, A. (2004) Neonatal mortality in roe deer. Unpublished doctoral thesis, Swedish University of Agricultural Sciences, Uppsala.

Jarmeno, A. and Liberg, O. (2005) Red fox removal and roe deer fawn survival – a 14 year study. *Journal of Wildlife Management* 69: 1090–1098.

Jefferies, D.J. and Arnold, H.R. (1977–1984) Mammal report for each year, 1976–1983. *Annual Reports of the Huntingdonshire Fauna & Flora Society*, volumes 29–36.

Jennings, N. (2008) Brown hare *Lepus europaeus*. In S. Harris and D.W. Yalden (eds) *Mammals of the British Isles*, 4th edn. Southampton: The Mammal Society. pp. 210–220.

Jiang, Z., Ma, Y., Wu, Y., Wang, Y., Zhou, K., Liu, S. and Fcng, Z. (2015) *China's Mammal Diversity and Geographic Distribution*. Beijing: Science Press.

Joint Nature Conservation Committee (2004) *Common Standards Monitoring Guidance for Woodland Habitats*. Accessed at: http://www.jncc.defra.gov.uk/pdf/CSM_woodland.pdf (3 November 2018).

Jung, J., Shimizu, Y., Omasu, K., Kim, S. and Lee, S. (2016) Developing and testing a habitat suitability index model for Korean water deer (*Hydropotes inermis argyropus*) and its potential for landscape management decisions in Korea. *Animal Cells and Systems* 20: 218–227.

Kay, S. (1993) Factors affecting severity of deer browsing damage within coppiced woodlands in the south of England. *Biological Conservation* 63: 217–222.

Keeling, J.G.M. (1995a) Ecological determinants of muntjac *Muntiacus reevesi* behaviour. Unpublished Ph.D. thesis, University of Bristol.

Keeling, J.G.M. (1995b) Resource utilization by muntjac deer (*Muntiacus reevesi*). In B.A. Mayle (ed.) *Muntjac Deer: Their Biology, Impact and Management in Britain*. Farnham: Forestry Commission and Trentham: British Deer Society. pp. 25–26.

Key, H.A.S. (1959) Mammals. *Bedfordshire Natural History Society and Field Club* 13: 45–46.

Kilgo, J.C., Labisky, R.F. and Fritzen, D.E. (1998) Influences of hunting on the behavior of white-tailed deer: implications for conservation of the Florida panther. *Conservation Biology* 12: 1359–1364.

Kim, B., Oh, D., Chun, S. and Lee, S. (2011) Distribution, density and habitat use of the Korean water deer (*Hydropotes inermis argyropus*) in Korea. *Landscape and Ecological Engineering* 7: 291–297.

Kirby, K.J. (2001a) The impact of deer on the ground flora of British broadleaved woodland. *Forestry* 74: 219–229.

Kirby, K.J. (2001b) Where have all the flowers gone? Are our woodland flowers disappearing? *Biologist* 48: 182–186.

Kirby, K.J. (2005) Monks Wood NNR – vegetation studies 1993–2003. In C. Gardiner and T. Sparks (eds) *Ten Years of Change: Woodland Research at Monks Wood NNR, 1993–2003*. English Nature Research Report 613. Peterborough: English Nature. pp. 27–33.

Kirby, K.J. and Morecroft, M.D. (2010) The flowers of the forest. In P.S. Savill, C.M. Perrins, K.J. Kirby and N. Fisher (eds) *Wytham Woods: Oxford's Ecological Observatory*. Oxford: Oxford University Press. pp. 75–89.

Kirby, K.J. and Thomas, R.C. (1999) *Changes in the Ground Flora in Wytham Woods, Southern England, 1974–1991 and Their Implications for Nature Conservation*. English Nature Research Report 320. Peterborough: English Nature.

Kirby, K.J., Thomas, R.C. and Dawkins, H.C. (1996) Monitoring of changes in tree and shrub layers in Wytham Woods (Oxfordshire), 1974–1991. *Forestry* 69: 319–334.

Knight, G.E. (1964) The factors affecting the distribution of *Endymion nonscriptus* (L.) Garcke in Warwickshire woods. *Journal of Ecology* 52: 405–421.

Koh, S., Lee, B., Wang, J., Heo, S. and Kyung, H. (2009) Two sympatric phylogroups of the Chinese water deer (*Hydropotes inermis*) identified by mitochondrial DNA control region and cytochrome *b* gene analyses. *Biochemical Genetics* 47: 11–12.

Lai, J. and Sheng, H. (1993) A comparative study on scent-marking behavior of captive forest musk deer and Reeves' muntjac. In N. Uhtalshl and H. Sheng (eds) *Deer of China*. Amsterdam: Elsevier. pp. 204–208.

Langbein, J. (2007) *National Deer–Vehicle Collisions Project: England (2003–2005)*. Deer Initiative Research Report 07/1. Wrexham: Deer Initiative.

Langbein, J. (2011) *Monitoring Reported Deer Road Casualties and Related Accidents in England to 2010*. Final report to the Highways Agency. Wrexham: Deer Initiative.

Langbein, J. and Chapman, N.G. (2003) *Fallow Deer*. London: Mammal Society and Fordingbridge: British Deer Society.

Langbein, J. and Rutter, S.M. (2003) Quantifying the damage wild deer cause to agricultural crops and pastures. In E. Goldberg (ed.) *Proceedings of the Future for Deer Conference*. English Nature Research Report 548. Peterborough: English Nature. pp. 32–39.

Lawrence, M.J. and Brown, R.W. (1967) *Mammals of Britain: Their Tracks, Trails and Signs*. London: Blandford Press.

Lawrence, R.P. (1982) Chinese water deer. *Shooting Times & Country Magazine*, 4–10 March 1982: 25.

Lawrence, R. (2009) Mammals 2008. *Bedfordshire Naturalist* 63(1): 12–18.

Lawson, R.E., Putman, R.J. and Fielding, A.H. (2000) Individual signatures in scent gland secretions of Eurasian deer. *Journal of Zoology, London* 251: 399–410.

Leadbeater, S. (2011) Deer management and biodiversity in England: the efficacy and ethics of culling. *Ecos* 32: 59–68.

Leech, D. (2008) *Norfolk Mammal Report 2007*. Norwich: Norfolk & Norwich Naturalists' Society.

Lever, C. (1977) *The Naturalized Animals of the British Isles*. London: Hutchinson.

Lever, C. (2009) *The Naturalized Animals of Britain and Ireland*. London: New Holland.

Linnell, J. and Panzacchi, M. (2006) Bambi meets the red devil. *Deer* 13(8): 34–37.

Lyme Disease Action (2013, 7 March) LDA questions the role of deer in the spread of Lyme disease. Press release. Accessed at: http://www.lymediseaseaction.org.uk/press-releases/lda-questions-the-role-of-deer-in-the-spread-of-lyme-disease/ (3 November 2018).

Ma, F., Yu, X., Chen, M., Liu, C., Zhang, Z. and Ye, J. (2013) Diurnal time budgets and activity rhythm of captive Chinese water deer (*Hydropotes inermis*) in spring and summer. *Acta Theriologica Sinica* 33: 1–7.

Mabey, R. (1996) *Flora Britannica*. London: Sinclair-Stevenson.

Macdonald, D., Tattersall, F., Johnson, P., Carbone, C., Reynolds, J., Langbein, J., Rushton, S. and Shirley, M. (2000) *Managing British Mammals: Case Studies from the Hunting Debate*. Oxford: University of Oxford Wildlife Conservation Research Unit.

Macklin, R.N. (1990) Chinese water deer at Minsmere, 1989. *Transactions of the Suffolk Naturalists' Society* 26: 5.

Malins, M. and Oliver, P. (2017) Alternatives to steel and lead. *Deer* 18(5): 26–28.

Manning, C.J. (2006) *Deer and Deer Parks of Lincolnshire*. Horncastle: Lincolnshire Naturalists' Union.

Manton, V.J.A. and Matthews, S. (1983) Chinese water deer. In A.J.B. Rudge (ed.) *The Capture and Handling of Deer*. Peterborough: Nature Conservancy Council. pp. 140–150.

Mårell, A., Archaux, F. and Korboulewsky, N. (2009) Floral herbivory of the wood anemone (*Anemone nemorosa* L.) by roe deer (*Capreolus capreolus* L.). *Plant Species Biology* 24: 209–214.

Massey, M.E. (1994) Managed change. In M.E. Massey and R.C. Welch (eds) *Monks Wood National Nature Reserve: The Experience of 40 Years 1953–93*. Peterborough: English Nature. pp. 13–17.

Massey, M.E. and Welch, R.C. (eds) (1994) *Monks Wood National Nature Reserve: The Experience of 40 Years 1953–93*. Peterborough: English Nature.

Mathews, F., Kubasiewicz, L.M., Gurnell, J., Harrower, C.A., McDonald, R.A. and Shore, R.F. (2018) *A Review of the Population and Conservation Status of British Mammals: Technical Summary*. Report by the Mammal Society. Peterborough: Natural England, Natural Resources Wales and Scottish Natural Heritage.

Mauget, R., Mauget, C., Dubost, G., Charron, F., Courcoul, A. and Rodier, A. (2007) Non-invasive assessment of reproductive status in Chinese water deer (*Hydropotes inermis*): correlation with sexual behaviour. *Mammalian Biology* 72: 14–26.

Mayle, B. (1999) *Managing Deer in the Countryside*. Forestry Commission Practice Note 6. Edinburgh: Forestry Commission.

Mayle, B.A., Peace, A.J. and Gill, R.M.A. (1999) *How Many Deer?* Edinburgh: Forestry Commission.

Mayle, B.A., Putman, R.J. and Wyllie, I. (2000) The use of trackway counts to establish an index of deer presence. *Mammal Review* 30: 233–237.

McCarrick, M. (2006) Mammals 2004. *Bedfordshire Naturalist* 59: 10–14.

McCarrick, M. (2007) Distribution of Bedfordshire mammal species 2000–2006. *Bedfordshire Naturalist* 61: 18–24.

McCullough, D.R., Pei, K.C.J. and Wang, Y. (2000) Home range, activity patterns, and habitat relations of Reeves' muntjacs in Taiwan. *Journal of Wildlife Management* 64: 430–441.

McShea, W.J. and Rappole, J.H. (1997) Herbivores and the ecology of forest understory birds. In W.J. McShea, H.B. Underwood and J.H. Rappole (eds) *The Science of Overabundance: Deer Ecology and Population Management*. Washington DC: Smithsonian Institution. pp. 298–309.

McShea, W.J., Underwood, H.B. and Rappole, J.H. (eds) (1997) *The Science of Overabundance: Deer Ecology and Population Management*. Washington DC: Smithsonian Institution.

Mellanby, K. (1973) *Mammalia*. In R.C. Steele and R.C. Welch (eds) *Monks Wood: A Nature Reserve Record*. Huntingdon: Nature Conservancy. pp. 289–295.

Middleton, A.D. (1937) Whipsnade ecological survey. *Proceedings of the Zoological Society of London* 107: 471–481.

Miller, S.G., Bratton, S.P. and Hadidian, J. (1992) Impacts of white-tailed deer on endangered plants. *Natural Areas Journal* 12: 67–74.

Mitchell-Jones, A. (2005) Introduction of dormice. In T. Collins, B. Dickerson, P.E.G. Walker and T.C.E. Wells (eds) *Brampton Wood: A Natural History*. N.p: Huntingdonshire Fauna and Flora Society. pp. 79–82.

Mitchell-Jones, A. and Kirby, K. (1997) *Deer Management and Woodland Conservation in England*. Peterborough: English Nature.

Morecroft, M.D., Taylor, M.E., Ellwood, S.A. and Quinn, S.A. (2001) Impacts of deer herbivory on ground vegetation at Wytham Woods, central England. *Forestry* 74: 251–257.

Morellet, N., Champely, S., Gaillard, J.-M., Ballon, P. and Boscardin, Y. (2001) The browsing index: new tool uses browsing pressure to monitor deer populations. *Wildlife Society Bulletin* 29: 1243–1252.

Mountford, E.P. and Peterken, G.F. (1998) *Monitoring Natural Stand Change in Monks Wood National Nature Reserve*. English Nature Research Report 270. Peterborough: English Nature.

Munro, R. (2002) *Report on the Deer Industry in Great Britain, 2002*. Unpublished report to Defra and the Food Standards Agency.

Mysterud, A., Easterday, W.R., Stigum, V.M., Aaas, A.B., Meisingset, E.L. and Viljugrein, H. (2016) Contrasting emergence of Lyme disease across ecosystems. *Nature Communications* 7: article number 11882. Accessed at: https://doi.org/10.1038/ncomms11882 (3 November 2018).

National Institute for Environmental Studies (2017) *Invasive Species of Japan: Muntiacus reevesi*. Ibaraki, Japan: National Institute for Environmental Studies.

Natural England (2013) *Natural England Standard. SSSI Monitoring and Reporting*. Accessed at: http://publications.naturalengland.org.uk/file/4564618932387840 (3 November 2018).

Natural Resources Wales (2011) *Wild Deer Management in Wales*. Strategy document, Welsh Assembly Government. Aberystwyth: Forestry Commission Wales.

Nau, B.S. (1992) Chinese water deer in Bedfordshire. *Bedfordshire Naturalist* 46: 17–27.

Nentwig, W., Bacher, S., Kumschick, S., Pyšek, P. and Vilà, M. (2018) More than '100 worst' alien species in Europe. *Biological Invasions* 20: 1611–1621.

Newson, S.E., Johnson, A., Renwick, A.R., Baillie, S.R. and Fuller, R.J. (2012) Modelling large-scale relationships between changes in woodland deer and bird populations. *Journal of Applied Ecology* 49: 278–286.

Niemann, D. (2016) *A Tale of Trees*. London: Short Books.

Nobbs, D. (2002) Mammals of a Broadland reserve – Wheatfen. In I. Keymer (ed.) *Norfolk Bird and Mammal Report for 2001*. Norwich: Norfolk & Norwich Naturalists' Society. pp. 295–300.

Non-Native Species Secretariat (2011) Risk assessment. Accessed at: http://www.nonnativespecies.org/index.cfm?pageid=143 (3 November 2018).

Nuttle, T., Ristau, T.E. and Royo, A.A. (2014) Long-term biological legacies of herbivore density in a landscape-scale experiment: forest understoreys reflect past deer density treatments for at least 20 years. *Journal of Ecology* 102: 221–228.

O'Connell, A.F., Nichols, J.D. and Karanth, K.U. (eds) (2011) *Camera Traps in Animal Ecology: Methods and Analyses*. Tokyo: Springer.

O'Connell, S. (2007) Lyme borreliosis and other tick-borne infections. *Deer* 14(4): 10–13.

Oliver, J. (2013) Muntjac and British plants. *BSBI News* 123: 35–37.

Ostfeld, R.S., Jones, C.G. and Wolff, J.O. (1996) Of mice and mast. *BioScience* 46: 323–330.

Palmer, G. (2014) Deer in Britain: population spread and the implications for biodiversity. Unpublished doctoral thesis, University of Durham. Accessed at: http://etheses.dur.ac.uk/10597/ (3 November 2018).

Palmer, R. (1947) Mammals of Bedfordshire. *Journal of the Bedfordshire Natural History Society and Field Club* 1: 47–50.

Parliamentary Office of Science and Technology (2009) *Wild Deer*. Postnote 325. London: Parliamentary Office of Science and Technology.

Patmore, A.J. (1995) Reeves' muntjac (*Muntiacus reevesi*) in Northamptonshire. In B.A. Mayle (ed.) *Muntjac Deer: Their Biology, Impact and Management in Britain.* Farnham: Forestry Commission and Trentham: British Deer Society. pp. 44–47.

Peace, T.R. and Gilmour, J.S.L. (1949) The effect of picking on the flowering of bluebell *Scilla non-scripta*. *New Phytologist* 48: 115–117.

Pearce, F. (2015) *The New Wild: Why Invasive Species Will Be Nature's Salvation.* London: Icon Books.

People's Trust for Endangered Species (2017) *Living with Mammals Update 2017.* London: PTES.

Pepper, H. (1999) *Recommendations for Fallow, Roe and Muntjac Deer Fencing: New Proposals for Temporary and Reusable Fencing.* Forestry Commission Practice Note 9. Edinburgh: Forestry Commission.

Pepper, H.W., Rowe, J.J. and Tee, L.A. (1985) *Individual Tree Protection.* Forestry Commission Arboricultural Leaflet 10. London: HMSO.

Perrins, C.M. and Overall, R. (2001) Effect of increasing numbers of deer on bird populations in Wytham Woods, central England. *Forestry* 74: 299–309.

Perrins, C.M. and Gosler, A.G. (2010) Birds. In P.S. Savill, C.M. Perrins, K.J. Kirby and N. Fisher (eds) *Wytham Woods: Oxford's Ecological Observatory.* Oxford: Oxford University Press. pp. 145–171.

Peterken, G.F. (1974) A method for assessing woodland flora for conservation using indicator species. *Biological Conservation* 6: 239–245.

Peterken, G.F. (1993) *Woodland Conservation and Management*, 2nd edn. London: Chapman & Hall.

Peterken, G.F. (1994) Natural change in unmanaged stands within Monks Wood NNR. In M.E. Massey and R.C. Welch (eds) *Monks Wood National Nature Reserve: The Experience of 40 Years 1953–93.* Peterborough: English Nature. pp. 1–8.

Peterken, G.F. and Mountford, E. (2005) Natural woodland reserves – 60 years of trying at Lady Park Wood. *British Wildlife* 17: 7–16.

Peterken, G. and Mountford, E. (2017) *Woodland Development: A Long Term Study of Lady Park Wood.* Wallingford: CABI.

Plantlife (2004) *Bluebells for Britain.* Salisbury: Plantlife.

Pocock, R.I. (1923) On the external characters of *Elaphurus, Hydropotes, Pudu* and other *Cervidae*. *Proceedings of the Zoological Society of London* 93: 181–207.

Pollard, E. (1979) Population ecology and change in range of the white admiral butterfly *Ladoga camilla* L. in England. *Ecological Entomology* 4: 61–74.

Pollard, E. and Cooke, A.S. (1994) Impact of muntjac deer *Muntiacus reevesi* on egg-laying sites of the white admiral butterfly *Ladoga camilla* in a Cambridgeshire wood. *Biological Conservation* 70: 189–191.

Pollard, E., Woiwod, I.P., Greatorex-Davies, J.N., Yates, T.J. and Welch, R.C. (1998) The spread of coarse grasses and changes in numbers of Lepidoptera in a woodland nature reserve. *Biological Conservation* 84: 17–24.

Prestt, I. (1973) *Amphibia* and *reptilia*. In R.C. Steele and R.C. Welch (eds) *Monks Wood: A Nature Reserve Record*. Huntingdon: Nature Conservancy. pp. 270–274.

Putman, R.J. (1984) Facts from faeces. *Mammal Review* 14: 79–97.

Putman, R.J. (1988) *The Natural History of Deer*. New York: Comstock.

Putman, R.J. (1994) Deer damage in coppice woodlands: an analysis of factors affecting the severity of damage and options for management. *Quarterly Journal of Forestry* 88: 45–54.

Putman, R.J. (1996) *Deer Management on National Nature Reserves: Problems and Practices*. English Nature Research Report 173. Peterborough: English Nature.

Putman, R.J. (1998a) Deer impact on conservation vegetation in England and Wales. In C.R. Goldspink, S. King and R.J. Putman (eds) *Population Ecology, Management and Welfare of Deer*. Manchester: Manchester Metropolitan University. pp. 61–66.

Putman, R.J. (1998b) The potential role of habitat manipulation in reducing deer impact. In C.R. Goldspink, S. King and R.J. Putman (eds) *Population Ecology, Management and Welfare of Deer*. Manchester: Manchester Metropolitan University. pp. 95–101.

Putman, R. (2000) *Sika Deer*. London: Mammal Society and Fordingbridge: British Deer Society.

Putman, R. (2003) *The Deer Manager's Companion*. Shrewsbury: Swan Hill.

Putman, R. and Moore, N.P. (1998) Impact of deer in lowland Britain on agriculture, forestry and conservation habitats. *Mammal Review* 28: 141–164.

Putman, R.J., Edwards, P.J., Mann, J.C.E., How, R.C. and Hill, S.D. (1989) Vegetational and fauna changes in an area of heavily grazed woodland following relief of grazing. *Biological Conservation* 47: 13–32.

Putman, R., Langbein, J., Green, P. and Watson, P. (2011a) Identifying threshold densities for wild deer in the UK above which negative impacts may occur. *Mammal Review* 41: 175–196.

Putman, R., Watson, P. and Langbein, J. (2011b) Assessing deer densities and impacts at the appropriate level for management: a review of methodologies for use beyond the site scale. *Mammal Review* 41: 197–219.

Rackham, O. (1975) *Hayley Wood: Its History and Ecology*. Cambridge: Cambridgeshire and Isle of Ely Naturalists' Trust.

Rackham, O. (1999) The woods 30 years on: where have the primroses gone? *Nature in Cambridgeshire* 41: 73–87.

Rackham, O. (2003) *Ancient Woodland: Its History, Vegetation and Uses in England*. Colvend: Castlepoint Press.

Rackham, O. (2006) *Woodlands*. New Naturalist Library 100. London: Collins.

Rackham, O. (2014) *The Ash Tree*. Toller Fratrum: Little Toller.

Randi, E., Mucci, N., Pierpaoli, M. and Douzery, E. (1998) New phylogenetic perspectives on the Cervidae (Artiodactyla) are provided by the mitochondrial cytochrome *b* gene. *Proceedings of the Royal Society of London B* 265: 793–801.

Ratcliffe, D.A. (1977) *A Nature Conservation Review*. Cambridge: Cambridge University Press.

Ratcliffe, P.R. (1992) The interaction of deer and vegetation in coppice woods. In G.P. Buckley (ed.) *Ecology and Management of Coppice Woodlands*. London: Chapman & Hall. pp. 233–245.

Revington, P.J. (1996) Age determination in deer: practical tips and literature review. *Deer* 10: 20–21.

Rhim, S. and Lee, W. (2007) Influence of forest fragmentation on the winter abundance of mammals in Mt Chirisan National Park, South Korea. *Journal of Wildlife Management* 71: 1404–1408.

Rooney, T.P. (2001) Deer impacts on forest ecosystems: a North American perspective. *Forestry* 74: 201–208.

Rooney, T.P. and Dress, W.J. (1997) Species loss over sixty six years in the ground-layer vegetation of Heart's Content, an old-growth forest in Pennsylvania, USA. *Natural Areas Journal* 17: 297–305.

Rowcliffe, J.M., Field, J., Turvey, S.T. and Carbone, C. (2008) Estimating animal density using camera traps without the need for individual recognition. *Journal of Applied Ecology* 45: 1228–1236.

Sale, G.N. and Archibald, J.F. (1957) Working plan for Monks Wood. Unpublished report for Nature Conservancy.

Scottish Natural Heritage (2016) *Deer Management in Scotland*. Report to the Scottish Government. Inverness: Scottish Natural Heritage.

Seo, H., Kim, J., Seomun, H., Hwang, J., Jeong, H., Kim, J., Kim, H. and Cho, S. (2017) Eruption of posterior teeth in the maxilla and mandible for age determination in water deer. *Archives of Oral Biology* 73: 237–242.

Sheng, H. (1992a) Reeves' muntjac *Muntiacus reevesi*. In H. Sheng (ed.) *The Deer in China*. Shanghai: East China Normal University. pp. 126–144.

Sheng, H. (1992b) Chinese water deer *Hydropotes inermis*. In H. Sheng (ed.) *The Deer in China*. Shanghai: East China Normal University. pp. 109–110.

Sheng, H. and Lu., H. (1984) A preliminary study on the river deer population of Zhoushan Island and adjacent islets. *Acta Theriologica Sinica* 4: 161–166.

Sheng, H. and Ohtaishi, N. (1993) The status of deer in China. In N. Ohtaishi and H. Sheng (eds) *Deer of China*. Amsterdam: Elsevier. pp. 1–11.

Sheng, H. and Zhang, E. (1992) The conservation of deer species in China. In H. Sheng (ed.) *The Deer in China*. Shanghai: East China Normal University. pp. 257–266.

Simpson, J.W., Morrison, L. and Stevenson, K. (2012) Johne's disease. *Deer* 16(5): 42–45.

Smith-Jones, C. (2004) *Muntjac: Managing an Alien Species*. Machynlleth: Coch-Y-Bonddu Books.

Smith-Jones, C. (2017) The 2016 BDS deer distribution survey. *Deer* 18(5): 15–18.

Son, S., Hwang, H., Lee, J., Eom, T., Park, C., Lee, E., Kang, J. and Rhim, S. (2017) Influence of tree thinning on the abundance of mammals in a Japanese larch *Larix kaempferi* plantation. *Animal Cells and Systems* 21: 70–75.

Sönnichsen, L., Bokje, M., Marchal, J., Hofer, H., Jędrzejewska, B., Kramer-Schadt, S. and Ortmann, S. (2013) Behavioural responses of European roe deer to temporal variation in predation risk. *Ethology* 119: 233–243.

Soper, E.A. (1969) *Muntjac*. London: Longmans.

Sparks, T., van Gaasbeek, F., Waasdorp, D. and Willi, J. (2005) Monks Wood NNR and its neighbours: a comparison with local woods. In C. Gardiner and T. Sparks (eds) *Ten Years of Change: Woodland Research at Monks Wood NNR, 1993–2003*. English Nature Research Report 613. Peterborough: English Nature. pp. 59–64.

Sparks, T., Garforth, J. and Lewthwaite, K. (2017) What was the effect of the warmest December on record? *British Wildlife* 28: 244–247.

Spencer, J. and Kirby, K. (1992) An inventory of ancient woodland for England and Wales. *Biological Conservation* 62: 77–93.

Stadler, S.G. (1991) Behaviour and social organisation of Chinese water deer (*Hydropotes inermis*) under semi-natural conditions. Unpublished Ph.D. thesis, Universität Bielefeld.

Staines, B. (1995) Discussion and summing up. In B.A. Mayle (ed.) *Muntjac Deer: Their Biology, Impact and Management in Britain.* Farnham: Forestry Commission and Trentham: British Deer Society. pp. 69–70.

Staines, B., Palmer, S.C.F., Wyllie, I., Gill, R. and Mayle, B. (1998) *Desk and Limited Field Studies to Analyse the Major Factors Influencing Regional Deer Populations and Ranging Behaviour.* MAFF Project No. VC 0314 (ITE Project No. T08093a5). Banchory: Institute of Terrestrial Ecology.

Staines, B.W., Langbein, J. and Burkitt, T.D. (2008) Red deer *Cervus elaphus*. In S. Harris and D.W. Yalden (eds) *Mammals of the British Isles*, 4th edn. Southampton: The Mammal Society. pp. 573–587.

Steele, R.C. (1973) Vegetation. In R.C. Steele and R.C. Welch (eds) *Monks Wood: A Nature Reserve Record.* Huntingdon: Nature Conservancy. pp. 43–53.

Steele, R.C. and Welch, R.C. (eds) (1973) *Monks Wood: A Nature Reserve Record.* Huntingdon: Nature Conservancy.

Stewart, A.J.A. (2001) The impact of deer on lowland woodland invertebrates: a review of the evidence and priorities for future research. *Forestry* 74: 259–270.

Sun, L. and Dai, N. (1995) Male and female association and mating system in the Chinese water deer (*Hydropotes inermis*). *Mammalia* 59: 171–178.

Sun, L. and Sheng, H. (1990) Chinese water deer at the areas of Poyang Lake. *Journal of East China Normal University (Mammalian Ecology Supplement)* 9: 21–26.

Sun, L. and Xiao, B. (1995) The effect of female distribution on male territoriality in Chinese water deer (*Hydropotes inermis*). *Zeitschrift für Säugetierkunde* 60: 33–40.

Sun, L., Xiao, B. and Dai, N. (1994) Scent marking behaviour in the male Chinese water deer. *Acta Theriologica* 39: 175–184.

Swinhoe, R. (1870) On a new deer from China. *Proceedings of the Zoological Society of London* (1870): 89–92.

Symonds, R.J. (1983) A survey of the distribution of deer in Cambridgeshire (vice-county 29). *Nature in Cambridgeshire* 26: 52–60.

Tabor, R.C.C. (1993) Control of deer in a managed coppice. *Quarterly Journal of Forestry* 87: 308–313.

Tabor, R.C.C. (1999) The effects of muntjac deer, *Muntiacus reevesi*, and fallow deer, *Dama dama*, on the oxlip, *Primula elatior*. *Deer* 11: 14–19.

Tabor, R.C.C. (2002) Notes on Essex specialities, 7: The distribution of the oxlip *Primula elatior* (L.) Hill in Essex. *Essex Naturalist* 19: 113–134.

Tabor, R.C.C. (2004) Assessing deer activity and damage in woodlands. *Deer* 13: 27–29.

Tabor, R.C.C. (2005) Woodland flora recovery after damage by deer, (1) Control of pendulous sedge *Carex pendula*. *Essex Naturalist* 22: 67–75.

Tabor, R.C.C. (2006) Woodland plants and deer – the potential for damage to plant biodiversity by deer and the use of plant indicators to monitor and predict this damage. *Essex Naturalist* 23: 65–76.

Tabor, R. (2009) On the fence. *Deer* 15(1): 40–42.

Tabor, R. (2011) Culling for conservation. *Deer* 15(10): 13–15.

Tack, C. (2000) Distribution of Bedfordshire mammal species 1995–1999. *Bedfordshire Naturalist* 54: 20–27.

Tanentzap, A.J., Burrows, L.E., Lee, W.G., Nugent, G., Maxwell, J.M. and Coomes, D.A. (2009) Landscape-level vegetation recovery from herbivory: progress after four decades of invasive red deer control. *Journal of Applied Ecology* 46: 1064–1072.

Tanentzap, A.J., Bazeley, D.R., Koh, S. and Coomes, D. (2011) Seeing the wood for the trees: do reductions in deer-disturbance lead to forest recovery? *Biological Conservation* 144: 376–382.

Tanentzap, A.J., Kirby, K.J. and Goldberg, E. (2012a) Slow responses of ecosystems to reductions in deer (Cervidae) populations and strategies for achieving recovery. *Forest Ecology and Management* 264: 159–166.

Tanentzap, A.J., Mountford, E.P., Cooke, A.S. and Coomes, D.A. (2012b) The more stems the merrier: advantages of multi-stemmed architecture for the demography of understorey trees in a temperate broadleaf woodland. *Journal of Ecology* 100: 171–183.

Tanton, M.T. (1969) The estimation and biology of populations of the bank vole (*Clethrionomys glareolus* (Schr.)) and wood mouse (*Apodemus sylvaticus* (L.)). *Journal of Animal Ecology* 38: 511–529.

Thomas, J. and Lewington, R. (2010) *The Butterflies of Britain and Ireland*. Milton on Stour: British Wildlife Publishing.

Thompson, K. (2014) *Where Do Camels Belong?* London: Profile Books.

Thornley, J. (2016) *Stalking Fallow*. Shrewsbury: Quiller.

Thurfjell, H., Ciuti, S. and Boyce, M.S. (2017) Learning from the mistakes of others: how female elk (*Cervus elaphus*) adjust behaviour with age to avoid hunters. *PLoS ONE* 12(6): e0178082. Accessed at: https://doi.org/10.1371/journal.pone.0178082 (2 December 2018).

Tilghman, N.G. (1989) Impacts of white-tailed deer on forest regeneration in north-western Pennsylvania. *Journal of Wildlife Management* 53: 524–532.

Timmins, J. and Chan, B. (2016) *Muntiacus reevesi*. IUCN Red List of Threatened Species 2016: e.T42191A22166608. Accessed at https://www.iucnredlist.org/species/42191/22166608 (3 November 2018).

Toms, M. and Leech, D. (2005) *Norfolk Mammal Report 2004*. Norwich: Norfolk & Norwich Naturalists' Society.

Trout, R. and Pepper, H. (2006) *Forest Fencing*. Technical Guide. Edinburgh: Forestry Commission.

Tubbs, C.R. (1986) *The New Forest*. New Naturalist Library 73. London: Collins.

van Gaasbeek, F., Waasdorp, D. and Sparks, T. (2000) The status of *Primula* and *Daphne laureola* in Monks Wood NNR in 1999. *Annual Report of the Huntingdonshire Fauna & Flora Society* 52: 8–12.

Vincent, J.-P., Gaillard, J.-M. and Bideau, E. (1991) Kilometric index as a biological indicator for monitoring forest roe deer populations. *Acta Theriologica* 36: 315–328.

Waeber, R., Spencer, J. and Dolman, P.M. (2013) Achieving landscape-scale deer management for biodiversity conservation: the need to consider sources and sinks. *Journal of Wildlife Management* 77: 726–736.

Walker, K. (2005) An extra 10 acres? Botanical research in the Monks Wood Wilderness. In C. Gardiner and T. Sparks (eds) *Ten Years of Change: Woodland Research at Monks Wood NNR, 1993–2003.* English Nature Research Report 613. Peterborough: English Nature. pp. 48–58.

Wang, H. and Sheng, H. (1990) Population density and habitat selection of Chinese water deer in Zhoushan Islands. *Journal of East China Normal University (Mammalian Ecology Supplement)* 9: 43–46.

Ward, A.I. (2005) Expanding ranges of wild and feral deer in Great Britain. *Mammal Review* 35: 165–173.

Ward, A., Etherington, T. and Ewald, J. (2008) Five years of change. *Deer* 14(8): 17–20.

Watson, P. (2017) Priority management. *Deer* 18(2): 14–16.

Watson, P., Putman, R. and Goulding, M. (2012) Management in modern Britain. *Deer* 16(2): 28–29.

Welch, R.C. (1975) The fauna of Bedford Purlieus: Amphibia, Reptilia and Mammalia. In G.F. Peterken and R.C. Welch (eds) *Bedford Purlieus: Its History, Ecology and Management.* Abbots Ripton: Institute of Terrestrial Ecology. p. 131.

Welch, R.C. (2005) Monks Wood Coleoptera – an update: 1973–2003. In C. Gardiner and T. Sparks (eds) *Ten Years of Change: Woodland Research at Monks Wood NNR, 1993–2003.* English Nature Research Report 613. Peterborough: English Nature. pp. 111–127.

Wells, T.C.E. (1973) Flowering plants and ferns. In R.C. Steele and R.C. Welch (eds) *Monks Wood: A Nature Reserve Record.* Huntingdon: Nature Conservancy. pp. 54–62.

Wells, T.C.E. (1994) Changes in vegetation and flora. In M.E. Massey and R.C. Welch (eds) *Monks Wood National Nature Reserve: The Experience of 40 Years 1953–93.* Peterborough: English Nature. pp. 19–28.

Wells, T.C.E. (2003) *The Flora of Huntingdonshire and the Soke of Peterborough.* N.p.: Huntingdonshire Fauna & Flora Society and T.C.E. Wells.

Wells, T.C.E. and Cox, R. (1991) Demographic and biological studies on *Ophrys apifera*: some results from a 10 year study. In T.C.E. Wells and J.H. Willems (eds) *Population Ecology of Terrestrial Orchids.* The Hague: SPB Academic. pp. 47–61.

White, P.C.L., Smart, J.C.R., Bohm, M., Langbein, J. and Ward, A. (2004a) *Economic Impacts of Wild Deer in the East of England.* Accessed at: http://www.woodlandforlife. net/PDFs/DEER%20studyExecutive_Summary[1].pdf (3 November 2018).

White, P.C.L., Ward, A.I., Smart, J.C.R. and Moore, N.P. (2004b) *Impacts of Deer and Deer Management on Woodland Biodiversity in the English Lowlands.* Unpublished final contract report for the Woodland Trust.

Whitehead, G.K. (1950) *Deer and Their Management in the Deer Parks of Great Britain and Ireland.* London: Country Life.

Whitehead, G.K. (1964) *The Deer of Great Britain and Ireland.* London: Routledge & Kegan Paul.

Willi, J. and Sparks, T. (2003) From Pingle to Perry West: the ground flora of twenty ancient woods in Huntingdonshire. *Annual Report of the Huntingdonshire Fauna & Flora Society* 55: 10–15.

Williams, E.S. and Young, S. (1980) Chronic wasting disease of captive mule deer: a spongiform encephalopathy. *Journal of Wildlife Diseases* 16: 89–98.

Williams, T., Harris, S., Chapman, N., Wayne, R., Beaumont, M. and Bruford, M. (1995) A molecular analysis of the introduced Reeves' muntjac in southern England: genetic variation in the mitochondrial genome. In B.A. Mayle (ed.) *Muntjac Deer: Their Biology, Impact and Management in Britain.* Farnham: Forestry Commission and Trentham: British Deer Society. pp. 6–22.

Wilson, C.J. (2003) *Current and Future Deer Management Options.* Report on behalf of Defra Wildlife Division. Exeter: Defra.

Wilson, K. (1995) The impact of muntjac in woodlands. In B.A. Mayle (ed.) *Muntjac Deer: Their Biology, Impact and Management in Britain.* Farnham: Forestry Commission and Trentham: British Deer Society. pp. 40–43.

Won, C. and Smith, K.G. (1999) History and current status of mammals of the Korean peninsula. *Mammal Review* 29: 3–36.

Worden, A.N. (1959) Mammals. *Annual Report of the Huntingdonshire Fauna & Flora Society* 11: 3.

Worden, A.N. (1960) Mammals. *Annual Report of the Huntingdonshire Fauna & Flora Society* 12: 3–4.

Wray, S. (1995) Competition between muntjac and other herbivores in a commercial coniferous forest. In B.A. Mayle (ed.) *Muntjac Deer: Their Biology, Impact and Management in Britain.* Farnham: Forestry Commission and Trentham: British Deer Society. pp. 27–39.

Wright, L.J., Newson, S.E. and Noble, D.G. (2014) The value of a random sampling design for annual monitoring of national populations of larger British terrestrial mammals. *European Journal of Wildlife Research* 60: 213–221.

Wyllie, I., Palmer, S.C.F. and Mayle, B. (1998) *Desk and Limited Field Studies to Analyse the Major Factors Influencing Regional Deer Populations and Ranging Behaviour. Annex 2: Radio-Telemetry Studies of Muntjac and Fallow Deer.* MAFF Project No. VC 0314 (ITE Project No. T08093a5). Banchory: Institute of Terrestrial Ecology.

Xiao, B. and Sheng, H. (1990) Home range and activity patterns of Chinese water deer (*Hydropotes inermis*) in Poyang Lake region. *Journal of East China Normal University (Mammalian Ecology Supplement)* 9: 27–36.

Xu, H. and Lu, H. (1996) A preliminary analysis of population viability for Chinese water deer (*Hydropotes inermis*) lived in Yancheng. *Acta Theriologica Sinica* 16: 81–88.

Xu, H., Lu, H. and Liu, X. (1996) The current status and habitat use of Chinese water deer in the coast of Jiangsu Province. *Zoological Research* 17: 217–224.

Xu, H., Zheng, X. and Lu, H. (1998) Impact of human activities and habitat changes on distribution of Chinese water deer along the coast area in northern Jiangsu. *Acta Theriologica Sinica* 18: 161–167.

Yahner, R.H. (1980) Activity patterns of captive Reeve's muntjacs. *Journal of Mammalogy* 61: 368–371.

Yalden, D. (1999) *The History of British Mammals.* London: Poyser.

Yu, X., Chen, M., Zhang, E., Ye, J., Pei, E. and Yuan, X. (2013) Correlation between seasonal changes in territory behaviour and fecal testosterone level in captive male Chinese water deer (*Hydropotes inermis*). *Chinese Journal of Zoology* 48: 43–48.

Zhang, E. (1994) Chinese water deer in Yancheng Nature Reserve. Unpublished report for the University of Cambridge.

Zhang, E. (1996) Behavioural ecology of the Chinese water deer at Whipsnade Wild Animal Park, England. Unpublished Ph.D. thesis, University of Cambridge.

Zhang, E. (1998) Uniparental female care in the Chinese water deer at Whipsnade Wild Animal Park, England. *Acta Theriologica Sinica* 18: 173–183.

Zhang, E. (2000a) Ingestive behaviour of the Chinese water deer. *Zoological Research Sinica* 21: 88–91.

Zhang, E. (2000b) Daytime activity budgets of the Chinese water deer. *Mammalia* 64: 163–172.

Zhang, E. and Guo, G. (2000) Poaching as a major threat to Chinese water deer in Zhoushan Archipelago, Zhejiang Province, PR China. *Deer* 11: 413–414.

Zhang, E., Teng, L. and Wu, Y. (2006) Habitat selection of the Chinese water deer (*Hydropotes inermis*) in Yancheng Reserve, Jiangsu Province. *Acta Theriologica Sinica* 26: 49–53.

Zhang, X. and Zhang, E. (2002) Distribution pattern of *Hydropotes inermis* in various habitats in Jiangsu Dafeng Pere David's Deer State Nature Reserve. *Sichuan Journal of Zoology* 21: 19–22.

Index

Figure numbers are in bold type and table numbers are in italics.